Toxic Hazards in Food

D. M. Conning and
A. B. G. Lansdown

CROOM HELM
London and Canberra

Croom Helm Ltd, Provident House, Burrell Row,
Beckenham, Kent BR3 1AT
Croom Helm Australia, PO Box 391
Manuka, ACT 2603, Australia

British Library Cataloguing in Publication Data
Toxic hazards in foods.
1. Food poisoning
I. Conning, David M. II. Lansdown, Alan B.G.
615.9'54 RC143
ISBN 0-7099-0253-0

Typeset by Mayhew Typesetting, Bristol
Printed and bound in Great Britain

CONTENTS

Foreword *D. V. Parke* 1

Contributors 4

1. **Systemic Toxicity Due to Foodstuffs** *D.M. Conning* 5
   Introduction 5
   Cardiovascular and Haematological Disease 5
   Central Nervous System 7
   Hepatotoxicity 12
   Renal Toxicity 16
   References 18

2. **Intolerance and Allergy to Foods and Food Additives: Its Relevance to Toxicology** *W.E. Parish* 22
   Allergy, Tolerance and Pseudo-allergy (Idiosyncrasy) 22
   Lymphocyte Stimulation and Antibody Formation 25
   Mucosal (Gastro-intestinal) Immune Systems 31
   Classification of Allergic Responses 35
   Anaphylaxis 37
   Arthus Reaction and Immune Complex Disorders 45
   Delayed Hypersensitivity (Cell-mediated Immune Response) 50
   Intolerance to Natural Food Chemicals and to Food Additives 57
   Reduced Immunocompetence (Immunosuppression) 61
   Immunological Tests to Predict Toxicological Allergic Hazards to Man 64
   References 68

3. **Teratogenicity and Reduced Fertility Resulting from Factors Present in Food** *A.B.G. Lansdown* 73
   Introduction 73
   Epidemiological and Experimental Aspects 75
   Food Constituents 78
   Minerals 84
   Food Additives 96
   Alcohol 100

Food Contaminants 102
Discussion 111
References 112

4. **Carcinogens in Food** *P. Grasso* 122
Carcinogens in Edible Plants 122
Fungal Carcinogens 124
Carcinogens Introduced During Food Processing and Preservation 127
Polycyclic Aromatic Hydrocarbons 127
Nitrosamines 130
Food Additives 133
Pesticides 136
Food and Cancer 138
Conclusion 140
References 140

5. **The Mutagenicity of Food** *D. Anderson and I.F.H. Purchase* 145
Introduction 145
Principles of Testing Food for Mutagenicity 146
Background to Mutagenicity Testing 147
Mutagenicity Testing 148
Testing Strategy 158
Mutagens in Food 158
Extrapolation of Results from Laboratory Assays to Man 165
Significance of Food-borne Mutagens to Man 166
References 168

6. **The Gastrointestinal Tract in Food Toxicology** *I.R. Rowlands and R. Walker* 183
Introduction 183
Role of Gut Microflora in Intestinal Metabolism 183
Role of Mammalian Enzymes in Intestinal Metabolism 208
Toxicological Aspects of Intestinal Metabolism 211
Pathology of the Gastrointestinal Tract in Safety Evaluation 246
References 250

7. **Food Allergies: Clinical Aspects and Natural Allergens** *J.R. Mansfield* 275
Introduction 275
Historical Development 276

Clinical Ecology 279
Clinical Conditions Related to Clinical Ecology 279
Investigation of Food Allergy 280
General Discussion 286
Summary 290
Further Reading 290
References 291

Index 292

CROOM HELM APPLIED BIOLOGY SERIES
Edited by Peter J. Baron, BSc. Hons, PhD

NUTRITION AND DISEASE
Edited by R.J. Jarrett

THE NATURE OF ENZYMOLOGY
R.L. Foster

NUTRITION AND FOOD PROCESSING
H.G. Muller and G. Tobin

THE MASTERPIECE OF NATURE:
The Evolution and Genetics of Sexuality
G. Bell

CELL LOCOMOTION IN VITRO
C.A. Middleton and J.A. Sharp

FOREWORD

In the world of plentiful, cheap food that so many of us in the Western World have come to accept as our birthright, it takes sensational journalism and horrific television coverage to make us realise that not all of the world's population shares this abundance. Visits to the Soviet Bloc countries, to China, most African states and various other countries of the Third World make one quickly realise that widespread shortage of food is just over the horizon and would be common experience to all were it not for the advanced technologies of Western agriculture, food production and food manufacture. Without doubt, pesticides and other agricultural chemicals have made enormous contributions to world food production – and indeed to world peace. The introduction of many food additives, especially antioxidants, has also contributed greatly to the amount and quality of food which is available, preventing early spoilage and waste, and possibly indirectly contributing to man's health, as these antioxidants have been shown to have beneficial properties in the prevention of experimental malignancy and cardiovascular disease. Nevertheless, despite the enormous benefits derived from the use of agrochemicals and food additives, it is essential to remember that these are all selectively toxic chemicals, with no absolute guarantee of safety, and where the benefit to risk ratio is always a compromise adjusted by the expediency of political and financial aspects of food production.

The major hazards from toxic chemicals in food have been generally recognised to comprise: (1) bacterial contamination; (2) naturally-occurring toxins and carcinogens; (3) accidental chemical contaminants; (4) pesticide contaminants and; (5) food additives, with the last two being regarded as carrying little or no real hazard. Nevertheless, at the interface of these two groups of chemicals, when pesticides are deliberately added to food, as in the treatment of cereal grains for the control of insect pests, it is essential that the pesticides added should be rigorously examined for their safety and health hazards. Recent developments in this area have seen the introduction of a very wide range of new chemicals for use in grain silos and these developments have to be watched carefully.

Constant vigil is necessary to ensure that imported food conforms with the accepted stringent standards for the absence of contaminant chemicals, such as aflatoxins and other mycotoxins and preformed nitrosamines. Food imported into North America and Western Europe from tropical countries generally contains only the smallest traces of these potent natural carcinogens, even though the same products on sale in the market places of the countries of origin may be several orders of magnitude greater in their contents of mycotoxins and nitrosamines. It is tempting to speculate that before the introduction of the rigorous chemical analyses, much imported tropical food was heavily contaminated with these natural toxic chemicals, and may have been responsible for as much human cancer in our Western civilisation as still occurs today in the tropical areas of production. Similarly, there is a real international need to prevent the accidental contamination of food with dangerous industrial chemicals, such as occurred in the Michigan disaster involving the polybromobiphenyl mixture, 'Firemaster'. The final chapters of this catastrophe have still to be written, but the persistence of this industrial chemical produced a magnitude of contamination that has never before been seen, and involved food production and manufacture on a world-wide scale.

The safety evaluation of food chemicals depends very largely on toxicity studies and the use of animal models. These widely accepted practices of toxicology are of value only provided that the animal data can be suitably translated to predict safety for man. More for convenience than for scientific reasons, small rodents are generally used in these toxicological studies, and in many cases there is little attempt to ascertain to what extent these animals are suitable models for man, for the specific chemicals under study. Carcinogenicity studies on pesticides, food additives and even natural food constituents, are generally conducted at extremely high dosage, and the smallest increase in occurrence of tumours has indicted as carcinogens such widely occurring food components as caffeine and saccharin.

This empirical testing of chemicals, often with scant regard for any scientific basis, and the use of unsuitable small animal species as models for man, will frequently grossly magnify the toxic and carcinogenic risks of many chemicals, and may generate spurious toxicity data that could lead toxicology into disrepute and ridicule. Politicians often prefer to use commonsense to the pseudo-science of some of their toxicology advisors, and international regulatory bodies such as the WHO and FAO, find it expedient not to pay too much attention to carcinogenicity data which indicate positive findings only in the mouse. There is an urgent need for more logical and scientific approaches to the choice of animal species for toxicity testing, to the selection of dose levels, and the consideration of species differences in the toxicokinetics, metabolism, and mechanisms of toxicity of chemicals. Several international bodies are now coming to grips with these problems and international experts are re-evaluating the most suitable techniques for determining and monitoring toxic hazards of chemicals, and the most appropriate scientific methods for interpreting the data generated.

Nevertheless, adulterant practices are still rife in many countries; lead chromate and carcinogenic yellow azo dyes are sometimes added illegally to curry pastes and powders in India, and well-meaning persons may send samples of these to relatives abroad. Of the numerous natural xenobiotics present in our food only a very few have been examined for safety, and of these perhaps too great a percentage has been found to be undesirably toxic (e.g. gossypol and ipomeanol) or potently carcinogenic (mycotoxins). The continuous and extended study of these natural components is well merited, but rarely funded. The safety evaluation of new pesticides and agrochemicals, and the re-evaluation of existing ones, is obviously a continuing and expanding need, and ways must be found to ensure prevention of gross contamination of food by industrial chemicals. Recent developments have again highlighted the the widespread occurrence of food allergies, which sometimes result in severely debilitating and incapacitating diseases. These and several facets of the toxic hzards of food are now reviewed in this most important new publication, by acknowledged experts in these many and diverse fields of study.

Dennis V. Parke
Department of Biochemistry,
University of Surrey,
Guildford.

CONTRIBUTORS

D. Anderson, PhD, British Industrial Biological Research Association, Woodmansterne Road, Carshalton, Surrey SM5 4DS.

D.M. Conning, MB, FRCPath, Director, British Industrial Biological Research Association, Woodmansterne Road, Carshalton, Surrey SM5 4DS.

P. Grasso, BSc, MD, DCP, DTM & H, FRCPath, British Petroleum Group Occupational Health Centre, Sunbury-on-Thames, Middlesex.

A.B.G. Lansdown, BSc, PhD, MRCPath, FIBiol, Department of Pathology, Life Science Research, Stock, Essex CM4 9PE.

J.R. Mansfield, MRCS, LRCP, DRCOG, Burgh Wood Allergy Clinic, Banstead, Surrey.

W.E. Parish, MA, PhD, BVSc, MRCVS, FRCPath, Environmental Safety Laboratory, Unilever Research, Colworth House, Sharnbrook, Bedford MK44 1LQ.

D.V. Parke, DSc, PhD, CChem, FRSC, FRCPath, FIBiol, Professor, Department of Biochemistry, University of Surrey, Guildford, Surrey GU2 5XH.

I.F.H. Purchase, PhD, BVSc, MRCVS, MRCPath, Imperial Chemical Industries Ltd., Alderley Park, Macclesfield, Cheshire SK10 4JJ.

I.R. Rowlands, PhD, British Industrial Biological Research Association, Woodmansterne Lane, Carshalton, Surrey SM5 4DS.

R. Walker, PhD, Reader in Food Science, Department of Biochemistry, University of Surrey, Guildford, Surrey GU2 5XH.

1 SYSTEMIC TOXICITY DUE TO FOODSTUFFS

D.M. Conning

Introduction

Throughout this book, the intention is to present the specific disease entities which have been associated with food and to define, where possible, the causative chemicals and the pathogenic process. It is the intention, also, to describe real entities that have been observed in man and attributed to materials ingested and to avoid the hypothetical diseases which would occur if the results of animal studies, often conducted with outlandish dosage, were to be directly extrapolated to man.

In this opening chapter toxic diseases which occur in the cardiovascular, central nervous, hepatic and renal systems, excluding carcinogenic and immunotoxic diseases, will be dealt with. It is clear that, in a volume of reasonable size, these types cannot be dealt with exhaustively, but it is hoped that the important features, as judged from the standpoint of human health, will have been included.

Cardiovascular and Haematological Disease

Cardiomyopathies

Alcohol. Excessive and prolonged consumption of alcohol can result, in a minority of cases, in cardiac failure associated with myocardial degeneration characterised by mitochondrial changes, fat deposition and diffuse fibrosis.[1,2,3] The condition is not due to coronary artery disease.[4] There appears to be no correlation with hepatic cirrhosis.[5] The mechanism remains obscure but there appears to be impairment of mitochondrial function in that isocitric and malic dehydrogenase activities are depressed, glycogen content is decreased and the mitochondrial respiratory control index (a measure of oxidative phosphorylation) decreases.[125,126,127]

Cobalt. The use of cobalt salts as antifoaming agents resulted in several cases of fatal heart disease among heavy drinkers of bottled beers. These occurrences ceased when the use of cobalt was discontinued. The condition is reproducible in the rat and is associated with accumulations of cobalt in the myocardium. It resembles thiamine deficiency and is probably due to the inhibition of α-lipoic acid, a co-factor in the activity of α-ketoglutarate dehydrogenase in myocardial mitochondria.

As large doses of cobalt used therapeutically have no such effects, it seems likely that alcohol or thiamine deficiency are predisposing factors.[6,7]

Vascular

Fats. The role of fats in the development of atherosclerosis has been the subject of substantial controversy, often due to the claims of many parties with vested interests. Atlhough there are innumerable associations between death from vascular disease and many effects which accompany obesity, there is no consistent causal relationship with any such as cholesterol, saturated fats, unsaturated fats or more specific lipids.[8] The latest correlations in which the so-called high density lipoprotein transport mechanisms show an inverse relationship with the incidences of vascular disease across a number of factors seen to be involved, show more promise. There is, as yet, no experimental support and much more information is required.[9,10]

Salt. There is increasing evidence that a small excess of salt consumption of the order of 12 grams per day is associated with an increase in systolic blood pressure in a proportion of the population.[11] It seems probable that there is a genetic susceptibility to essential hypertension which can be prevented by reduced salt consumption.[12] In the absence of the predisposition, very large dosage approaching 40 grams per day is required to effect an increase in blood pressure but it is possible to achieve such levels of consumption in Chinese and Japanese diets. There is experimental support for the genetic factor.[12]

Potassium. Depletion of potassium is not likely in the absence of predisposing factors such as treatment with certain diuretics or mineral corticoids. A case has been reported of cardiac arrest associated with hypokalaemia, due to liquorice. The aldosterone effect resulted from a large dosage (1.8 kg per week of liquorice) of glycyrrhizinic acid.

Cadmium. A number of studies have claimed a relationship between dietary cadmium, especially from drinking water (up to 3 μg/litre) and hypertensive cardiovascular disease in man[65] and there is experimental evidence of such a relationship.[66] Studies of populations where the environmental exposure to cadmium has been high, however, have shown no consistent evidence of hypertension.[67]

Miscellaneous Effects

Erucic Acid. Myocardial damage following massive infiltration of fat after feeding rape seed oil has been attributed to the erucic acid content.[13] Similar effects with other oils only occur if erucic acid is present. Although most studies have been done in the rat, other species are susceptible, but there is little evidence of effects in man. In a study of the extraction of erucic acid by human myocardium,[14] no preferential extraction was detected 3 hours after ingestion of rape seed oil containing 49 per cent erucic acid. Nevertheless a study in healthy Italian males[15] demonstrated serum levels on 40 subjects ranging from 0.3 to 3.8 per cent of total fatty acids.

Haematological Diseases

Favism. A sporadic and capricious haemolytic anaemia prevalent in the Mediterranean area, due to consumption of broad beans, especially raw or partially cooked.[16] It is associated with genetically determined glucose-6-phosphate dehydrogenase deficiency.[17]

Lead. A microcytic hypochromic anaemia due to two effects, to the inhibition of haem synthesis by inhibition of δ-aminolaevulinic dehydratase and to the inhibition of ferrochelatase thereby reducing the incorporation of iron into protoporphyrin.[18] There is also evidence of a direct haemolytic effect.[19] Impedance of iron incorporation might occur at blood levels as low as 15 μg/100 ml in children and women,[20] but chemically detected anaemia is unlikely below levels of about 40 μg/100 ml.

Other trace elements have been associated with anaemia of various types but only in cases of excessive dosage or deficiency.

Central Nervous System

Hitherto the study of neurotoxicity has been largely confined to systemic poisons associated with specific neurological disorders or with the various components which affect mood and which are almost invariably addictive. In recent years increasing attention has been paid to behavioural disorders and their possible toxicological origin. This account will deal with the variety of neurotoxins, both systemic and behavioural, for which the effects have been substantiated. It is not intended as an exhaustive account of the topic, much of which remains conjectural.

Natural Constituents

Neurolathyrism. A disease characterised by spastic paralysis of the lower limbs due to sclerosis of the dorsolumbar spinal cord which results from the prolonged consumption of the chickpea (*Lathyrus sativa*) as a major item of the diet.[21] It is probably due to oxalyl-diaminobutyric and oxalyl-diaminopropionic acids,[22] but in the absence of a definitive animal model, the precise causative agent remains uncertain.

Cycasin. This has been associated with the development of amyotrophic lateral sclerosis where cycad starch flour is consumed. Although the active principle, methylazoxymethanol, causes abnormalities of the developing brain in rats and hamsters[23] there is no evidence to support the occurrence of human disease.[24]

Locoweed. Consumption of locoweed (*Astragalus*) by domestic animals and the rat is associated with neurological (central and autonomic) and teratological effects. The active principle is unknown but appears to be water soluble and excreted unchanged.[25] A human counterpart has not been described as yet.

Fish Poisons. Shellfish and ciguatera fish poisoning results from the ingestion by these species of dinoflagellates and the storage and concentration of specific toxins. Consumption of the affected shellfish produces tingling sensations of the tongue, lips and fingers followed by ascending paralysis, culminating in respiratory failure. Death occurs within 12 hours in fatal cases. Ciguatera poisoning shows similar symptomatology but is rarely fatal.

Puffer fish poisoning which shows the same symptoms with additional gastro-intestinal effects may also result in respiratory failure. The toxin is located in the liver, testes and ovaries of the affected fish. If these viscera are removed, undamaged, the fish is safe to eat. The poisons concerned, saxitoxin and tetrodotoxin, block diffusion of sodium into nerve and muscle cells. They are lethal at doses of 4 μg/kg bodyweight.[26]

Natural Contaminants

Mycotoxins. The discovery of the potent and varied toxic effects of fungal toxins has opened a new and exciting aspect of toxicology which promises real benefits for man in the prevention of disease. Mycotoxins have been associated with several different disease entities. The elucidation of the respective mechanisms will bring new insights into aetiology.

Tremorgens. Diseases of grazing livestock characterised by staggering, tremors and convulsions resulted in the isolation from *Penicillium crustosum* of penitrem A, the first of four compounds to be isolated from this organism.[27] Earlier a tremor-producing agent had been detected as a metabolite of *Aspergillus flavus.*[28] This has resulted in the isolation of a range of complex molecules from several fungal strains capable of inducing severe tremors of a protracted kind in mice. Some are active by mouth and occur in human foodstuffs. No documented examples of human disease exist but several species are susceptible and there is no reason to believe that man is immune. The mechanism of action is not known.[29]

Citreoviridin. Isolated from *Penicillium citreoviride* contamination of rice,[31] citreoviridin causes convulsions and ascending paralysis following intraperitoneal injection in mice. Human disease has not been reported.

Verruculotoxin. Isolated from *Penicillium verruculosum* induces ataxia in cockerels.[31]

Roquefortine from *Penicillium roqueforti*, an indole derivative causing paralysis in mice.[32]

Ergotism. An ancient disease, not seen in England since 1928, which is due to lysergic acid derivatives produced by *Claviceps purpurea* infestation of wheat, barley and rye. It is associated with vasoconstricture, neurohumoral and cortical effects.[33]

Metals

Lead. An EEC Commission has reported on levels of lead in the blood of the European population.[34] Eighteen thousand subjects were examined and found to have a median concentration of 13 μg/100 ml. About 2 per cent exceeded 30 μg/100 ml and about 1 per cent exceeded

35 μg/100 ml. The survey did not discover any risk of lead poisoning except where specific sources of lead such as industrial emission or lead water pipes were present. In the UK the DHSS Working Party Report[35] recommended that mean air concentrations should be kept below 2 $\mu g/m^3$ and where blood levels were found to exceed 35 μg/100 ml specific sources should be identified and dealt with. The report identifies lead paint, lead water pipes and some manufactured goods as main sources. It states that as airborne lead can account for no more than 10 per cent of blood lead (and usually much less), this could not be regarded as a major problem, though it did recommend the gradual reduction of emission from traffic using leaded petrols. There is evidence that airborne lead might be a more significant source for children. Absorption of lead is approximately 8 per cent in adults but may be as high as 50 per cent in children.[36] Organic lead is more readily absorbed and is distributed in brain rather than in bone.[37]

Poisoning in children results in encephalopathy characterised by ataxia, stupor, coma and convulsions. Adults more commonly show a peripheral neuropathy. The level at which toxicity occurs is not certain and is subject to controversy, but there is evidence of peripheral nerve damage in adults with blood levels in the range 40 μg to 120 μg/100 ml.[38] In one study, children with blood lead concentrations greater than 99 μg/100 ml all had encephalopathy.[39] The controversy relates to behavioural effects in children with low but elevated blood concentrations. Such studies are difficult to control because of the interaction of intelligence, social status, parental attitudes and bias introduced by expectations of compensation. Of more value are similar studies in adult lead workers. In such studies there is good correlation between impaired intelligence, psychomotor ability and memory and raised blood concentrations in matched tests and control groups.[40–42] In general blood values ranged up to 90 μg/100 ml and the worst effects were seen at concentrations above 50 μg/100 ml. Comparable findings have been achieved in animal studies[43–45] and it seems reasonable to conclude that an elimination of lead from the environment is desirable.

Mercury. This is widespread in the human environment and the average human contains about 13 mg.[46] Apart from the metal, it is present as inorganic or organic ions. Man absorbs from the gut 95 per cent of an administered dose of organic mercury[47] and 7 per cent of the dose of inorganic mercury.[48] Organic mercury in the form of alkyl and phenyl mercurics are extensively used as fungicidal seed dressings and it has been demonstrated that inorganic forms can be converted into organic forms by microbial activity in estuary and bay bottom sediments.[49] In this form it readily enters the food chain. Accidental poisonings have occurred through the consumption of treated grain in Iraq,[50] pork reared on treated grain in New Mexico[51] and contaminated fish from Minamata Bay, Japan.[52] The syndrome is characterised by ataxia, blindness and coma and is due to degeneration of the cells of the granular layer in the cerebellum.[53] In rats there is evidence of spinal neuropathy in addition to a cerebellar lesion.[54]

Other metals. Lithium,[55] magnesium[56] and manganese[57] have all been associated with neurogenic toxicity but none has been dietary in origin.

Direct Additives

Monosodium glutamate. Used as an agent to enhance flavour, this amino acid was associated with subjective symptoms of numbness and tingling of mouth and tongue, and some peripheral weakness following the ingestion of Chinese-style food. Double-blind, crossover studies failed to confirm the effect at an oral dosage of 3 g.[58] Doses as high as 45 g/day have been used in the treatment of mental retardation and certain encephalopathies, with no adverse effects and no change in brain glutamate levels.[59]

Animal studies, however, demonstrated hypothalamic lesions in neonatal mice given 0.5 g/kg orally,[60] although similar changes could not be reproduced in the rat by subcutaneous injection.[61]

Adult human diets contain approximately 20 g glutamic acid per day, of which about 2 g is free acid and 0.7 g is added material. Breast-fed infants receive approximately 0.4 g/kg/day, of which 10 per cent is free. The amount may be greater from cow's milk.

It is conceivable therefore, that hypothalamic lesions or impaired development could occur if man was as sensitive to glutamic acid as the newborn mouse. It is extremely unlikely that dietary glutamate constitutes a neurogenic hazard to individuals over the age of six months.

Tartrazine. The facile association of 'hypersensitivity' in the immunological sense (allergy) and 'hypersensitivity' in the behavioural sense has led to the assertion that colours such as tartrazine, used among others, in foodstuffs might be the cause of the hyperactivity syndrome.[62,63] The evidence is equivocal and not supported by animal studies.[64]

In general the attempts to relate the consumption of food colours and the presence of hyperkinesis in children is fraught with the difficulties in measuring and controlling small behavioural changes in children. The effect of increased parental, teacher and physician attention and of uncertainties about the consistency and natural course of the syndrome all influence the child's responses. All studies so far have given equivocal results and at present, despite the conviction of many parents, the role of food colours is uncertain.[68,69]

Hepatotoxicity

The liver is the main site of metabolism of food constituents and is naturally vulnerable to excessive intake of toxic components. This section will deal with only those components of the diet which might be consumed in sufficient quantity to induce hepatotoxicity and will not include circumstances in which absurdly high dosage is involved.

Natural Constituents

Cycasin. This material is notable as an extremely potent liver carcinogen which requires metabolism by intestinal flora before it becomes active – a conversion of β-D-glucosyloxyazoxy-methane into methylazoxymethanol. At low concentrations, however, this material causes decreased protein synthesis, centrilobular necrosis and haemorrhage in the liver.[23]

Senecio alkaloids. These compounds have been responsible for many deaths amongst livestock throughout the world from hepatic cirrhosis.[68] Human deaths have also been reported due to the consumption of bread made from wheat contaminated with *Senecio illicifolius*. Death was due to hepatic necrosis and hepatic vein thrombosis.[69] It is also thought that the one-time prevalence of cirrhosis in Jamaican children was due to 'bush tea' made from several plants which included *Senecio*.[70]

The active moiety is a pyrrolizidine alkaloid of which there are many. The structural configuration which determines hepatotoxicity has been determined.[71] Monocrotaline, another pyrrolizidine alkaloid of the toxic configuration occurs in *Crotalacia*, and is also associated with liver disease in cattle but of a more chronic variety.

Hypervitaminosis. Although unlikely to result from the consumption of normal diets, excessive consumption of vitamin A and of niacin arising from therapy for deficiency or self-administration can occur. Fish-eating, fat-storing animals (polar bear, seal, shark) have very high concentrations of vitamin A in the liver, which, as a consequence, is poisonous to other animals.[72] Chronic over-consumption of vitamin A (above 100,000 i.u. daily) results in toxicity, mainly manifested by skin lesions (dermatitis and eczema), but is also accompanied by an enlarged liver with diffuse fibrosis, congestion and gross fatty degeneration,[73] which may progress to cirrhosis.[74] The mechanism is uncertain, but the fatty change may be an attempt to store excess and unbound vitamin A, with the consequent hepatic toxicity.[73]

Niacin (nicotinic acid), essential for the production of coenzymes NAD and NADP, may cause hepatic toxicity at doses as low as 750 mg/day.[75] It causes cholestatic hepatitis, hepatic cell necrosis and periportal fibrosis. There is evidence that nicotinamide may be even more toxic.[76]

Natural Contaminants

Mycotoxins

Aflatoxin. Fluorescent furanocoumarins are produced by *Aspergillus flavus* which infects ground nuts and maize under certain conditions of storage. Since their discovery, they have been studied extensively and there is an enormous literature.[77–79]

Most of the work relates to aflatoxin B_1, the first to be identified and responsible for the outbreak of 'Turkey X' disease which led to the discovery of the hepatotoxic effect. All animals tested have been susceptible, with the duckling being the most sensitive. The LD-50 in the duckling is approximately 0.4 mg/kg, whereas in the rat it is approximately 7 mg/kg (18 mg/kg in the female).[79] Hepatotoxicity is characterised by periportal necrosis and bile duct proliferation in the rat.[80] Aflatoxin is lethal to human liver cells in culture at 10 μg B_1/ml.[81]

There is considerable circumstantial evidence linking the presence of aflatoxins to the prevalence of hepatic disease in geographic areas where the climatic conditions predispose to mould infestation. Populations in such areas also show the predisposing factors such as malnutrition, exposure to other toxins such as the pyrrolizidine alkaloids, parasites and hepatitis B virus infections. The multiplicity of factors makes absolute proof of aflatoxin B_1 as the causal material very difficult but for practical purposes the hepatic morbidity should be regarded as of multiple origin and the remedies designed accordingly.[79] The mechanisms of action of aflatoxin are not understood. It inhibits

DNA and RNA and protein synthesis and causes degenerative changes in rough endoplasmic reticulum. There are several enzymatic consequences of these effects which vary from species to species. There is, as yet, no coordinated concept of the sequence of events.[79]

Ochratoxin A. Although currently regarded as a nephrotoxin, ochratoxin A was first isolated as a hepatotoxin from *Aspergillus ochraceus* and *Penicillium viridicatum*.[82,83] In the weanling rat, oral doses of about 2 mg/kg cause hyaline degeneration, periportal necrosis and fatty change. The hyaline areas are due to massive accumulations of smooth endoplasmic reticulum. Hepatic lesions were also described in ducklings and trout.[82,84] The LD-50 in the adult is 22 mg/kg. Although found in moulds on cereals and vegetables, there are as yet, no instances of human hepatic disease attributable to ochratoxins.

Rubratoxins. Penicillium rubrum is widespread and has been shown experimentally to produce toxins which cause lesions of liver, kidney and the central nervous system, at doses in the region of 4 mg/kg after intraperitoneal injection.[85,86]

Sporidesmins. A widespread disease of sheep 'facial eczema' which results in the loss of large numbers of carcasses due to jaundice and cirrhosis, was shown to be due to the fungus *Pithomyces chartarum* after extensive investigations.[87] The hepatic lesions are characterised by bile duct necrosis, portal tract fibrosis, hepatocellular necrosis (and infarct) and cholestasis.[88]

Direct and Indirect Additives

Modern food processing requires the use of large numbers of food additives covering all aspects of the physicochemical performance, preservation and colour. Such is the legislative control of the quality of food, that enormous amounts of money and human effort are used to ensure that materials of dubious safety are excluded from the diet. The methods used still leave a good deal of room for improvement and control is exerted largely by reducing the conceivable ingestion to a fraction of the dose which is known to have any biological effect. Safety thus depends upon the largely uncharted defence mechanisms of the human organism. The consequence is that the study of the toxicology of food additives is a study of the long-term administration of large doses to large numbers of experimental animals and the somewhat empirical extrapolation of the results to man, except in those few instances where accidental overdoses have occurred. In this section only those compounds where there is information concerning human exposure, and thus potential adverse effects, will be considered.

Nitrites. Nitrites are permitted in cured meats and some cheese up to a maximum of 200 ppb. Dietary nitrates may be converted into nitrite by bacterial action in the gut, an effect which may be increased by dietary fibre or other carbon-contributing compounds.[89] Nitrates are present in water and a variety of vegetables.[90] The nitrite produced can result in increased levels of methaemaglobinaemia.[89] There has been much recent speculation on the possible interaction of nitrite with dietary amines to produce *N*-nitrosamines. There is evidence that conditions in the human stomach would favour such a transformation[91] and several studies have produced circumstantial evidence that there may be a correlation between dietary nitrate and gastric cancer.[92–94] Notwithstanding that nitrosamines are primarily hepatic carcinogens and that hepatocarcinoma is preceded by hepatotoxicity, none of these studies included evidence of hepatic dysfunction. This is an omission which should be rectified as soon as possible.

Pesticides. Considerable concern has been expressed at the residues of many chlorinated chemicals identified in human tissues following the widespread use of chlorinated hydrocarbon pesticides. This concern is related to the prolonged existence of such compounds, rather than any demonstrable toxicity.[95,96]

Packaging chemicals. There is no evidence that the chemicals which may migrate from packaging materials into packaged foodstuffs have caused hepatotoxicity through such use.

Conclusion

It is clear that much of the anxiety which has been expressed concerning the hazards to human health which might arise from the presence of food additives in the diet results from a misconception of the amounts used, their qualitative relationships to other food components and a mistaken belief that because they are present, they are poisonous. This mistaken belief arises from the attempts to regulate the use of these substances by the empirical reduction, by a factor of a hundred or a thousand, of those quantities known to have produced any biological effect in any experimental system except man himself. It must now be obvious that such a practice cannot guarantee safety and might actually increase danger by restricting the number of materials which can be used in food processing, thus increasing the relative consumption of those that remain.

There appears to be, therefore, a strong case that the safety of compounds, which are deemed of value in maintaining the integrity and quality of man's food supplies, should be judged on the basis of mechanisms of toxicity which pertain to man, and that our regulatory system should be adjusted, increasingly, to that objective. The one exception to this concept might be that group of compounds which resist biodegradation and thus tend to accumulate in the biosphere. The synthesis of such materials should be discontinued as soon as practicable.

Practicability has to be judged with some intellectual subtlety. Thus the horrifying impact on world health of a premature discontinuation of use of chlorinated hydrocarbon pesticides reflects no credit on any of the regulatory machines involved.

Renal Toxicity

Natural Constituents

Oxalic Acid. Certain plant foods may contain as much as 1 per cent oxalate and there have been reports of renal damage due to the deposition of oxalate crystals in the kidney and bladder.[97,98] Although there are probably several reports of such effects in man, there is doubt as to whether the consumption of dietary oxalates is the sole factor.[99] Oxalic acid represents an end point of ascorbate degradation and there is concern that the current fad for swallowing large quantities of vitamin C may predispose to renal and bladder problems. Microcrystallisation increases after ascorbate by mouth[100] and oxalate stones occurred in rats treated with vitamin C.[101] A similar effect is seen with uric acid excretion due to the effect of ascorbate on renal tubular function.[102] This has not, as yet, been associated with urinary stones.

Natural Contaminants

Mycotoxins

Ochratoxin A. As methods for the detection of ochratoxin A became available, it was recognised to occur widely on wheat and barley and was particularly prevalent in Denmark.[103] It is thought to be the cause of porcine nephrosis and experimentally has been shown to induce tubular necrosis.[83] There is no evidence of a definite association with human disease, but it is strongly suspected that Balkan nephropathy may be, in part, due to this mycotoxin.[104–107]

Citrinin. This occurs widely on wheat and barley, often in the company of ochratoxin, as a result of infestation by *Penicillium viridicatum*. It causes, among other effects, proximal tubular necrosis and interstitial fibrosis with dilatation of collecting tubules in a number of species at oral doses ranging up to about 70 mg/kg bodyweight.[108] These findings have been confirmed and extended in a series of studies in dogs.[109]

Aflatoxins. In keeping with their general toxic capabilities, aflatoxins induce proximal tubular necrosis in rodents.[109]

Other Contaminants

Metals

Cadmium. It has been estimated that the human consumption of cadmium from the diet ranges from 50 μg up to 300 μg/day.[111–113] The variability probably relates to the proportion of fish in the diet.[112] There is a significant uptake of cadmium by the kidney[114] and this may persist when ingestion ceases.[115] This is thought to be due to cadmium being held in bound form to metallothionein.[116] Cadmium causes necrosis of the renal cortex beginning in the proximal tubules and extending to the distal tubules.[117] Dietary concentrations of about 200 ppm appear necessary to produce renal lesions in the absence of other dietary manipulations.[118] In a study of Japanese farmers, there was evidence of proteinuria indicative of tubular dysfunction in 14 per cent of the test group which had faecal cadmium levels of 146 μg compared with unexposed control levels of 41 μg.[119]

Given that dietary consumption for most people is less than 10 μg/kg/day, there is little likelihood of an adverse effect. This appears to be supported by studies of the population of Shipton, UK, where soil content of cadmium is high, ranging up to about 1000 ppm compared to control levels of below 10 ppm. Several of the inhabitants showed high tissue levels of cadmium and some showed hypertension and renal tubular damage. An analysis of morbidity and mortality later showed that there was an increased incidence of renal, hypertensive and cardiovascular diseases, but this was significant only for cardiovascular disease. The mortality rates were well below the national average.[120,121]

Lead. Chronic absorption of lead gives rise to nephrosclerosis in man akin to the appearance associated with chronic glomerulonephritis.[122] This is in contrast to the rat where the lesion is primarily tubular and tumours develop.[123]

Mercury. Mercuric chloride is well documented as a proximal tubular poison in dosage exceeding 10 mg/kg by injection.[124] Human risk of renal disease due to mercury relates to organic mercurials except in cases of accidental poisoning.[125] Dietary poisoning is dominated by the neurotoxic rather than nephrotoxic effects.

References

1. W. Laurie, *Med. J. Aust., 1* (1971), 1224–6.
2. G.E. Burch and T.D. Giles, *Am. J. Med., 50,* (1971), 141–5.
3. L. Gould, *Am. Heart J., 84,* (1972), 13.
4. J. Hognestard and P. Teisberg, *Acta. Path. miciobiol Scand., 81A,* (1973), 315–22.
5. L.A. Schenk and J. Cohen, *Pathologia Microbiol. (Basel), 35,* (1970), 96–104.
6. G.S. Wiberg, *Can. J. Biochem. 46,* (1968), 549–54.
7. H.A. Heggtviet, H.C. Grice and G.S. Wiberg, *Pathologia Microbiol. 35,* (1970), 110–13.
8. G.J. Brissom, *Lipids in Human Nutrition – An appraisal of some dietary concepts,* MTP Press Ltd. 1982.
9. T. Gordon, W.P. Castelli, M.C. Hjortland, W.B. Kannel and T. Drawber, *Am. J. Med. 62,* (1977), 707–14.
10. E.D. Freis in (eds) Y. Yaman, W. Lovenberg, E.D. Freis, *Prophylactic Approach to Hypertensive Disease*, Rowen, New York.
11. K.D. Knudson, L.K. Dahl, K. Thompson, J. Iwai, M. Heine and G. Leitl, *J. Exp. Med., 132,* (1970), 976–1000.
12. B. Rocquelin, B. Martin and R. Cluzan, *Proc. Ste-Adèle,* (1970), 405.
13. E. Jaillard, *Can. J. Animal Sci., 53,* (1973), 53.
14. R. Del Carmine, G.L. Gatti, H. Michalek and L. Boniforti, *Proc. Envr. Soc. Toxicol., 16,* (1975), 209.
15. W.H. Crosby, *Blood, II,* (1956), 91.
16. G. Sansone, G. Segui, *Boll. Soc. Ital Biol. Sper., 33,* (1958), 327.
17. N. Ortzonsek, *Int. Arch. Gewerbepathol., 24,* (1967), 66–73.
18. P.D. Berk, D.P. Tschudy, L.A. Shepley, J.G. Waggoner, and N.I. Berlin, *Am. J. Med., 48,* (1970), 137–44.
19. *Fed. Reg., 42,* (1977), 63076.
20. P.S. Sarma, and G. Padmanaban, in I.E. Liener (ed.), *Toxic Constituents of Plant Foodstuffs*, 267–91, (1969), Academic Press.
21. E.A. Bell and J.P. O'Donovan, *Phytochemistry, 5,* (1966), 1211–19.
22. M. Spatz, *Ann. N.Y. Acad. Sci, 163,* (1969), 848–59.
23. M.G. Yang, and O. Michelson, in I.E. Liener (ed.), *Cycads in Toxic Constituents of Plant Foodstuffs*, 159–67, 1969, Academic Press.
24. L.F. James and K.R. Van Kampen, *Am. J. Vet. Res., 37,* (1976), 845.
25. E.J. Schantz, in I.E. Leiner (ed.), *Toxic Constituents of Animal Foodstuffs*, p.111–30, 1974 Academic Press.

26. B.J. Wilson, T.C. Campbell, A.W. Hayes and R.T. Hankin, *Appl. Microbiol., 16,* (1968), 819.
27. B.J. Wilson and C.H. Wilson, *Science N.Y., 144,* (1964), 177–8.
28. B.J. Wilson in J.N. Hathcock (ed.), *Nutritional Toxicology, 1,* (1982), 239.
29. N. Sakabe, T. Goto and Y. Huata, *Tetrah. Lett., 27,* (1964), 1825–30.
30. R.J. Cole, J.W. Kirksey and G. Morgan-Jones, *Toxicol and Appl. Pharmacol., 31*, (1975), 465–8.
31. K. Kanota in M. Herzeberg (ed.), *Toxic Micro-organisms*, 129–32 (1970).
32. S.J. Van Rensburg and B. Altenkirk in I.F.H. Purchase (ed.), *Mycotoxins* 1974, 69–96, Elsevier.
33. EEC Report to Council COM (81), 88.
34. *Lead and Health*, HMSO, London, 1980.
35. L. Karhausen, *Proc. Int. Symp. Environ. Health aspects. Lead.* C.E.C., 427–40, Luxembourg.
36. K.R. Mahaffey, in J.O. Nriagu (ed.), *The Biogeochemistry of Lead in the Environment*, 1–36, 1978, Elsevier.
37. H.J. Catton, H.J.G. Harrison, P.H. Fullerton, G. Kazantzis, *Br Med. J., ii,* (1970), 80.
38. P.R. Betts, R. Astley and D.N. Raine, *Br. Med. J., i*, (1973), 402.
39. P. Grandjean, E. Arnvig and J. Beckman, *Scand. J. Work Envir. Hlth., 4,* (1978), 295–303.
40. J.A. Valciukas, R. Lilis, J. Eislinger, W.E. Blumberg, A. Fishbein and I.J. Selikoff, *Int. Archs. occup. envr. Hlth., 41*, (1978), 217.
41. H. Haenninen, S. Hernberg, P. Mantere, R. Vesanto, M. Jalkenon, *J. Occup. Med., 20*, (1978), 683–9.
42. P.J. Bushnell, and A.E. Bowman, *Neurobehav. Toxicol., 1,* (1979), 207.
43. D. Cory-Slechta and T. Thompson, *Toxicol. Appl. Pharmacol., 47*, (1979), 151–9.
44. J. Hastings, *Neurobehav. Toxicol., 1*, (1979), 227.
45. B. Venugopal, T.D. Luckey, *Metal Toxicity in Mammals*, (1978), Plenum, N.Y.
46. T.K. Miettinen, *Proc XVII Int. Congr. Occup. Hlth,* (1972).
47. T. Rahola, T. Hanula, A. Korolamen, J.K. Mettinen, *Ann. Clin. Res., 5,* (1973), 214–19.
48. S. Jensen and A. Jernolöv, *Nature, 223*, (1969), 753–4.
49. T.W. Clarkson, L. Amin-Zaki and S.K. Al-Tikriti, *Fed. Proc. Fed. Am. Soc. Exp. Biol, 35,* (1976), 2395–9.
50. A. Hinman in R. Hartnung and B. Dinman (eds.), *Environmental Mercury Contamination*, (1972), Am. Arber Sci. Publishers.
51. L. Friberg and J. Vostal (1972), *Mercury in the Environment*, CRC Press.
52. Myakawa and Deshimaru, *Acta. Neuropath, 14,* (1969), 126.
53. J.B. Cavanagh, F.C.K. Chen, *Acta. Neuropath,* (Berlin), *19*, (1971), 208–15.
54. G.P. Ellis and G.B. West (eds.), *Progress in Medicinal Chemistry*, (1975), *11* North Holland Publishing Co., Amsterdam.
55. W.E.C. Wacker, A.F. Parisi, *New Eng. J. Med., 278*, (1968), 658.
56. S. Tanaka, J. Lieben, *Arch. Envr. Hlth., 19,* (1969), 674.
57. P.L. Morselli, S. Garattini, *Nature* (Lon.), *227,* (1970), 611–12.
58. Lajtha, *J. Neurochem., 3*, (1959), 222.
59. J.N. Olney and O-L Ho, *Nature (London), 227*, (1970), 609–11.
60. N.J. Adano and A. Ratner, *Science N.Y., 169*, (1970), 673–4.
61. L.K. Salzmar, *Med. J. Aust., 2*, (1976), 248.
62. P.S. Cook and J.H. Woodhill, *Med. J. Aust., 2,* (1976), 85.
63. T.J. Sobotha, R.E. Brodie and S.L. Spard, *J. Toxicol envr. Hlth., 2,*

(1977), 1211.
64. H.A. Schroeder, *Med. Clin. N. Am., 58*, (1974), 381-96.
65. H.A. Schroeder, *Am. J. Physiol., 207*, (1964), 62-6.
66. K. Tsuchiya, *Fed. Proc., Fed. Am. Soc. Exp. Biol., 35*, (1976), 2412-18.
67. R.J. Garner, *Veterinary Toxicol*, Williams & Wilkins, New York, 1961.
68. G. Selzer and R.G.F. Parker, *Am. J. Pathol., 27*, (1951), 885-907.
69. K.R. Hill, K. Rhodes, J.L. Stafford and R. Aub, *Br. Med. J., i*, (1953), 117-22.
70. R. Schoental, *Bull. World Hlth. Org., 29*, (1963), 823-33.
71. F.W. Cleland and R.V. Southcost, *Med. J. Aust., i*, (1969), 760-3.
72. R.M. Russell, J.L. Boyer, S.A. Baghors and Z. Hinborn, *N. Eng. J. Med., 291*, (1974), 435.
73. R.R. Babb and J.H. Kieraldo, *J. West. Med., 128*, (1978), 244.
74. A.A. Sugarman and C.G. Clark, *J. Am. Med. Assoc., 228*, (1974).
75. L.R. Mosher, *Am. J. Psychiatry, 126*, (1970), 1290.
76. I.F.H. Purchase (ed.), *Mycotoxins*, (1974), Elsevier Scientific Publishing, Amsterdam.
77. B.J. Wilson in J. Hathcock (ed.), *Nutritional Toxicology, 1*, (1982), Academic Press.
78. C. Moreau, *Moulds, Toxins and Food* (John Wiley & Sons Ltd., 1979).
79. W.H. Butler, *Brit. J. Cancer, 18*, (1964), 756-62.
80. S.F. Sullman, S.J. Armstrong, A.J. Zuckerman and K.R. Rees, *Brit. J. Exp. Path., 51*, (1970), 314-16.
81. J.J. Theron, K.J. van der Merwe, N. Liebeuberg, H.J. Joubert and W. Nel, *J. Path Bacteriol., 91*, (1966), 521.
82. I.F.H. Purchase and J.J. Theron, *Food Cosmetic Toxicol, 6*, (1968), 479-83.
83. R.C. Doster, R.O. Sinnhuber and J.H. Wales, *Food Cosmetic Toxicol, 10*, (1972), 85.
84. G.N. Wogan, G.S. Edwards and P.M. Newberne, *Toxicol Appl. Pharmacol., 19*, (1971), 712-20.
85. A.W. Hayes in J.V. Rodick, C.W. Hesse Hine, M.A. Mehlman (eds.), *Mycotoxins in Human and Animal Hlth.*, 507-28, (Pathotox Publications Inc., 1977).
86. R.H. Thornton and J.C. Percival, *Nature* (Lond), *183*, (1959), 63.
87. L.G. Atheston, D. Brewer and A. Taylor in I.F.H. Purchase (ed.), *Pithomyces Chartarium Mycotoxins*, p.29-68, (Elsevier, Amsterdam, 1974).
88. A. Wise, A.K. Mallet and I.R. Rowland, *Xenobiotica, 12*, (1982), 111-18.
89. J.W. White, *J. Agric. Food Chem., 23*, (1952), 886-91.
90. N.P. Sen, D.L. Smith, H. Schwinghamer, *Food Cosmetic Toxicol., 7*, (1969), 301-7.
91. R. Tannenbaum, D. Moran, W. Rand, C. Cuello and P. Correa, *J. Natl. Cancer Inst., 62*, (1979), 9-12.
92. R. Zaldivar and W.H. Wetterstrand, *Experientia, 31*, (1975), 1354-5.
93. R. Sato, *Bull Inst. Public Hlth, 8*, (1959), 187-98.
94. D.E. Coffin and W.P. McKinley in H.D. Graham (ed.), *Safety of Foods*, p. 498-514, (AVI Publishing Co. Inc., 1980).
95. I.C. Munro and S.M. Charbonneau in H.R. Roberts (ed.), *Food Safety*, p. 141-80, (John Wiley & Sons N.Y., 1981).
96. P.M. Zarembski and A. Hodgkinson, *Brit. J. Nutr., 16*, (1962), 627-34.
97. R.F. Crampton and F.A. Charlesworth, *Brit. Med. Bull., 31*, (1975), 209-13.
98. D.W. Fassett, 'Toxicants occurring naturally in food', *Nat. Acad. Sciences*, (1973), 346-62.
99. von Sengbusch and Trimmerman, *Urol. int., 5*, (1957), 218.

100. M.O. Keith, B.G. Shah, O. Pelletier, *Fed. proc. Fed. Am. Soc. Exp. Biol.*, *33*, (1974), 665.
101. A. Pena, J.L. del Arbol and J.A. Garcia-Tones, *Nut. Abst. Rev.*, *34*, (1964), 195.
102. P. Krogh in *Mycotoxins in Human and Animal Hlth.*, p. 489–98, (Pathotox Publications, 1977).
103. M. Dimitrov in A. Puchler (ed.), *Endemic nephritis in Bulgaria*, (1960), 201–7 Sofia.
104. J.M. Barnes in G.E.W. Wolstenholme and J. Knight (eds.), *Balkan Nephropathy*, Ciba Geigy Study Group., *30*, 1967, Churchill, Lond.
105. P.K.C. Austwick, *Proc. Roy. Soc. Med.*, *68* (1975), 219–21.
106. P.W. Hall and G.J. Danmir, *Nephron*, *22*, (1978), 281.
107. W.W. Carlton, J. Tuite in *Mycotoxins in Human and Animal Hlth.*, (1977), 532, Pathotox Publications.
108. D.N. Kitchen, W.W. Carlton, J. Tuite and E.J. Hinsman, *Vet. Pathol.*, *14*, (1977), 154, 261 and 392.
109. W.H. Butler and W. Lijinsky, *J. Pathol.*, *102*, (1970), 209–12.
110. Kropf and Goldmacher, *Arch. Hyg. Bakt.*, *152*, (1968), 218.
111. Rantu and Sporn, *Nahrung*, *14*, 1970, 25.
112. I.H. Tipton, P.L. Stewart and J. Dickson, *Hlth. Phys.*, *16*, (1969), 455.
113. S.A. Gunn, T.C. Gould and W.A.D. Anderson, *J. Path Bact.*, *96*, (1968), 89.
114. H. Piscator and B. Axelsson, *Arch. Envr. Hlth.*, *21*, (1970), 604.
115. Wisniewska-Knypt Jn. and J. Jablenska, *Bull. Acad. Polon. Sci.*, *18*, (1970), 321.
116. B. Axelsson, S.E. Dahlgren and H. Piscator, *Arch. Envr. Hlth.*, *17*, (1968), 24–8.
117. T. Muto and H. Omori, *J. Nutr. Sci. Vitaminol*, *23*, (1977), 349.
118. S. Kojima, Y. Haga, T. Kurihara, T. Yumawaki and T. Kjellstrom, *Envr. Res.*, *14*, (1977), 436.
119. H. Carruthers and B. Smith, *Lancet*, *1*, (1979), 845–7.
120. H. Inskip, V. Beral and M. McDawell, *Lancet*, *1*, (1982), 896.
121. J.A. Inglis, D.A. Henderson and B.T. Emmerson, *J. Path.*, *124*, (1978) 65–76.
122. R.A. Goyer, D.L. Leonard, J.E. Moore, B. Rhyne and H.R. Kugman, *Arch. Envr. Hlth.*, *20*, (1970), 705–11.
123. C.W. Janssen, *Scand. J. Urol Nephrol.*, *4*, (1970), 157.
124. P.L. Bidstrup, *Toxicity of Mercury and its Compounds*, (Elsevier, Amsterdam, 1974).
125. L Gould, *Am. Heart. J.*, *79* (1970), 422.
126. P. Grozdjak, *Biochemical Pharmacol.*, *22*, (1973), 1807.
127. J. Parkinger, *J. Clin. Invest.*, *52*, (1973), 2690.

2 INTOLERANCE AND ALLERGY TO FOODS AND FOOD ADDITIVES: ITS RELEVANCE TO TOXICOLOGY

W.E. Parish

There is no clear distinction between food allergy and immunological toxic reactions to food substances. Allergy is one of the possible adverse reactions to natural substances or additives in the diet. The immunology of toxicology, or immunotoxicology, is the study of allergy, or of reduced immunocompetence, resulting from exposure to chemicals, drugs and natural substances modified by man, and the studies include the development of experimental models to predict the potential immunological effects of such substances on man. Though there has been much recent interest by toxicologists in immunological adverse reactions to man-made or natural substances, the studies do not differ from decades of investigations of allergies.

The subject is too complex to present in one chapter more than a general account of immunological responses to antigen, of the different types of allergies and food allergic disorders, and of toxicological predictive tests.

Allergy, Tolerance and Pseudo-Allergy (Idiosyncrasy)

Everyone is continually exposed in his environment to antigenic substances, which may enter the body by several routes, including the gastrointestinal tract. Apart from environmental antigens, including pathogenic organisms and parasites, the tissues and body fluids may be altered by combination with absorbed antigens, or by autologous degradation or aggregation, becoming auto-antigenic to the individual.

Antigens stimulate sensitisation/activation of lymphocytes with or without formation of antibody. The resulting cells or antibody react with chemical groups (determinants) specific for the antigen. Reaction of the antigen with the cells and antibodies, *in vivo* or *in vitro*, induces observable changes. Some substances, among which are chemicals of low molecular weight, are unable to stimulate formation of antibodies unless they first combine with protein, e.g. the host tissues.

Such substances are known as haptens. The hapten-protein complexes induce sensitisation of lymphocytes or formation of antibody specific for the hapten. Haptens are of particular importance in allergy/toxicology as they are frequently potent sensitising agents inducing contact dermatitis.

Immunological stimulation and response depends upon the nature of the antigen, and route, amount, duration and frequency of exposure. The result may be a negligible or weak response, being a state of awareness of the antigen but manifested clinically as a state of tolerance, or result in hypersensitivity, which may be beneficial when limiting infection by micro-organisms, or be harmful in clinical allergic reactions.[1]

The term allergy designates a specific altered reactivity of the tissues to antigens compared with the response on first exposure to the same substances. Re-exposure to an antigen is not necessary if the antigen persists in the blood or tissues sufficiently long to react with antibody formed against it. In this instance allergic change follows the first but prolonged exposure to the antigen. Allergic reactions are mediated by activated lymphocytes or by antibody specific for the antigen, or for another antigen with similar determinants (antigenic chemical groups). This distinguishes allergies from pseudo-allergies or idiosyncrasies not mediated by antigen-specific immune responses, discussed later. It is useful to restrict the term allergy or altered reactivity to mean obvious, i.e. clinically observable, allergic signs. Antigens inducing allergy are commonly known as allergens.

Tolerance is the opposite of allergy. In strict immunological usage, tolerance designates complete non-reactivity to an antigen. The phenomenon is best illustrated experimentally by treating fetuses of animals *in utero* with tissues or chemicals during the period that the fetus becomes tolerant to its own normal tissues. During this period the clones of lymphoid cells capable of responding to 'self' antigens are inactivated or destroyed. When the treated young animals grow to maturity, they are unable to respond immunologically to the antigens with which they were treated early in life. In more general usage, however, tolerance also refers to the state in which an individual has formed specifically sensitised lymphocytes and small amounts of antibody, but who shows no outward changes when exposed to the relevant antigen. Thus normal persons are considered tolerant to the many foods that they eat. Discriminating tests show few sensitised lymphocytes and low titres of antibody to nearly every food eaten. Usually small fluctuations occur in lymphocyte and antibody responses with changes in the diet.

However, whether small or large amounts of any particular food are eaten, or if applied to the skin, in these 'tolerant' persons there are no obvious signs of an immunological response. These persons are considered tolerant even though they have been immunologically 'primed' or 'activated' to the food antigens.

In both normal and allergic persons, small amounts of food protein are absorbed in the circulation, and for a short time afterwards complexes of the protein with IgA, IgG, and occasionally IgM antibody may be found in the blood. In normal persons the complexes are harmless and are quickly removed by macrophages. However, immune complexes containing IgG or IgM become harmful if for some reason they are not quickly removed from the blood, and become deposited in susceptible tissues, e.g. walls of blood vessels, the kidney or joints, where they induce the inflammatory changes known as immune complex-mediated disorders.

Allergic or otherwise immunotoxicological responses to foods are a manifestation of the breakdown of tolerance.

Pseudo-allergy mimics allergy in clinical manifestations but is not due to antigen-specific immunological reactions, and therefore not initiated by antigen combining with specific antibody, or with specifically sensitised lymphocytes.

The clinical signs of pseudo-allergies are induced by the same endogenous pharmacological substances, e.g. histamine or prostaglandins, that are released or synthesised following an antibody or lymphocyte-mediated immune response, but in pseudo-allergy the inducing substance, e.g. drug or food metabolite, has a direct action to release the pharmacological substances without participation by immunological mediators.

The close similarities in clinical signs make difficult the distinction between allergy and pseudo-allergy, especially when it may be difficult to identify allergens which are complexes of an antigen with a cell, protein or glycoprotein, or which are degradation products of exogenous substances.

The differentiation between allergy and pseudo-allergy is important for the toxicologist, particularly in studies of food intolerance, but the distinction is difficult to make, and at present cannot be examined by animal tests. The differences between the two phenomena are that allergic susceptibility usually follows repeated or prolonged contact with antigen (though immune complex-mediated 'serum sickness' is an exception), is mediated by antigen-specific antibody or sensitised lymphocytes, is confined to the specific antigen or to closely related chemicals with the same antigenic determinants (epitopes), and susceptibility is transferable to normal individuals of the same species by antibody or lymphocytes. Pseudo-allergy may occur on first exposure to the eliciting substance (although fluctuations in susceptibility may occur, and the first adverse response may follow several harmless exposures), intolerance to other drugs or foods unrelated chemically may occur in the same patient, and the intolerance cannot be transferred to normal individuals.

Many substances induce pseudo-allergies by release of histamine from mast cells and basophils of man, *in vivo*, among which are many basic compounds, drugs including several antibiotics, venoms, bacterial toxins and metals.[2] A very full account of pseudo-allergic anaphylactoid reactions is available.[3]

Some pseudo-allergies are referred to as idiosyncrasies. If this term is retained in current usage it is best confined to those pseudo-allergies affecting only a proportion of the population who have a selective inability to process a food substance normally. Examples are lactose intolerance due to a deficiency of the enzyme lactose, or fat intolerance due to bile salt deficiency.

Pseudo-allergic food intolerance is discussed in more detail later.

Lymphocyte Stimulation and Antibody Formation

Lymphocytes and Antigenic Stimulation

Lymphocytes mediate and monitor immunological competence and responses. They are derived from stem cells of the bone marrow and released into the blood. A large proportion of these enter the thymus where they are modified by thymic hormones, and divide rapidly, before being released into the blood and distributed throughout the immunologically active tissues (Figure 2.1). These cells form about 64 per cent of the circulating lymphocytes and are termed T lymphocytes or cells (T indicating thymus modification). Other lymphocytes are not modified by the thymus, and it is not known if they have any special site of maturation after leaving the bone marrow. These are known as B cells (B designates Bursa of Fabricius of the chick where they were first identified). B cells form approximately 17 per cent of the circulating lymphocytes and differ greatly in properties and function from T cells. B cells synthesise immunoglobulin which is bound to the cell surface; on appropriate activation B cells become antibody-secreting plasma cells, each B cell secreting one immunoglobulin at any one time. In normal persons there tends to be a consistent proportion of

Figure 2.1: Diagrammatic Representation of the Origin and Differentiation of Lymphyocytes Resulting in Delayed Hypersensitivity (DH).

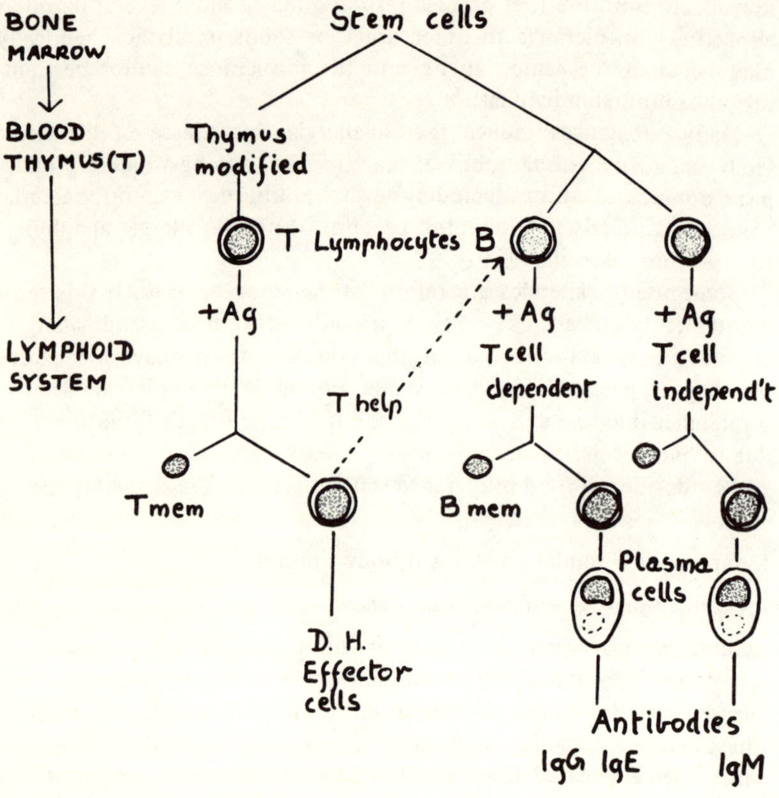

T effector cells, T helper cells co-operating with B cells, T and B memory cells, and plasma cells synthesising immunoglobulins. The activities of T suppressor cells are omitted. Source: From Parish.[1]

cells secreting each immunoglobulin. However, the number of circulating B cells synthesising a particular immunoglobulin does not reflect the amount of that immunoglobulin in the blood. Variations from the normal pattern of subsets or subpopulations of B cells occur regularly in some disorders.

Some features used to identify human and rodent T and B cells are presented in Table 2.4.

Subsets of T cells also occur and are differentiated by their membrane receptors or intrinsic antigens. The significance of the relation between subset features and biological activity is still to be determined. Helper T cells which activate B cells to form antibody are found in the T subset with receptors for Fc of IgM designated Tμ cells. Suppressor T cells that regulate immunological stimulation occur in the subset with receptors for Fc or IgG designated Tγ. However not all Tμ cells are helper, or Tγ suppressor, cells.

This classification by membrane antigens and receptors, and by biological activity may be summarised as follows:

Lymphocytes: general classification

T effector cells	*functional activity*
T helper cells	K (killer) cells (antibody dependent)
T suppressor cells	NK (natural killer) cells (antibody independent)
T memory cells	
B cells	
B memory cells	
N (null) cells	

Null cells are those without the characteristic criteria of T or B cells, though some appear to be immature T cells. K and NK cells are functional activities and tend to have the properties of T cells.

An antigen entering the tissues is ingested and modified by macrophages, or in the skin by the Langerhans cells. The macrophages may migrate to a draining lymph node, or react with lymphocytes in the tissue into which the antigen penetrated. There is direct contact between the macrophage and receptive lymphoid cells, and there follows a complex series of events illustrated diagrammatically in Figure 2.1. T cells develop special functions, becoming *effector* cells (Te) mediating delayed hypersensitivity reactions, *helper* cells, which activate B cells to form antibody (Th), *suppressor* cells which regulate the immunological response (Ts) and *memory* cells (Tm) which are long-lived and respond immediately on renewed contact with the specific antigen without the need for macrophage intervention. Thus after every exposure to an antigen to which the individual is not tolerant, there is a greater number of specific memory cells, and an increased susceptibility to hyper-responsiveness and allergy.

The majority of antigens inducing antibody formation are T cell dependent. That is, the B cell stimulated by the macrophage-modified antigen is further activated by a T helper cell before it divides to form B cell precursors of plasma cells and B memory cells. The resulting plasma cells form one of each of the classes of immunoglobulin, but mainly IgG (Figure 2.1). Some antigens do not require helper cell intervention to complete the maturation of B cells to plasma cells; these antigens are usually large molecules, often with replicating chemical groupings, as in long-chain polysaccharides, and which usually enter the tissues in relatively large amounts. After the T cell-independent stimulation, the plasma cells tend to secrete IgM, but some switch later to synthesis of IgG.

T effector cells mediate the changes of delayed hypersensitivity reactions, and on stimulation by the specific antigen synthesise substances with a wide range of biological activities, as discussed later.

Suppressor cells, and some humoral substances, control the activities of effector and antibody-forming cells. They regulate the intensity of the response to each immunological stimulus and prevent most of them developing into an allergic reaction. Exposure to many environmental antigens, including foods, results in immunological stimulation, but the activities of T effector cells, and of B cell precursors of antibody-forming plasma cells, is controlled by suppressor cells. Thus immunological stimulation and memory for these antigens is manifested as a weak immunological response without allergy. This state of tolerance is believed to be regulated by suppressor cells.[4,5]

Stimulation by inoculation to induce protective immunity against infection results in a greater T effector and T helper response than T suppressor activity, and consequently confers variable degrees of protective immunity. Inherent or acquired deficiencies in the numbers or activity of suppressor cells predisposes the individual to several disorders, including auto-immune diseases,[6] inflammatory bowel disease,[7,8] and atopic disorders, e.g. asthma.[9]

In man, suppressor cells form part of the heterogeneous population in the subset of T lymphocytes, and probably induce inactivation by direct contact with effector or helper cells in any one of the following stages of an immune response. (i) Inhibit activation of Te cells; (ii) inhibit activation of Th cells; (iii) inhibit differentiation of B cells after they have been activated by Th cells; (iv) suppression of antibody formation by plasma cells. Most suppressor cell activity is antigen-specific, but there are lymphoid cells with antigen-nonspecific regulating activity.

Antibodies (Immunoglobulins)

Immunoglobulins (designated Ig) are proteins able to combine with specific determinants of antigens, the results of which depend upon the biological properties of the particular class of immunoglobulin. In man, five classes of immunoglobulin have been identified according to their chemical and physical properties, and their own antigenic determinants on the Fc portion. Some of their properties are summarised in Table 2.1. A more detailed description is available elsewhere.[1,10]

Table 2.1: Human Serum Immunoglobulins

Ig Class	Subclasses	H chain antigen	Concentration (mg/ml)	Complement fixation
IgG	4	γ	12.5	+
IgA (serum)	2	α	2.0	C3
IgM	2	μ	0.9	+ +
IgD	2	δ	0.03	–
IgE (reagin)	1 (D1 & D2)	ϵ	0.0005	–

IgE: No definite subclass but two 'domain' variants known.
Complement fixation: Aggregated IgA may act through the alternative C3 component sequence.
Concentration: There is considerable variation in normal persons. The amounts are usually expressed as units as determined by international standards.

IgG Antibodies. These are responsible for the greater part of antibody activity. The four subclasses, each with variable biological activity, result in a wide range of reactivity. They form complexes with antigen, and subclasses 1 and 3 are potent activators of complement. IgG_1 not infrequently occurs in immune complexes associated with allergic disorders.

IgG antibodies, being present in large amounts, and able to activate complement (with the exception of IgG_4) are important in limiting infection, inactivating toxins, and removing foreign antigenic substances. Macrophages with receptors for IgG bind the antibody to their surface and potentiate their activity to ingest the relevant antigens. Similarly lymphocytes with receptors for IgG become antibody-dependent cytotoxic cells known as killer or K cells.

Some human IgG has been found that sensitises human tissue, anaphylactically, for a few hours. This has been designated IgG S-TS (short term sensitising) and probably participates in immediate-type responses to foods; its properties are described in the later section on anaphylaxis. In experimental animals, guinea pigs, rats and mice, anaphylactic IgG antibodies are formed readily.

IgA Antibodies. In serum these are probably not important in limiting infection or removing foreign antigens. They are most important in the protection of mucosal surfaces as described later. They do not fix and activate complement, though aggregated IgA appears to activate complement through the alternative pathway starting with the C_3 component. Complexes of IgA with food antigens have been detected in serum, and the cutaneous deposits of IgA in dermatitis herpetiformis probably represent antibody complexes with wheat gluten antigens. (Secretory or mucosal IgA is discussed later.)

IgM Antibodies. These are large molecules made up of five basic antibody units, each resembling IgG linked by Junction or J chains. They combine strongly with antigen, and most IgM activates complement.

IgM occurs in the blood soon after first antigenic stimulation, and during infections it is a quickly formed, potent, protective antibody, but as the antigenic stimulation continues, the amount of specific IgM antibody decreases and is replaced by IgG. IgM, like IgG, forms tissue-damaging immune complexes with antigen and complement.

Most rheumatoid factors are IgM anti-IgG and form immune complexes with autologous degraded IgG, or with IgG in a changed form bound to antigen. Rheumatoid factors are associated with chronic connective tissue disorders.

IgD Antibody. The function of IgD antibody is not known, though it often occurs bound to the lymphocyte membrane and may be functioning as an antigen-recognition molecule for the cell. It does not fix complement or sensitise tissues for anaphylaxis.

IgE Antibodies (Reagins).[1,10-12] These are the classical anaphylactic sensitising antibodies of man, mediating allergic rhinitis, asthma, immediate-type gastro-intestinal disturbances and other atopic disorders. IgE occurs in very small amounts in normal serum, 10 to 70 μg IgE/100 ml serum, but occurs in larger amounts in anaphylactically-sensitised persons, particularly those with atopic eczema together with another allergic manifestation, e.g. asthma. The amounts of IgE in serum are too small to be detected by conventional *in vitro* techniques, and special methods were developed to detect them. The total amount in serum or body fluids may be assayed by the paper disc radio-immunoassay (PRIST), and assay of IgE specific for antigen by the radio-allergosorbent test (RAST). The total amount is usually expressed in international units, each i.u. being equivalent to about 2.2 ng IgE.

The amounts of total or antigen-specific IgE do not always reflect the degree of allergic sensitivity. It is possible for some persons to have large amounts of total IgE, 800 to 100 i.u., without evidence of allergy. Others may have very small amounts of IgE in the serum and a non-significant RAST 'score' and yet be allergic. Such sensitisation occurs with the small amounts of IgE binding to the mast cells and basophils, but with no excess in the circulation. IgE is also synthesised locally in several mucosal and other tissues, e.g. tonsils, adenoids, bronchial mucosa and in lesser numbers in the gastro-intestinal mucosa. Examples of this variability are shown in the section on Anaphylaxis (Table 2.3).

The cells with receptors for the Fc of IgE are mast cells and basophils, and when antigen links these specific antibody molecules on the cell surface, there is release of the pharmacological mediators of the cell (see section on Anaphylactic sensitivity). The tissue-sensitising properties of IgE are destroyed by heating at 56°C for 30 minutes to 4 hours; a feature sometimes used to detect its presence in serum containing several other classes of antibody.

Recently, receptors for IgE have also been found on macrophages, and in experiments on rats stimulation of the macrophage-bound IgE resulted in increased secretion of lysosomal enzymes and generation of antibacterial superoxide.[13]

Mucosal (Gastro-intestinal) Immune Systems

Mucosal tissue, possibly because it is more exposed than the skin to penetration by antigen, has special immuno-active cells and products. The essential features are IgA and IgM antibodies adapted to persist and function on exposed mucous surfaces, and lymphoid cells which have at least three functions, (1) continual formation of antibody, especially IgA, to antigens penetrating the mucosa, (2) dissemination of these cells to other mucosal tissue throughout the body, (3) immunoregulatory or suppressor activity that in normal persons prevents the large amount of harmless antigenic material which penetrates the intestine from inducing excessive antibody formation, lymphocyte stimulation and tissue damaging immune responses.

These activities are illustrated diagrammatically (Figure 2.2) omitting

Figure 2.2: Diagrammatic Representation of Local Antibody Synthesis and Lymphocyte Activities in the Intestine.

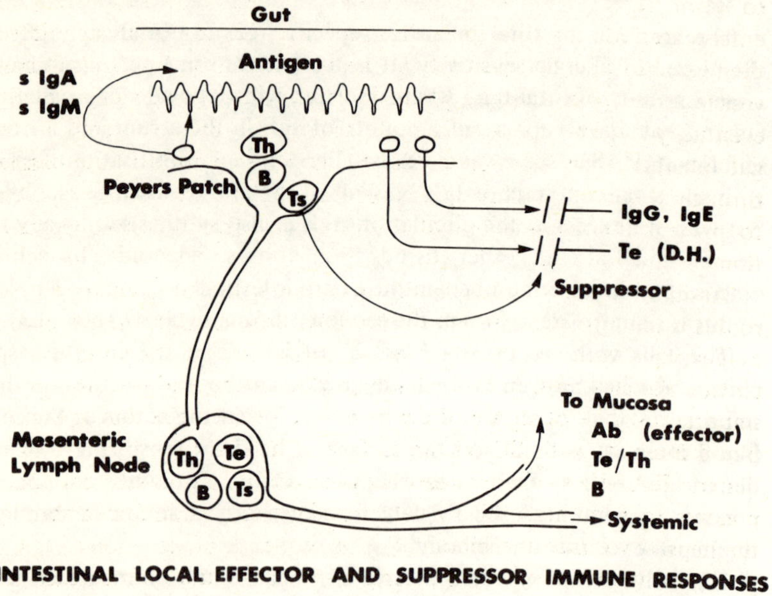

Secretory IgA and sIgM are formed beneath the epithelium and pass through the cells to the lumen. Peyer's patches give rise to Ts and Th cells and contain B precursor cells. The suppressor and helper T cells control formation of IgA and other Ig in the lamina propria. Mesenteric lymph nodes give rise to further helper cells and numerous IgA-secretory B lymphoblasts. These, together with cells from the systemic circulation may pass back to the bowel mucosa or enter the blood and settle in other mucosae.

other contributory features, the intra-epithelial leucocytes or lymphocytes, the mucosal mast cells which are thymus dependent, and antigen-nonspecific cell and humoral substances with immunoregulatory properties.

Secretory Antibodies

Mucosal tissues synthesise and are protected by two classes of immunoglobulin which are modified forms of those found in the blood. The most important is IgA; the other IgM. These are known as secretory immunoglobulins designated sIgA and sIgM. Secretory of IgA is a dimer of two monomer units linked by a J (Junction of joining) chain and incorporating a protein known as the secretory component (SC). The IgA and IgM are formed by plasma cells in the lamina propria, and the SC by epithelial cells. It is believed that SC combines with the globulin to form sIgA or sIgM on the basement membrane supporting the enterocytes, i.e. the epithelial cells lining the villi, and the new molecule then is absorbed by the cell and passes through it in a cytoplasmic vesicle to be passed to the gland lumen together with some free SC, eventually to line the exposed surface of the epithelial cells in the gut lumen.[14] The function of the SC is believed to facilitate passage through the cell, possibly to aid adherence to the epithelial cells, and to protect the Fc portions of the molecules from enzymic degradation.[14,15]

Another source of sIgA in the intestine is the bile. There are two routes leading to accumulation in the bile. In one, sIgA on the enterocyte surface becomes absorbed and recycled through the portal system to the bile and back to the gut lumen. The other, and probably more important source of sIgA in the bile, as seen in the rat, is that IgA in the blood interacts with SC on the surface of hepatocytes in the form of dimeric IgA (dIgA). Only dimeric IgA can combine with secretory components to form sIgA. The newly formed sIgA is transferred through the hepatocyte into the bile duct. This sequence of secretion of IgA in plasma cells of the gut wall, passage to lymph and blood to reach the hepatocytes, combination with SC and transfer to the bile, ensures an additional source of sIgA within the gut lumen. In man, the liver does not appear to have such an important role as in the rat, as dIgA from the thoracic duct passes first to the gut where most combines with SC of the enterocytes and passes to the lumen, leaving little to reach the blood.

Secretory IgA reduces adherence of micro-organisms and their colonisation of the mucosa and, with sIgM, prevents infection, and contributes to resistance against helminth parasites. Antigen-specific sIgA also reduces absorption of food antigens (immune exclusion) and thereby has an important function in controlling absorption of substances able to stimulate immune responses possibly resulting in allergy.

Persons with an inherent partial deficiency of IgA tend to absorb more food substance, e.g. milk protein, into the circulation, than persons with normal amounts of sIgA. This absorption results in synthesis of IgG which forms immune complexes with the circulating antigen. It is believed that a transient deficiency of IgA in the first three months of life predisposes an infant later in life to synthesise IgE and develop atopic disorders (e.g. hay fever or asthma).

Gut Associated Lymphoid Tissue

Protection of mucosae by sIgA is dependent upon the continual infiltration of the tissue by cells synthesising IgA.[15–17] These cells are lymphocytes, some of which mature to plasma cells, and are derived from the gut-associated lymphoid tissue (known as GALT).

GALT is composed of the lymphoid tissue of the Peyer's patches, the appendix, and the many lymphoid cells found throughout the depth of the gut wall. The cells have variable biological activities. Lymphocytes in the Peyer's patches appear to be mainly T suppressor and T helper cells, and also precursor B cells, large lymphocytes with IgA determinants. The wall of the intestine contains macrophages, T helper, T effector and B cells, and is a complete immunological organ secreting large amounts of immunoglobulin (Figure 2.2).

The essential features of the lymphocyte circulation or lymphocyte 'traffic' are as follows

(1) Lymphocytes become primed with antigen specificity in the gut wall.
(2) The sensitised lymphocytes including T helper cells and large lymphocytes with α chain determinants which are precursors of IgA-synthesising cells, emerge from the site of sensitisation.
(3) Some spread via the lymphatic vessels within the gut lamina propria. Others enter the blood to return to the gut or to other mucosal tissue, e.g. the lung. There is a hormone-mediated accumulation of these cells in the mammary gland, particularly during lactation, so that large amounts of protective sIgA is available to the infant.

Lymphocytes stimulated in one mucosal tissue, e.g. the gut wall, may migrate to other mucosal tissues, e.g. the lung or genital tract, and secrete antibody to an antigen that has not entered the second organ. If there is no antigenic stimulus in this organ, the cells probably degenerate. If there is an antigenic stimulus the lymphocytes proliferate, secrete more antibody, and some continue the cycle by returning to the gut. Most cells derived from the gut usually return to the gut. Similarly those activated in the lung tend to remain in or return to the lung.

Mesenteric lymph nodes contain many large lymphocytes secreting IgA, B cells secreting other immunoglobulins, Th and Ts cells. The population of Te cells is less than that of the spleen or peripheral lymph nodes, and there are almost no cells capable of spontaneous cytotoxicity. After antigenic stimulation of the mucosa, antibody-forming cells appear first in the lamina propria then in the mesenteric lymph nodes, to be followed by the spleen.[18,19] Blast cells from the mesenteric lymph nodes tend to return to the intestinal wall (Figure 2.2).[20]

An immunoregulatory system controls antigenic stimulation within the intestinal mucosa. Allergens that are very potent sensitising substances when applied to the skin may induce tolerance when ingested. Such tolerance has been shown to occur in guinea pigs, rats and mice. Investigations in which skin-sensitising chemicals were administered orally, and the properties of the lymphocytes examined by transfer to other animals, showed that tolerance was associated with (1) suppressor B cells which inhibited skin reactivity (in rats and mice there are suppressor B as well as T cells); and (2) suppressor T cells which inhibited DNA synthesis and development of blast cells in the lymph nodes.[21]

It would appear that antigen penetrating the intestinal mucosa of the rodents quickly induced sufficient numbers of suppressor B cells to prevent a significant, observable clinical response to the antigen. On contact with skin few suppressor B cells are formed, and allergic sensitisation results. The relevance of this to man is still to be ascertained, but subsets of human T suppressor cells may have similar activity to both the suppressor B and T cells of rodents.

Animals fed protein antigens develop, in the intestinal Peyer's patches, T cells which are antigen-specific in suppressing formation of IgG and IgE antibodies, T effector cell activity in delayed hypersensitivity responses, and suppressing T cell proliferation (probably Te cells). Later, similar T suppressor cells occur in the spleen.[22]

Apart from antigen-specific suppressor cells there are antigen non-specific cells with suppressor activities, as seen in the mucosae of the rabbit and the lamina propria of man. These probably participate in the control of antigenic stimulation, possibly in the first phases of response to a new antigen, but appear much weaker than the antigen-specific systems.

Classification of Allergic Responses

There have been several classifications of allergies, by descriptive names or by the numerical classification of Coombs and Gell.[23] The classification used here (Table 2.2) retains the old nomenclature as being less confusing than type numbering, and is in a different order being a

Table 2.2: Characters of Types of Allergic Reactions and Some Clinical Examples

	Anaphylaxis	Arthus reactions (immune complex mediated)	Cytotoxic reactions	Delayed (cell-mediated immune (cmi) response)
Response mediated by	Tissue passive sensitising antibody	Precipitating antibody	Antibody or cell mediated	Cell mediated Controlled by T lympho-cytes
Skin test	Immediate weal and flare	Arthus reaction with polymorph infiltration, appearing in 2–4 h but may progress for hours or days to necrosis	Immediate weal and flare. Also granulomatous lesion with or without poly-morphs, first appearing in 2–6 h	Delayed tuberculin-type response. Maximum about 48 h
Clinical examples	Allergic rhinitis Asthma Urticaria	Serum sickness (immune complexes) Cutaneous or systemic vasculitis Glomerulonephritis	Eczema Purpura ? Systemic lupus erythematosus	Eczema Contact dermatitis Some granulomatous disorders
	Immediate allergic intestinal dysfunction	Arthritis ? Colitis ? Coeliac disease maladsorption	Auto-allergies e.g. allergic thyroiditis	? Coeliac disease maladsorption ? Crohn's disease and ulcerative colitis

logical progression from anaphylactic and Arthus (immune complex mediated) responses which are antibody mediated, to cytotoxic which is antibody or cell-mediated, to delayed hypersensitivity which is entirely cell mediated. The type numbering system for anaphylaxis is I, Arthus III cytotoxic II and Delayed Sensitivity IV.[23]

The changes occurring in anaphylactic, Arthus and delayed hypersensitivities are described below. The clinical examples have been selected for their toxicological rather than their immunological interest.

Food allergic disorders may be manifested in all susceptible organs, e.g. lung or skin, without clinical involvement of the intestine. In a sensitised person food antigens may penetrate the intestine to pass by the blood to the susceptible organ of that individual, resulting in asthma, or rhinitis, or urticaria, or very mild drop in blood pressure and lassitude, without any evidence of intestinal dysfunction. The normal physiological barriers of the intestine, and mucosal immune exclusion, fail to reduce to an essential minimum, penetration by the antigen which induces an allergic reaction in the most susceptible organ of that individual.

The following descriptions of three types of allergic response refer to their pure form. In any one patient more than one type may co-exist, and it may be difficult to ascertain which is mediating the illness. This complex of reactions is observed in some skin test responses. An immediate (10 min) anaphylactic weal and flare is followed by an Arthus-type induration at 2 hours and by a delayed hypersensitivity response at 48 hours. However, in food allergies, the most frequent skin test responses are anaphylactic (immediate) and delayed hypersensitivity.

Anaphylaxis

Anaphylaxis occurs when mast cells or basophils *in vivo* or *in vitro*, sensitised passively with antibody, are exposed to specific antigen. This used to be termed immediate type hypersensitivity because the signs appeared within minutes of reaction with the antigen. The weal and flare response to a skin prick test in an anaphylactically sensitised person, which occurs within minutes of test and reaches a maximum intensity in 10 to 15 minutes, is a good example of this response.

The term anaphylaxis in clinical usage refers to generalised shock, with or without bronchial and other complications. In immunological usage the term refers to those conditions resulting from antigen reacting with antibody passively sensitising mast cells and basophils, which apart

Figure 2.3: Diagrammatic Representation of the Anaphylactic Response

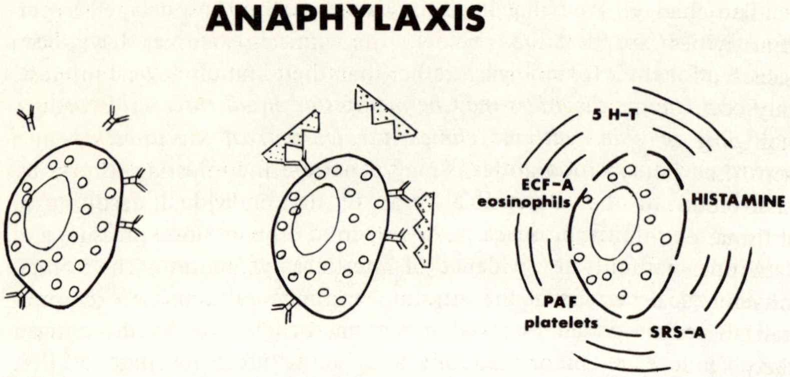

Anaphylactic antibody Fc binds to specific receptors on mast cell or basophil membranes. Antigen reacts with free Fab (antigen binding) sites to bridge or link two antibody molecules. This initiates release or generation of pharmacological mediators of the clinical signs: histamine, 5-hydroxytryptamine in rodents. Slow reacting substance of anaphylaxis (SRS-A), platelet activating factor (PAF) and several others. Source: From Parish.[1]

from anaphylactic shock includes allergic rhinitis (hay fever), asthma, some acute urticaria and angio-oedema, and some acute gastro-intestinal disorders which are among the atopic group of conditions.

The features of the anaphylactic response are illustrated diagrammatically in Figure 2.3.

Mast cells and basophils are the reactive cells. There is a specific reaction between antigen combining with antibody bound to the cell through receptors specific for Fc of IgE. The union of antigen with two antibody molecules draws together the Fc receptors to a critical distance which initiates the release of preformed stored mediators, e.g. histamine, and generation of others, e.g. slow reacting substance (or SRS-A). Several substances of diverse inflammatory activity are released.[24] It is these pharmacological mediators released locally or systemically which induce the clinical signs.

The anaphylactic antibody of man is IgE, whose properties have already been described. The amount of IgE specific antibody in serum may be assayed by RAST, by ELISA (enzyme-linked immunosorbent assay) or similar discriminating assays detecting very small amounts of antibody. There are also several biological tests for IgE antibody, the first described being the Prausnitz-Küstner transfer test in which about 0.1 ml of the patient's serum is injected into the skin of a normal recipient and 24 to 48 hours later a prick test with a dilute solution of antigen is made over the site. In a positive response a weal and flare occurs at the site. This test is rarely used and for research purposes only, due to the danger of transfer of viral diseases. Among the other biological tests are passive cutaneous anaphylaxis in monkeys, and *in vitro* challenge of the patient's leucocytes (basophils) with antigen, or passive transfer of antibody to sections of human lung or to suspensions of basophils from non-allergic persons (see reference 10).

A proportion of the population, known as atopic subjects, have a tendency to form significant amounts of IgE spontaneously to environmental antigens, to show tissue hypersensitivity on skin tests, and are likely to develop one or more of the anaphylactic, also termed atopic, disorders. Included in the group of disorders is atopic dermatitis which is an eczematous condition and appears to be associated with dysfunction of some subsets of lymphocytes. However, the condition is accompanied by increased synthesis of IgE particularly if there is another concomitant allergy, e.g. asthma. Though the IgE does not mediate the eczema, the anaphylactic sensitivity may exacerbate the condition. Allergy to foods has long been considered to induce the eczema but though most patients have relevant IgE antibodies, and give prick test responses, special or 'avoidance' diets have seldom effected improvement.

Atopic sensitivity tends to be familial, with a history of atopic susceptibility in the parents or near relatives. Criteria for judging a person to be atopic vary, and a person may be susceptible without showing illness. Positive prick test responses to at least three common environmental antigens, e.g. house dust mite, animal dander, *Aspergillus*, milk or egg, is reasonable evidence. According to the criteria selected about 15 per cent of the UK population has atopic susceptibility, though the proportion with atopic illness is considerably less.

Man also forms an anaphylactic IgG antibody, though this is less potent and occurs less frequently than does IgE. This is the reverse of the response of laboratory animals in which IgG anaphylactic antibodies are those most frequently observed.

The subclass of the anaphylactic IgG has not been determined, but it is designated IgG S-TS (short-term sensitising).[25,26] It sensitises human or monkey skin anaphylactically for about 4 hours, and differs in several other properties from IgE. IgG S-TS blocks sensitisation by IgE for about 4 hours. There is some evidence that the basophil receptors for IgG S-TS and IgE differ, but are sufficiently close for one to hinder sensitisation by the other.

Anaphylactic Histological Changes[1]

The chief features are oedema, congestion of venules and capillaries with variable vacuolation of endothelium, infiltration by eosinophils, and in more severe responses by neutrophils. Mast cells, if they can still be identified by conventional light microscopy, show degranulation, and metachromania by the Romanowsky stains.

Experiments on guinea pigs and rats show that in parenteral sensitisation and anaphylaxis there is no change in the intestines, or mild congestion and oedema with occasional desquamation of the epithelium.

Young guinea pigs sensitised by feeding β-lactoglobulin were found to be anaphylactically sensitive on intravenous or intradermal challenge, but tolerated the antigen given in the diet (discussed later). If no antigen was fed for 14 days, and then the β-lactoglobulin was given with egg albumin in the diet, little or no clinical change was seen; at most there was reddening of the ears, and indications of itching in the nose and paws. On post-mortem examination very mild or mild histological changes were seen in the intestine, the chief features being congestion, oedema of the villi, patchy desquamation of the epithelium lining the small intestine, and a slight increase in the numbers of neutrophils in the lamina propria and submucosa (Figure 2.4). Examination of the blood at short intervals after feeding, for absorbed ovalbumin, or prior injection of the skin with anti-ovalbumin antibody, followed by intravenous blue dye after feeding which showed an anaphylactic response at the site, revealed that ovalbumin was readily absorbed, and in greater amounts than in unsensitised controls fed both proteins, or in β-lactoglobulin-sensitised animals fed ovalbumin only (Figure 2.4). The ovalbumin absorbed from the one feed in the sensitised, antigen-challenged animals, which were kept, induced formation of IgG_1 and IgG_2 antibodies and on later re-exposure to the antigen, circulating complexes were formed with activation of complement. Apart from the evidence of anaphylactic change in the intestine, in some instances associated with only mild clinical signs, the results illustrate the phenomenon of an allergic 'leaky' gut predisposing to immune complex disease.

Figure 2.4: Diagrammatic Representation of Changes in Villi Following Anaphylactic Reaction in the Intestine

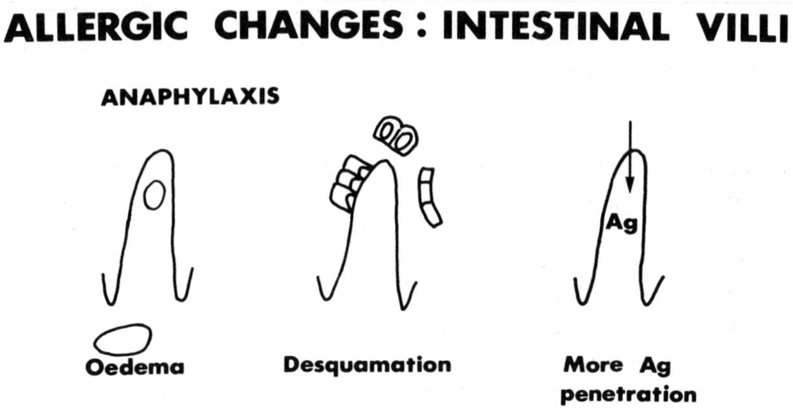

Food Anaphylactic Disorders

Several disorders result from hypersensitivity to food antigens, and the response may appear in the lung or skin without obvious intestinal involvement. Though an anaphylactic reaction is usually apparent within 15 minutes, e.g. in provocation inhalation tests, or skin prick tests, clinical signs may be delayed for hours or even two days after food challenge tests. This may be due to changes in the intestine, or slow absorption until a critical amount of antigen reaches the susceptible organ. Thus far, there is little evidence that protein metabolites are the antigens rather than the antigenic determinants of the original protein. Variations in time of appearance in clinical signs are illustrated in Figure 2.5 from the work of Professor Lessof and colleagues,[27] and in Figure 2.6 from samples submitted to the author; these are related to the classes of antibody to the relevant food antigens. However, apart from the immediate reactions attributable to IgE, it cannot be assumed that all these reactions are anaphylactic, particularly eczema, which is probably cell mediated, and the IgE antibody a concomitant phenomenon. Detection of IgE antibody is not certain evidence of food allergy. IgE antibodies to food antigens occur in asymptomatic persons,[28] in tolerant adults who were sensitive as children,[29] and even in normal adults with no history of allergy. Fluctuations in the amount of IgE antibody to foods occur with changes in diet.[30]

Figure 2.5: Time Course of Onset and Possible Duration of Anaphylactic Signs of Allergic Food Intolerance.

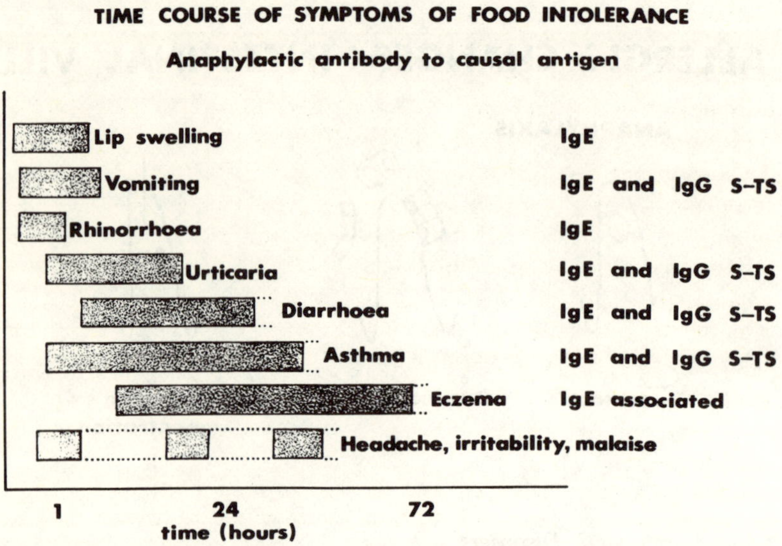

Superimposed are the two types of antibody found associated with each sign.
Source: From Lessof *et al.*[27]

Figure 2.6: Time of Onset of Clinical Signs of Anaphylactic Changes in Food Allergy to the Type of Anaphylactic Antibody Associated with Each Group of Signs.

FOOD ALLERGY

Responses varying with time of onset

5 to 60 mins	3 to 6 hrs	12 to 36hrs	2 to 3 days
rhinitis	asthma	(breathlessness)	eczema
asthma	abdominal pain	abdominal pain	
lip oedema	diarrhoea	diarrhoea	
vomiting	erythema	urticaria	
abdominal pain	urticaria	eczema	
diarrhoea			
erythema			
urticaria			
shock			
IgE	IgE	(IgE)	IgE associated
IgG S–TS	IgG S–TS	IgG S–TS	

The changes occurring at 12 to 36 hours are more frequently associated with IgG S-TS than IgE; sometimes in the absence of significant amounts of specific IgE.

Foods which are potential allergens, in order of frequency, though the incidence varies slightly in studies from different laboratories are:

1	Milk	7	Rice
2	Egg white	8	Nuts
3	Fish	9	Chicken
4	Wheat	10	Peas
5	Beans	11	Yeasts
6	Mushrooms	12	Tomatoes

Of the milk proteins inducing IgE antibody formation the frequency, and apparent potency in descending order is β-lactoglobulin, α-lactalbumin, casein, albumin, γ-globulin, and a few instances of bovine IgM.[29] Among the nuts, peanuts (ground nuts) appear to be the most allergenic.

Figures 2.5 and 2.6 indicate the clinical signs for which there is most evidence of an allergic cause. (For more detailed information see references 27, 31–34.) Diagnosis is best made on the results of ingestion tests and critically observed elimination diets. Other aids to diagnosis are skin prick tests; RAST or ELISA for detection of antibodies, or assay of histamine and other mediators released from the patient's basophils on treatment with antigen *in vitro*.

RAST to detect specific IgE antibody has done much to reveal the identity of food allergens not readily apparent from the clinical history. However, the correlation of food allergy and IgE antibody titres is not absolute as discussed in the previous section on IgE.

Examples of the variability in amounts of specific antibody in non-allergic persons is evident from Table 2.3, comparing results with egg white in examination of sera of four persons, two of whom were not clinically allergic and yet had more IgE than a person proven to have vomiting and diarrhoea within one hour of eating the antigen on every 'blind' exposure. The total concentration of IgA was within normal limits for all four persons and the amount of IgG anti-egg in the patient with an immediate response was not much greater than in the other three persons.

Immediate-type Pseudo-allergic Idiosyncrasies

Pseudo-allergic reactions, discussed earlier, result not infrequently from foods, several of which may induce anaphylactoid (anaphylaxis-like) reactions not mediated by specific IgE. Among these are shellfish, salmon and some other fish, egg white, strawberries, chocolate and

Table 2.3: Amounts of Immunoglobulin and Specific Antibody for Egg Albumin in Two Allergic and Two Non-allergic Persons

Clinical state	Prick test	Anti egg RAST Sorb-units	Total IgE Units	Anti globulin IgG	Antibody to egg[a] haem. aggln. IgA	IgM	IgA concn. (mg/ml)
Normal	+	2.05	68	32	<8	<8	1.6
Normal	+	1.45	137	32	16	16	1.9
Eczema[b]	+	3.75	598	64	16	32	2.3
Vomiting/Diarrhoea[c]	+	0.45	29	128	<16	32	2.2

a. Detection of antibody to egg by coupling egg antigen to red cells, treating with patient's serum to bind any antibody, followed by test with class specific antiglobulin sera to detect bound antibody by agglutination.
b. Atopic dermatitis, but egg apparently did not exacerbate condition.
c. Proven allergic to egg on history and 'blind' provocation tests.

possibly tomatoes. These usually have to be eaten in fairly large amounts, or are absorbed in abnormally large amounts.

Histamine-releasing substances can be extracted from some foods, e.g. protamine sulphate from fish, especially salmon, tyramine from cheese and chocolate, and at least two arginine-lysine containing peptides from egg white. Basic polypeptide metabolites induce release of histamine and other pharmacological mediators. Some of these foods may induce both antigen-specific allergy and pseudo-allergic food intolerance.

The pseudo-allergic reactions probably have some relevance to abnormal responses to food additives, discussed later, and share with them a range of conditions, e.g. hyperactivity, enuresis, mouth ulcers, attributed to both.

Arthus Reaction and Immune Complex Disorders

The Arthus reaction, named after the man first describing it, results from formation of complexes of antigen with antibody, in particular ratios of the two, which activate complement, and on deposition in susceptible tissues, attract neutrophils. The neutrophils ingest the complexes and release their lysosomal enzymes, including acidic proteases, collagenase, elastase and desoxyribonuclease, which degrade the tissues, and are responsible for much of the ensuing tissue change. The sequence of changes is illustrated in Figure 2.7 and a more detailed description is available in reference 1.

Complexes vary in their ability to induce tissue damage. Those with antigen in slight to moderate excess of equivalence of antibody[1] are much more harmful than those with a gross excess of antigen or of antibody. To induce damage, the complexes must be deposited within a tissue, e.g. beneath the endothelium of venules, or in the synovium of joints, where they act as a focus for neutrophil infiltration.

After the acute phase of neutrophil infiltration, during which the complexes are degraded, there is a mononuclear cell infiltration, containing in some instances an accumulation of plasma cells. Severe reactions progress to necrosis.

Potentially harmful complexes are formed by IgG or IgM antibodies activating complement. Complexes formed of IgA are usually harmless, and are a means of removing, without tissue damage, foreign antigens penetrating mucosae.

Many substances may form the antigen component of immune

Figure 2.7: Diagrammatic Representation of the Arthus (Immune Complex Mediated) Response.

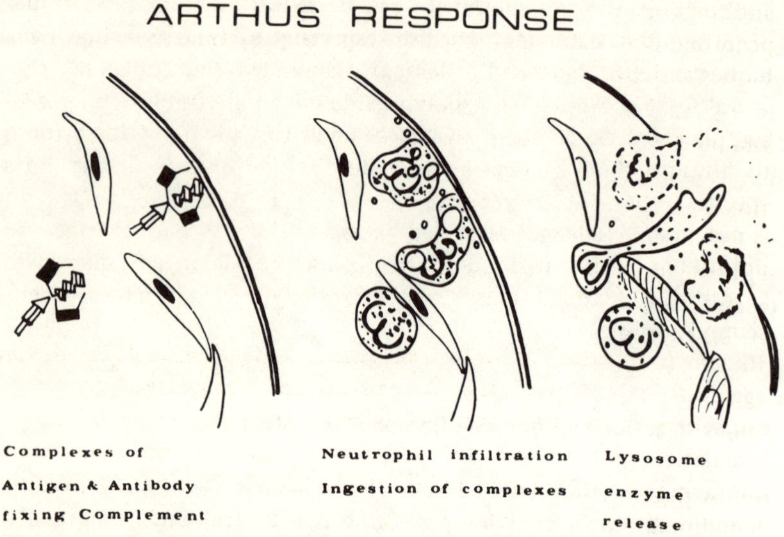

Complexes of antigen and antibody that activate complement are deposited beneath vascular endothelium. Neutrophils attracted by the activated complement, ingest the complexes and release tissue-damaging lysosomal enzymes. Source: From Parish.[1]

complexes, e.g. bacterial and fungal somatic and secreted products, viruses, food substances and additives, drugs, and autologous cell and humoral substances. Drugs and chemicals that covalently bind to proteins are of particular interest to the toxicologist. Though each chemical molecule in the free state may be too small to provide sufficient epitopes for antibody molecules to bind in sufficient number to form immune complexes, if they become bound or coupled to proteins, the chemicals provide the epitopes, and the protein provides the mass of the antigen. Such chemicals are known as haptens. Effective protein-chemical hapten formation requires firm, usually covalent, binding of the hapten, and sufficient hapten molecules appropriately spaced, to enable antibodies to cross link the antigen molecules to form aggregates or complexes. However, for some haptens, e.g. penicillin, only three to four molecules may be required to bind to a protein molecule to form antigen.

As drugs, e.g. penicillins, streptomycin and sulphonamides, are given in relatively large amounts for several days or weeks, there is a steady supply of chemical hapten to stimulate antibody formation, and to combine with protein to form antigens, which will bind to antibody resulting in immune complexes. Chemicals in foods which penetrate the intestine could have similar activity as the drugs; antibiotics and sulphonamides are administered orally.

In some instances the haptens combine with membranes of cells, e.g. platelets, erythrocytes or epidermal cells, making them susceptible to the potential cytotoxic activities of antibody or lymphocytes specific for the cell-bound chemical Transient thrombocytopenia is not an uncommon feature of acute allergic reactions due to chemicals or proteins.

Autologous tissue substances, e.g. DNA from damaged cells, or serum proteins will also form the antigen of immune complexes. Rheumatoid factor is an IgM antibody to IgG which has become aggregated or denatured; thus first-formed complexes containing virus or a food antigen and IgG antibody, may be cleared from the circulation, but are replaced by complexes of IgM anti-IgG. Persons forming these rheumatoid factor antibodies are particularly prone to immune complex-mediated disorders, e.g. rheumatoid arthritis, the course of which is associated with the presence of these complexes.

Complexes are continually being formed in the blood and tissue fluids as a normal physiological process in the removal of penetrating foreign antigens. It is only when the clearance system fails that the complexes may become deposited in tissues susceptible to damage. It is not the person who forms much antibody, but the person who has formed insufficient antibody or who forms antibody of weak affinity (binding-power) for antigen who appears most at risk. In such complexes there is little IgG Fc or complement to bind with the receptors on the phagocytic cells that remove and degrade them.

Antigens penetrate the intestine after every meal, and complexes appear in the blood, usually in a biphasic sequence. Firstly in small amounts for a short time, within 30 minutes after ingesting the food. Secondly in larger amounts for a longer period, about 2 hours later. These complexes usually contain IgA, and those that are composed of IgG and complement are removed harmlessly.

Histology of Immune Complex Mediated Responses

The histological changes of immune complex mediated reactions vary much according to the susceptibility of the individual and the nature and amount of antigen. Acute reactions are infiltrated by neutrophils; these progress to mainly mononuclear cell infiltrations and resolve, or become granulomatous.[1] There is evidence that some responses associated with the deposition of complexes start as infiltration of mononuclear cells and have been termed 'lymphocytic' in classifications of vasculitis, but the identity of the cells is still to be determined, and in some lesions the author has found a predominance of monocytes.

The essential lesion is a vasculitis. Neutrophils infiltrate the media of blood vessels, particularly venules in which the immune complexes are most likely to be deposited. In mild lesions the endothelium is shed, the elastic lamina and basement membrane degraded and fragmented, and nuclear debris accumulates as basophilic 'dust'. In severe lesions there is necrosis.

Histopathological changes in the intestine are described from a compilation of results of systemic dissemination of immune complexes, or case histories from human and veterinary practice.

The essential features (Figure 2.8) are deposition of antigen-antibody complexes detected by immunofluorescence or immunoperoxidase techniques, infiltration by neutrophils and their subsequent degeneration, oedema of the stroma of the villi, degeneration of the capillaries with erosion of the tips of the villi.

Figure 2.8: Diagrammatic Representation of Changes in Villi Following an Immune Complex Mediated Reaction in the Intestine

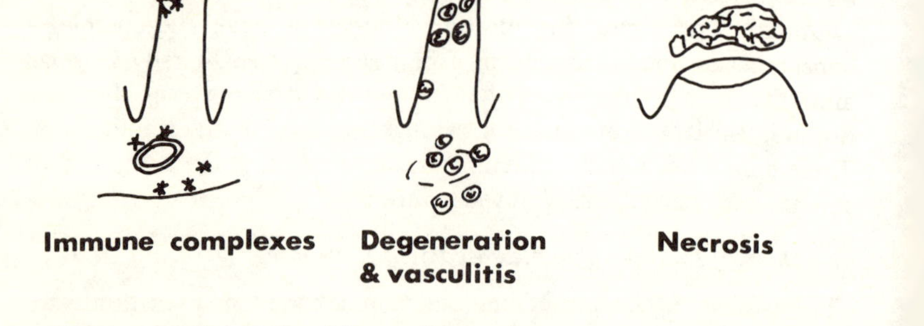

Most experimental evidence from tests on rabbits results from intramucosal injection into sensitised animals[35] or antigenic challenge of sensitised animals after chemical irritation of the colon,[36] neither of which are truly representative of the human disorders.

Immune Complex Mediated Disorders

Many disorders are associated with the presence of circulating immune complexes, but in many instances they reflect the clearance of foreign substances or damaged tissue, and not the causal or a significant exacerbating stimulus.

The classical example of an immune complex mediated disorder is serum sickness following the injection of therapeutic antisera of animal origin. The injected foreign protein persists for many days in the blood of the recipient, stimulating formation of antibody which combines with the persistent antigen to form complexes of antigen in excess of equivalence of antibody. Deposition of these complexes damages susceptible tissues. With further formation of antibody which binds to the residual antigen, the remaining complexes are harmlessly removed, and the lesions heal. The chief pathological features are arthritis, vasculitis, acute glomerulonephritis, lymphadenopathy and leucocytosis. In the experimental condition in rabbits, myocarditis and endocarditis are sometimes observed.

Though serum sickness due to therapeutic antisera is now rarely seen, a similar condition induced by drugs occurs. However, immune complex mediated disorders associated with microbial antigens, chemicals and some food antigens, may affect particular organs without the generalised signs of serum sickness, affecting the kidneys, joints, blood vessels and the lungs.

Though all food proteins are capable of forming complexes with antibody, few foods regularly induce formation of IgG antibody in sufficient amounts to be detected by precipitation or agglutination tests, which require large amounts of antibody for positive results compared to the very discriminating tests, e.g. ELISA detecting the small amount of antibody formed against nearly all foods. The foods most frequently stimulating formation of IgG detected by precipitation or agglutination are milk, wheat and similar cereals, and egg. These antibodies occur in normal persons, but in greater amounts in persons with gastro-intestinal disease.[29,32,37-40] IgG antibody to cow's milk is frequently present in sera of infants who are bottle-fed, but declines in amount as the infants are weaned onto a mixed diet.[37]

Immune complexes have been demonstrated in the blood of persons with IgG antibody after a meal containing the relevant food. These complexes are quickly and apparently harmlessly removed. There are few disorders in which foods appear to induce tissue-damaging immune complexes. Among the intestinal disorders is coeliac disease in which complexes of gluten and possibly other wheat antigens are found, and the associated condition of dermatitis herpetiformis in which complexes of IgA with possibly wheat antigen are deposited beneath the dermo-epidermal basement membrane. Increased amounts of IgG and circulating complexes are found in inflammatory bowel disease (ulcerative colitis; Crohn's disease). However, food antigens more readily penetrate during intestinal disorders, and though complexes of wheat or similar antigen may contribute to the aetiology of the condition, complexes of other food antigens, which for example occur in coeliac disease, probably reflect the harmless clearance of substances penetrating the disordered bowel.

There is good evidence that chronic pneumonitis in children is related to immune complexes containing milk antigens, but recent studies do not support the contention that foods induce or exacerbate arthritis.

Delayed Hypersensitivity (Cell-Mediated Immune Response)

Some lymphocyte stimulation occurs in all immunological responses, but delayed hypersensitivity refers to an immunologically specific reaction which takes several hours to reach its maximum intensity and occurs in the absence of demonstrable antibody in the serum.

In a sensitised person the reaction is initiated by antigen, which for a few hours becomes fixed or retained at a tissue site, reacting with a few randomly migrating T lymphocytes which are sensitised to that antigen. The lymphocytes enlarge to form blasts and divide; at the same time they synthesise substances with many biological activities, known as products of activated lymphocytes (PALs) or lymphokines. These substances attract and stimulate other leucocytes, monocytes, neutrophils, basophils and other lymphocytes not sensitised to the eliciting antigen (Figure 2.9). These in turn contribute to the inflammatory changes. If the antigenic stimulus is eliminated and the inflammation subsides, there remains an increased population of T effector cells, T memory cells and possibly T helper and activated B cells (Figure 2.1) so that the individual is further sensitised.

Lymphokines synthesised by T cells stimulated by antigen have a wide range of activities on other leucocytes, attracting them to the site of antigen deposition, preventing movement away from the site, and stimulating their function.[41–43] Some lymphokines promote

Figure 2.9: Diagrammatic Representation of Delayed Hypersensitivity (Cell-Mediated Immune Response).

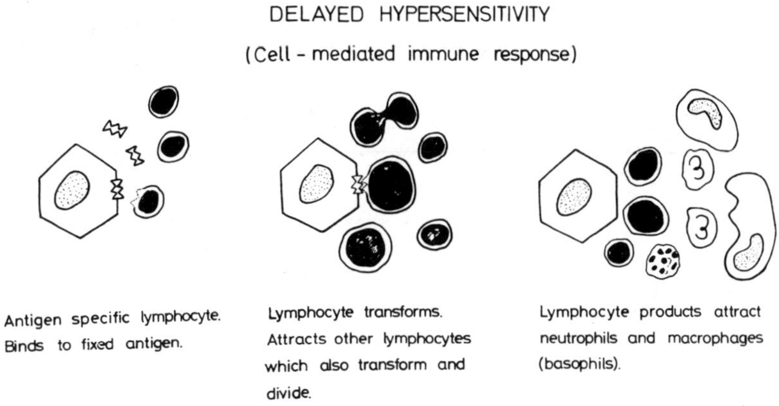

Antigen reacting with specifically sensitised lymphocytes stimulate them to enlarge, divide, and to synthesise products (lymphokines) that attract other cells and stimulate their activity, particularly that of macrophages. Source: From Parish.[1]

inflammation; others reduce or suppress inflammatory change. Thus lymphocytes control the nature and intensity of the response induced by other lymphocyte products. Though many biological activities induced by T cells have been observed, it is not known if there is a corresponding diversity of lymphokine substances, or if some substances have several biological activities.

The macrophage is particularly responsive to T cell stimulation, which also influences maturation of monocytes to macrophages, and subsequent modification of macrophages to epithelioid cells and giant cells. Hypersensitivity granulomata, e.g. the miliary tubercule, are mainly allergic responses to the specific antigen mediated by activated T lymphocytes.

Determination of Lymphocyte Activity

The examination of lymphocytes for their responsiveness to detect deviations from normal properties, and examination for reactivity to specific antigens to assess delayed hypersensitivity or cytotoxic activity, is becoming routine in the study of allergies, and in toxicology. The

Table 2.4: Features Used to Distinguish Between T and B Lymphocytes of Man, Mouse and Rat, and Substances Used to Examine T and B Lymphocyte Responsiveness

	Man		Mouse		Rat	
	T	B	T	B	T	B
Identification of suspensions of blood or tissue lymphocytes						
E rosettes (sheep)	+	–	–	–	–	–
C3b receptors (EAC rosettes)	–	+	–	+	–	+
Immunoglobulin synthesis[a]	–	+	–	+	–	+
Antisera to mouse theta (ϕ) Ag	–	–	+	–	b	b
Identification of lymphocytes in tissue sections						
E rosettes (sheep)	+	–	–	–	–	–
EAC rosettes	–	+	–	+	–	+
Ig bearing cells (also specific antisera)	–	+	–	+	–	+
Mitogen induced transformation or thymidine incorporation						
Concanavalin A (Con A)	+	–	+	–	+	–
Phythaemagglutinin (PHA)	+	–	+	–	+	–
Pokeweed mitogen (PWM)	+	+	+	+	+	+
E. coli lipopolysaccharide (LPS)[c]	–	–	–	+	–	+
General antigens to which most persons are sensitised						
Tuberculin						
Candidin						
Measles						
Homologous lymphocytes (the mixed lymphocyte reaction (MLR) in which lymphocytes of one individual, inactivated by mitomycin C, are mixed with the test lymphocytes to determine the intensity of reaction to the foreign lymphocyte antigens).						

a. Detection of synthesis: not cell binding as observed in some T cells.
b. Not mouse theta, but specific antibody prepared to rat T and B.
c. In man B cell response to LPS appears to be antigenic stimulus, not mitogenic.

tests commonly used are summarised in Table 2.4. The identification of particular subsets or subpopulations of lymphocytes requires monoclonal antibodies not readily available. For details of test procedures see references 44–46.

Counts are made of absolute numbers of lymphocytes and of relative numbers to other leucocytes in the blood. The numbers of T and B cells should be determined. These never account for the total number of lymphocytes, partly because of variations in, and deficiencies of, the techniques, and partly because some cells do not have the typical characters of T or B cells. These are termed null cells or the third population. Some of these appear to be immature T lymphocytes;

Figure 2.10: Results of Adding β-lactoglobulin Antigen (+ column) to Suspensions of Lymphocytes and Determining Susceptibility to Transformation by the Amount of Radio-isotope Labelled Thymidine Incorporated

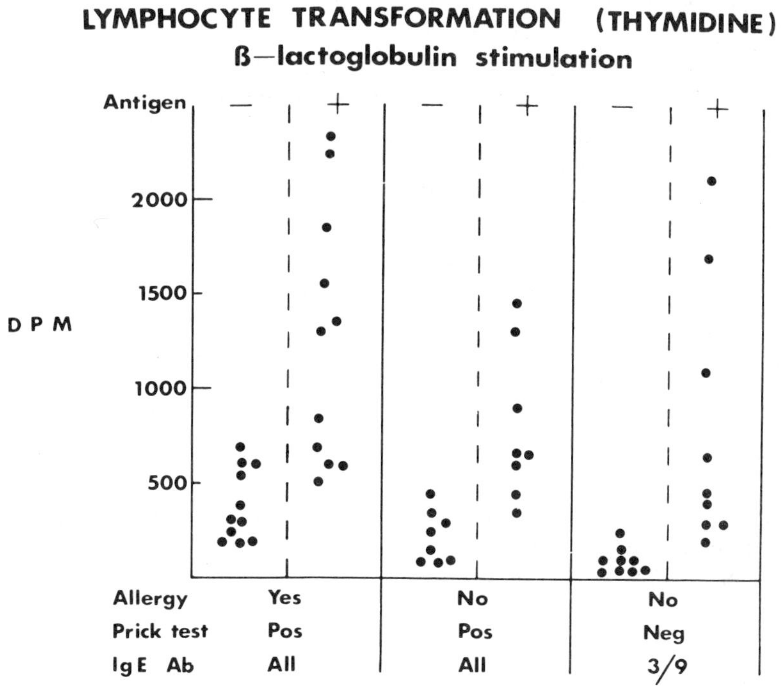

Milk allergic persons showed the greatest responsiveness, but lymphocytes of non-allergic persons, with or without specific IgE antibody, also responded, indicating immunological awareness to antigen. Lymphocytes of some allergic and non-allergic persons with specific antibody did not respond to the antigenic stimulation, indicating the limitation of this test as an aid to diagnosis of anaphylactic food allergy.

others have special properties. Simple tests are available to identify subsets (Table 2.4), but eventually very specific antisera will become available. These will greatly facilitate identification of lymphocytes in tissue sections and differentiate them from monocytes or macrophages. Rosette adherence tests on tissue sections have limited applications.

The activity of lymphocytes is examined by the use of various mitogens. Mitogens are antigen non-specific substances stimulating transformation to lymphoblasts with synthesis of DNA and of lymphokines relevant to the particular subset, in a manner similar to that induced by antigen. Different mitogens stimulate different subsets of lymphocytes, and there is some species variation in susceptibility (Table 2.4). The responsiveness to general environmental antigen, or to other specific antigens may be examined in lymphocyte cultures. As mitogens stimulate most of a subset of lymphocytes, and antigens initially only stimulate specifically sensitised cells which form only a small proportion of the total cells, responses induced by mitogens tend to be much greater than those induced by antigen.

Mitogens detect normal or decreased responsiveness of lymphocytes, and by choosing the appropriate mitogens (Table 2.4) it is possible to determine dysfunction in T or B cells. Antigen challenge of lymphocytes from sensitised man or animals may indicate hyper-reactivity as evidence of greater than normal responsiveness indicative of allergy, though some normal subjects may show similar lymphocyte responsiveness (Figure 2.10).

However, responsiveness to antigens to which the person is allergic may be below normal in some persons. In the atopic diseases, particularly if associated with atopic dermatitis, there tends to be decreased reactivity to the antigens to which the subject has formed antibody, particularly IgE.

Histology of the Delayed Hypersensitivity Response

Antigen-activated T effector cells mediate or participate in many acute and chronic inflammatory conditions which may present as induration or cell necrosis.

The classical example of a delayed hypersensitivity response is the tuberculin reaction in which in the dermis there is a generalised infiltration by mononuclear cells and characteristic multifocal accumulations round thin-walled blood vessels.[1] Selective immunological and histochemical reagents enable identification of the cells, and those round the blood vessels are composed of monocytes and lymphocytes. The presence of neutrophils infiltrating the tissue indicates the severity of the reaction, as few are found in weak responses. Much of the induration is due to deposition of fibrin rather than infiltration by large numbers of cells. The patch test of contact dermatitis sensitive patients shows similar changes in the dermis, with overlying spongiosis and vesicles in the epidermis.

The miliary tubercle is a hypersensitivity granuloma and another example of a delayed hypersensitivity response. It is composed of lymphocytes, macrophages, epithelioid cells and occasional giant cells, with deposition of strands of reticulin and collagen. Some chemicals, e.g. beryllium, induce similar but not identical granulomata.

Delayed hypersensitivity responses have also been implicated in destruction of tissue, as in chronic intestinal disorders, e.g. coeliac disease, Crohn's disease and ulcerative colitis. It has not yet been demonstrated conclusively that antigen-sensitised lymphocytes are primarily responsible for the tissue degeneration.

Cutaneous Basophil Hypersensitivity

An allergic skin response delayed in onset, but differing in several features from the classical delayed hypersensitivity response is cutaneous basophil hypersensitivity. This response is less indurated and sites contain less fibrin than seen in delayed hypersensitivity tuberculin reactions, but there is a similar perivascular accumulation of mononuclear cells. The characteristic feature is the large number of basophils that infiltrate the dermis, being most numerous at 24 to 48 hours after antigen challenge. Usually only a small proportion of the basophils shed their granules, but after 48 hours there tends to be a decreased affinity for Giemsa stain.[47]

The basophil response is mediated by lymphocytes but this hypersensitivity is transient, at least in guinea pigs, and is considered to be a phase in the change of delayed hypersensitivity to formation of specific antibody and susceptibility to an Arthus response. In other test systems the cutaneous basophil hypersensitivity persists.[47]

Some tuberculin or contact dermatitis skin test reactions in man are infiltrated by basophils.[48]

Delayed Hypersensitivity in Allergy

Manifestations of delayed hypersensitivity occur in many infections, bacterial, fungal, viral and by helminth parasites. They are found in many persons sensitive to insect bites. Some chemicals, e.g. beryllium and zirconium induce lymphocyte-mediated granulomatous lesions. Auto-allergic cytotoxic changes, in thyroiditis, encephalomyelitis and aspermatogenesis are mainly delayed hypersensitivity reactions.

Contact dermatitis, or allergic eczema, is a classical example of delayed hypersensitivity, and is a common response to natural or man-made chemical haptens that bind to proteins to form antigens. Sensitised persons may have exacerbations of their eczema if the allergen is ingested, though repeated ingestion may lead to tolerance.

Persons frequently handling foods, particularly fish, shellfish, onion and garlic, but also many other foods, may develop cutaneous sensitisation resulting in contact dermatitis. This is a potential occupational hazard in the food industry.

Lymphocytes participate in gastro-intestinal allergic tissue change apart from promoting synthesis of antibody, but the nature of their activity, whether dependent upon release of inflammatory lymphokines or upon cell damage induced by cytotoxic T or K cells, and the identity of the antigens, whether foreign, as in foods or microorganisms, or autologous as in epithelial cell or basement membrane antigens, is largely unknown. An example of the complexity is briefly considered, but for detailed reviews see references 49 and 50.

Coeliac disease (gluten-sensitive enteropathy) and some related malabsorption syndromes are the best studied examples of allergic change in the intestinal tract, but the stimuli to change, and the nature of the pharmacological or cellular mediators are not well understood. The association with particular HLA (human leucocyte antigen) groups indicates a genetic susceptibility.

The jejunal villi become flattened, the epithelial cells of the villi are reduced from columnar to distorted cuboidal forms, and the lamina propria is infiltrated by mononuclear cells including plasma cells. There is a relative and absolute reduction in the total number of circulating T lymphocytes which returns to normal when gluten is excluded from the diet. There is, however, much conflicting evidence about abnormal responsiveness of the lymphocytes. Cultured lymphocytes from coeliac patients showed decreased thymidine incorporation compared to cells from normal persons. When, however, gluten is added to the culture medium, there is some increase in responsiveness of lymphocytes from coeliac patients compared to the controls, particularly if the lower initial activity is accepted as the base line for calculation.[51] In an examination of a small group of patients, lymphocytes of the small intestine of untreated cases showed evidence of cell migration inhibition factor on challenge with alpha-gliadin.[52]

Another defect is the reduced ability of lymphocytes from coeliac disease patients to be stimulated *in vitro* to become cytotoxic to Epstein Barr lymphoma cells; a model system to assess lymphocyte cytotoxic function.

There is much evidence of allergic change in coeliac disease. Antibodies are formed to one or more antigens of wheat; circulating immune complexes may contribute to the tissue changes, and there is some evidence of reduced IgA secretion which could result in impaired immune exclusion of antigen. The decrease below normal number of T cells and their reduced activity may reflect a fundamental immune deficiency rather than a T cell-mediated cytotoxic allergy.

Maladsorption syndromes are not uncommon. The possibility that they are induced by antigens other than wheat gluten, and their relevance to toxicology, is still to be ascertained.

Intolerance to Natural Food Chemicals and to Food Additives

Misuse of terminology tends to make many clinical reports of intolerance misleading. The term allergy, which indicates specific immunological reactivity, is applied to cases in the absence of proven antigenic stimulation, and failure to do critical 'double blind' provocation tests and diet elimination tests on patients (withdrawal of a coloured drink removes the other constituents of the drink together with the colour) results in much confusion. Only a brief summary and discussion of the main issue is possible.

It is difficult to distinguish between the clinical signs of food intolerance induced by allergy from those resulting from pseudo-allergic stimulation. The same pharmacological mediators inducing inflammatory change are released or generated in both forms of intolerance, but in pseudo-allergy the eliciting substance acts directly on the cell or on the mediator precursor substrate, without initiation by an immune response to a specific antigen.

It is noteworthy that most allergic responses can be reproduced in appropriate animal models; several of the common pseudo-allergies cannot. Other differences between allergy and pseudo-allergy are stated in the Introduction.

The following five points summarise our present understanding of intolerance to food chemical additives.

(1) Several chemicals have been incriminated as causal agents of a wide variety of conditions, but tartrazine is usually cited as representative of these chemicals.

(2) The adverse clinical signs should be described as food or chemical *intolerance* and not allergies, as there is very little proven immunological stimulation. Very few immunological changes have been detected and these have never been shown to be responsible for the clinical signs in the patients, and they are likely to be a concomitant immunological awareness of the substance, like the weak reactivity

to many environmental antigens including foods.

(3) In the majority of persons with a proven intolerance, the clinical signs may prove to be pseudo-allergies or idiosyncrasies, in which the individual has changes in cell receptors for the chemicals or their metabolites, a hypersusceptibility to release or generate pharmacological, or hyper-reactivity of the tissues to mediators (as occurs in a different context in atopic dermatitis).

(4) The intolerance disorders, supported by strong evidence, are urticaria, angio-oedema, purpura, asthma, rhinitis and gastro-intestinal discomfort or colic, nausea and diarrhoea. Other less well supported manifestations of intolerance are headache, urinary incontinence or urgency, hyperactivity syndromes and psychological disturbances.

(5) In pseudo-allergies to chemicals, particularly food additives, only a small proportion of persons is susceptible, and the incidence of susceptibility appears unrelated to atopic allergic status, though in experimental models IgE enhances spontaneous release of mediators from basophils and mast cells. In persons with cutaneous disorders, e.g. urticaria, food additives, if they have any effect, appear to provoke or exacerbate an existing condition, and do not induce it.

Several food additives have been claimed to induce intolerance, the more common in incidence being tartrazine, and other azo dyes, Sunset Yellow, New Coccine, Patent Blue V; also salicylates; Annatto, benzoates, and butylated hydroxyanisole (BHA).[53-55] Other substances claimed to have similar activities are sulphur dioxide, metabisulphite, and monosodium glutamate. In a critical study of 330 patients with recurrent urticaria, food was considered to be an exacerbating agent in 30 per cent of cases with fruits, vegetables and nuts the most common inducing substances. Provocation tests with a wide range of food additives showed one or more substances to provoke reactions in one third of the patients.[56] In these cases the food additives appeared to exacerbate an existing condition rather than induce a disorder.

Exacerbation or recrudescence of urticaria by salicylates and other additives has been reported several times, and the susceptibility is dose dependent;[56,57] in some cases a response is obtained early when large amounts are given.

An associated feature of urticaria,[56,57] and to a lesser degree in asthma,[54] in aspirin and azo compound intolerance tests, is gastro-intestinal dysfunction, described variously as epigastric discomfort, gastritis or intestinal colic, and may be accompanied by nausea, vomiting or diarrhoea.

There is a tendency to confine attention to food additives, but several natural foods provoke skin disorders without evidence of allergy, including nuts, eggs, fish, chocolate, milk, fruit and yeasts.[57] There may be some evidence of a concomitant immunological responsiveness with formation of antibody, but in these cases of intolerance there is no certain evidence of allergy. This also applies to migraine which may be induced by similar foods, e.g. chocolate, cheese, meat extracts containing yeasts, wine, citrus fruits and strawberries.

Some foods contain salicylates and benzoates as natural constituents, e.g. apples, citrus fruits, strawberries, cabbage, beans, grapes, wine or coffee, and in persons in whom small amounts of the pure chemical will induce urticaria or asthma, similar amounts of chemicals may be ingested with natural foods.

Immunological Investigations

There is very little evidence that salicylates and benzoates, as natural constituents of food or as additives, induce antigen-specific allergy, though the clinical disorders, urticaria, asthma, rhinitis, are identical with those of proven allergic reactions.

The intolerance is not usually antigen-specific, even though there are close chemical similarities among the salicylates or between the benzoates. A person susceptible to one tends to be susceptible to others (see Schlumberger[3]). In patients with asthma induced by aspirin, concomitant susceptibility to indomethacin was observed in 71 per cent, to sodium benzoate in 14 per cent and to tartrazine in 17 per cent, and in patients sensitive to tartrazine, asthma could be induced in 100 per cent by the azo dye Sunset Yellow, in 42 per cent by aspirin and in 39 per cent by sodium benzoate.[54] Similar susceptibility to several different additives also occurs in individuals with urticaria.[55] The variations in chemical side chains, the usual antigenic determinants, indicate a remarkable cross-reactivity if all the reactions were antigen induced. The term cross-reactivity, frequently used, may be a misnomer, It is more likely that there is a concomitant susceptibility to substances with similar pharmacological activity.

Though the clinical signs, asthma or urticaria, are those of atopic (anaphylactic) allergies, susceptibility to these chemicals is not selectively confined to atopic persons, and is not related to the amount of serum IgE.[54,58,59] The assertion by Juhlin[56] that many of the patients with urticaria were atopic is inconsistent with his Table 10 of laboratory abnormalities in which only 4 per cent of the patients had above normal amounts of IgE. Though not all atopic persons have increased amounts of IgE, a high proportion do.

In another study of aspirin and tartrazine intolerance, no increase in total IgE was found, and no IgE antibody to aspirin or tartrazine, with the possible exception of one person with antibody to aspirin, but this showed no correlation with clinical intolerance.[60]

In another investigation there were no local prick or patch test reactions to tartrazine or its two metabolites, sulphanilic acid and 1-sulphophenyl-3-carboxy-4 amino 5-oxypyrazole (phenyl sulphonic acid); no histamine release from the patient's leucocytes tested with these substances *in vitro*, and no IgE antibody to these substances by several conjugation systems by RAST.

No IgE antibody to tartrazine or its metabolites was found in another study,[60] but there was IgD antibody to the phenyl sulphonic acid metabolite, though not to the complete tartrazine molecule. IgD is usually considered to be a lymphocyte-associated antibody and part of the antigen-recognition system. IgD antibody to a tartrazine metabolite indicates immunological 'awareness', but its role in allergy is not known.

Another example of immunological awareness, but not of allergy, is the observation that lymphocyte transformation (thymidine incorporation) occurred in cells of 18.4 per cent of 258 patients with chronic urticaria, to at least one of three substances, acetyl salicylic acid, tartrazine and hydroxybenzoic acid, though in the same study 35.2 per cent of the patients had cells reacting to common foods particularly milk, chocolate and egg.[61] However, chronic urticaria does not usually resolve by eliminating items of diet to which patients are considered allergic.

It is not possible to review the numerous reports, especially on adverse responses to aspirin, but thus far there is no certain evidence that intolerance to aspirin, tartrazine and similar chemicals is immunological.

Pharmacological Intolerance

Intolerance to aspirin is associated with intolerance to several salicylates and benzoates. It is logical to consider that the natural chemicals in food and food additives have pharmacological activities in common with aspirin.

For the majority of asthmatic patients, aspirin is contra-indicated, as it provokes severe attacks. Similarly aspirin, and azo compounds, induce exacerbations of recurrent urticaria and angio-oedema, for which there are indications of other initiating causes. These substances appear to stimulate pharmacological changes in organs already in an excited or irritated state. Nevertheless, asthma, for example, is a clinical sign which may be mediated by different pharmacological systems, and in some patients, aspirin give relief.

Among the pharmacological activities of aspirin is inhibition of prostaglandin synthesis, which has been attributed to the provocation of asthma in asthma-sensitive patients.[62] However, it appears unlikely that tartrazine, or sulphanilic acid, have similar activity, and in two studies there was no inhibition of cyclo-oxygenase activity in animal cell micro-organelles or human platelets,[63,64] indicating that clinical signs were not due to inhibition of prostaglandin synthesis. Some anaphylactoid responses are due to histamine-releasing activity of substances, but no significant release of histamine was observed from mast cells of normal rats, or of rats fed tartrazine, Sunset Yellow and other approved colours, when suspensions of mast cells were treated with the colours *in vitro*.[65] No significant release of histamine from leucocytes of tartrazine-sensitive patients, or from cells of normal persons has been observed.

Other suggestions of the aetiology of the intolerance to additives are activation of kinins and inhibition of platelet aggregation, but these are likely to be associated phenomena. The nature of intolerance to foods containing natural salicylates and benzoates or to similar drugs and additives remains a problem for physicians and toxicologists.

Reduced Immunocompetence (Immunosuppression)

There is much interest in the possibility that some foods, food additives or chemicals reduce immunological competence, and by inference are potential toxicological hazards. By describing the activity of such substances as immunosuppression, it is inferred that the effects are as severe as therapeutic treatment for transplantation surgery.

The term immunosuppression should be restricted to the use of drugs, or radiation treatment administered with the intent to completely or greatly reduce all immunological responsiveness.

Reduced immunological competence or reduced immunocompetence designates the impedance of function of lymphoid tissue (the immunocompetent cells) including macrophages resulting from exposure to environmental substances, including foods. Such accidental reduced responsiveness to antigens may be manifested as reduced formation of antibodies or reduced cell-mediated immune responses and, in theory, could predispose an individual to infection, impaired removal of foreign substances and tissue debris, to auto-allergy, and the extreme possibility to neoplasia.

Regrettably, recent claims that several foods or food additives are 'immunosuppressant' and deemed to be harmful are based solely on the results of *in vitro* culture tests, using animal tissues. It is likely to be misleading to interpret results of cell culture systems as evidence of similar activity *in vivo*, or that reduced antibody formation or lymphocyte responses *in vivo* are certain evidence that the individual becomes susceptible to disease. Furthermore tests on animal, e.g. rat or mouse, lymphocytes, dependent upon responses to mitogens, do not exactly reflect the responses of human lymphocytes to the same mitogens (Table 2.4). The animal tests indicate that the test substance has potential activity for lymphoid tissue, but its significance for man has to be confirmed on human lymphocytes.

The sequence of tests to examine the potential of a substance to reduce immunological competence, and the significance to health, is discussed elsewhere.[66] The logical programme of tests is as follows.

(i) Examine the activity of the test substance on animal lymphocytes *in vitro*. This provides limited but indicative evidence.

(ii) Show that reasonable amounts of the test substance have a selective effect on lymphoid tissue *in vivo*, preferably on the same species whose cells were tested in (i), and that it reduces lymphocyte responsiveness and or antibody formation, by *in vitro* and *in vivo* examination of the cells of the treated animal. Any effect must be selective for the lymphoid tissue and not the result of general systemic toxic activity resulting in poor condition or damage to one organ, e.g. the liver.

(iii) If immunological competence is much reduced *in vivo*, its significance for health should be assessed, e.g. resistance to infection, or susceptibility to auto-allergy. Some immune responses may be much reduced without the animal becoming susceptible to any disease. Tests properly designed to determine significance of reduced immunological competence are complex and costly.

(iv) Human blood lymphocytes, in suspensions of leucocytes, or as purified suspensions of total lymphocytes or their subsets may be examined to assess their susceptibility to the test substance. The results of these tests will be compared with those obtained in (i), and will form the basis of judgement in significance of any activity obtained in the subsequent experiments.

(v) If the test substance is already widely used by man, a study of records of use by the population in relation to health, particularly susceptibility to infection, may indicate any likelihood of significant adverse effects.

This regimen of investigation is a demanding study. It should not be undertaken without good evidence that the test substance may have significant activity on the lymphoid system, and the results of *in vitro* tests need to be carefully assessed before commitment to detailed *in vivo* studies.

Several chemicals have been reported to reduce immunocompetence. Among the food additives or their metabolites are butylated hydroxyanisole, gallic acid and propyl gallate. The claim is based almost entirely on the results of tests using the Mishell-Dutton technique which assays *in vitro* the number of plaques representing each plasma cell secreting haemolytic antibody for the sheep red cell antigen.[67,68] The relevance of this decreased formation of antibody to any significant adverse effect *in vivo* is unknown. An examination of calcium cyclamate in rabbits showed that those fed 5 per cent formed less antibody to antigen (bovine serum albumin) than those fed 3 per cent or the controls not fed cyclamate.[69] However, as there was a concomitant reduction in the number of red cells, and there was no significant change in the total number of white cells (apparently selective counts for lymphocytes were not made) and no change in the lymphoid tissue was reported, the effect could well be due to a systemic toxic effect and not to selective activity on the lymphoid tissue.

More extensive and well-controlled studies have shown that several chemicals (e.g. salts of tin, lead, cadmium, arsenic) and some pesticides, reduce immune responses in animals *in vivo*,[70,71] but in most instances the amounts used to induce an effect are far in excess of the likely exposure to the general population (excluding factory staff who are always at risk if exposed to large amounts of chemicals without appropriate care) indicate that reduced immunocompetence is not a serious risk compared to the other more potent toxic activities of these substances.

An example of the need to assess laboratory data very critically before concluding that substances mediating reduced immunocompetence are likely to be harmful to man is evident from the current studies on polyunsaturated fatty acids (PUFA). PUFA undoubtedly reduces immune responsiveness *in vitro* and *in vivo* in animals, but there is no evidence at all that this activity is in any way harmful. PUFA reduces responsiveness of lymphocytes to mitogens *in vitro*[72] and depresses lymphocyte responses *in vivo* in mice sufficiently to prolong the duration of skin allografts.[73] These laboratory findings in animals exposed to large amounts of PUFA do not indicate that PUFA induces or predisposes man to disease.

In a study in which pregnant guineapigs were fed large amounts of PUFA and the offspring were weaned on large amounts of PUFA, there was decreased formation of antibody to a thymus-dependent antigen and decreased delayed hypersensitivity responsiveness to tuberculin.[74] Three important findings become evident. Firstly, on repeated antigen challenge of the PUFA-fed animals the immunological activities returned to normal. Secondly, animals fed saturated fats also showed decreased immunological responsiveness. Thirdly, the animals fed PUFA or saturated fat and showing decreased immunocompetence showed no disorder related to the diet and remained healthy.[74]

In further studies, mice fed PUFA also showed reduced immune responsiveness, but when exposed to a Gram positive coccus, a Gram negative salmonella or a yeast, they were no more susceptible to infection than control mice on a standard diet.

Evidence is accumulating that the activity of fats, unsaturated or saturated, and of other food substances and additives claimed to reduce immunological competence, is a laboratory phenomenon and of no toxicological significance in man eating even large amounts of the substances.

Immunological Tests to Predict Toxicological Allergic Hazards to Man

This summary is taken from a more detailed description of procedures and assessment of their validity as predictive tests for immunological hazards to man.[66]

Tests to detect immunological activity of substances are designed to:

(1) Determine their potency as antigens or haptens, usually compared with the results from an equal weight, and equal number of molecules of standard substances.
(2) Determine their potency as allergens or allergenic haptens, particularly for anaphylactic and for delayed hypersensitivities.
(3) Examine their immunological activity by particular routes of exposure, e.g. by inhalation or by contact with skin.
(4) Examine their potential to induce photo-allergy.

(5) Detect any activity likely to result in significant reduced immunological competence, as previously outlined in this chapter.

The routine tests to determine the potential of substances as contact or photo-allergens have some relevance to food allergy, but are not predictive of the effects of ingested food. Persons frequently handling raw foods may, however, develop contact dermatitis, and possibly contact urticaria.

In consideration of a substance, e.g. a foodstuff for its possible allergenic potency, it is first desirable to assess its antigenic potential. This is best achieved by injection into a guineapig or rat and comparing the amount and class of antibody formed, or tissue responses, with results of suitable control tests. This provides evidence of the potential activity of the substance should it enter the body. However, food allergy in man results from failure of tolerance of the ingested material; this is an individual susceptibility, and there is at present no animal model that reflects the individual human susceptibility enabling a predictive test to be made for human responsiveness.

There are laboratory animal models to predict possible allergic responses in man to substances contacting the skin or inhaled. There is no true model in rodents by ingestion, that mimics human allergic susceptibility to injected allergens. The human alimentary tract regularly permits penetration of food antigens soon after meals, which can be found in the blood and even in the breast milk of lactating mothers. Though in the majority of persons no allergy results, some antibody is formed to many common food substances, and immune complexes occur soon after ingestion of sufficient amounts of the food antigens. If rodents are fed natural diets mainly of cereals, or even substances containing potent allergens for man, e.g. nuts or beans, it is rare to find antibody, and immune complexes are not detected. The intestine does not appear to permit passage of these antigens. If guineapigs are fed cow's milk or hen's egg, particularly early in life, they develop anaphylactic sensitivity but no allergy (see below). There are findings, inconsistent, that milk or egg allergens pass the intestinal barrier in guineapigs or rats, but some tests are subject to artefact, because injection of passive sensiting antibody in a local skin site as an indication of antigen penetration into the blood from the gut, or, gavage of the test animals, induces stress that may reduce the sensitivity of the skin site, or abnormally increase penetration from the gut.

Antigenic or Allergenic Potency

For reasons stated above, this is best determined by injection into guineapigs or rats, and not by ingestion. Details are available in reference 66 from which Table 2.5 and the following extract is taken.

Table 2.5: Procedures to Determine the Potential of a Substance to Induce Anaphylactic Sensitivity

---

Active sensitisation by injection

(1) Serum samples from the sensitised animals examined for the total amount of antibody in each immunoglobulin class, especially for anaphylactic antibody. RAST* is available for rat IgE. Titres of anaphylactic antibody can be determined by passive cutaneous anaphylaxis, or by passive sensitisation of susceptible tissues or cells *in vitro*.

(2) Challenge by the same routes to observe generalised responses, or by skin test.

(3) Tests *in vitro* on tissues from sensitised animals.
  - (a) Challenge with antigen of perfused lung or slices of lung or skin, and assay of the pharmacological mediators released;
  - (b) Challenge with antigen on whole blood or leucocyte buffy coat, and assay of mediators released;
  - (c) Challenge with antigen of the ileum of guineapigs in Schultz-Dale tests.

---

* RAST: Radio-allergosorbent test.

The essentials for any protocol to examine a substance for anaphylactic sensitivity (Table 2.5) are as follows.

(a) Sensitise the selected species by injection, or by other routes if more appropriate.

(b) Take samples of blood from the sensitised animals and determine the amount of anaphylactic antibody (IgG or IgE) formed.

(c) Challenge the sensitised animals for generalised anaphylaxis, or by skin tests. Alternatively, remove lungs, skin, mesentery, peritoneal mast cells or blood leucocytes (as appropriate for the species used) and challenge the tissues *in vitro* with antigen, for release of anaphylactic mediators.

(d) Always compare the anaphylactic potential of the test substance with that of a standard control, e.g. egg albumin on an equal weight and molar basis, because the number of allergenic determinants is influenced by the number of molecules. Small molecules, more widely dispersed in the tissue, induce a different response than the same number of determinants on fewer, but larger, molecules.

The results of these tests indicate the antigenic potency, and anaphylactic potential of the new substance compared to the activity of known allergens for man, e.g. egg or β-lactoglobulin. They reflect what would happen should sufficient antigen penetrate the intestine.

Animal Models to Predict Human Allergy by Ingestion

The evidence presented in several sections of this chapter indicate that food allergy results from impairment of the normal process of immune (antigen) exclusion, so that antigens penetrating the gut mucosa first sensitise, and then elicit an allergic response if the 'fail-safe' process of forming antigen-antibody complexes does not remove the antigen harmlessly.

Ingestion tests using guineapigs have been reported by Coombs and his colleague to compare the anaphylactic sensitising potential of milk products intended for infant feeds, with the sensitising potential of untreated cow's milk. The guineapigs are fed the milk for varying periods, and after an interval of about 14 days are challenged by the appropriate milk antigen intravenously. Susceptible sensitised animals react anaphylactically.[75,76] This procedure has been shown to distinguish between allergenic potency of milk preparations, like that achieved by the injection test. However, the method is a good example of the difference in response between the guineapig and man. The guineapigs remain unaffected by continually drinking milk even though they have formed circulating anaphylactic antibody, and are exquisitely sensitive to small amounts of milk injected intravenously. Such large amounts of anaphylactic antibody (though of a different immunoglobulin class) are not usually found in normal, i.e. non-allergic, persons, and penetration of milk antigen from the gut, as regularly occurs in man, could be fatal. Though, as described earlier, some persons with significant amounts of anaphylactic antibody are not allergic to the relevant food allergen.

Tolerance in Laboratory Rodents Fed Food Antigens

Laboratory rodents tend to develop tolerance, or refractoriness to anaphylaxis, after repeated exposure to foods allergenic to man. Guineapigs drinking milk for a short period, as described above, become anaphylactically sensitised. If, however, they continue to drink milk for 70 days or longer, animals of at least one strain become tolerant, and do not even respond to an intravenous injection. They cease to make anaphylactic antibody, but some non-anaphylactic antibody is still present.[77] Similar results were obtained with guineapigs fed ovalbumin, though tolerance was age dependent.[78]

Small amounts of ingested serum albumin[78] or egg albumin are absorbed by rats, but this greatly decreases once the rats form specific antibody. Furthermore, rats fed extremely small amounts of ovalbumin, about 1 μg, and also injected with Bordetella pertussis vaccine as adjuvant, form IgE antibody, but if larger amounts of ovalbumin are fed there is a tendency to develop a refractory or tolerant state.[79,80] Rats which have formed IgE antibody do not show significant clinical changes when fed the antigen again.

Though such studies will become important in understanding tolerance to foods, they do not provide test procedures to examine allergenic potencies of foodstuffs.

Tests on Man

Once the antigenic or allergenic potency of a new food substance has been determined by the animal injection tests, and compared with the potency of known human food allergens, the test substance is best fed to man to determine its acceptability. Antibody responses (which for a new food will show an early peak and then decline) for each immunoglobulin class are easily determined, and a variety of other tests on leucocytes can be made for supporting evidence, e.g. histamine release from leucocyte suspensions containing basophils.

References

1. W.E. Parish, 'Clinical Immunology and Allergy' in A. Rook, D.S. Wilkinson and F.J.G. Ebling (eds.), *Textbook of Dermatology*, 3rd edn. (Blackwell Scientific Publications, Oxford, 1979), pp. 249–287.

2. W. Kazimierczak and B. Diamant, 'Mechanisms of Histamine Release in Anaphylactic and Anaphylactoid Reactions', *Progr. Allergy, 24* (1978), p. 295.

3. P. Dukor, P. Kallos, H.D. Schlumberger and G.B. West (eds.), 'Genetic Aspects and Anaphylactoid Reactions', in *Pseudo-Allergic Reactions. Involvement of Drugs and Chemicals*, (Karger Press, Basel, 1980), vol. 1.

4. T.Y. Ha and B.H. Waksman, 'Role of the Thymus in Tolerance Suppressor Activity of Antigen-stimulated Rat Thymocytes Transferred to Normal Recipients', *J. Immunol., 110* (1973), pp. 1290–1299.

5. A. Basten, F.A.P. Miller, J. Sprent and C. Cheers, 'Cell-to-cell Interaction in the Immune Response T-cell-dependent Suppression in Tolerant Mice', *J. Exp. Med., 140* (1974), pp. 199–217.

6. A.J. Strelkauska, R.T. Callery, J. McDowell, Y. Borel and S.F. Schlossman, 'Direct Evidence for Less of Human Suppressor Cells During Active Auto-immune Disease', *Proc. Natl Acad. Sci. USA, 75* (1978), pp. 5150–5154.

7. H.J.F. Hodgson, J.R. Wards and K.J. Isselbacher, 'Decreased Suppressor Cell Activity in Inflammatory Bowel Disease', *Clin. Exp. Immunol., 32* (1978), pp. 451–458.

8. R.M.M. Victorino and H.J.F. Hodgson, 'Alteration in T Lymphocyte Populations in Inflammatory Bowel Disease', *Clin. Exp. Immunol. 41* (1980), pp. 156–165.

9. I.L. Strannegard, L. Lindholm and O. Strannegard, 'T Lymphocytes in Atopic Children', *Int. Arch Allergy appl. Immunol., 50* (1976), pp. 684–692.

10. D.R. Stanworth and M.W. Turner, 'Immunochemical Analysis of Immunoglobulins and Their Subunits', in D.M. Weir (ed.), *Handbook of Experimental Immunology. Immunochemistry*, 3rd edn. (Blackwell Scientific Publications, Oxford, 1978), vol. 1, pp. 6.1–6.102.

11. K. Ishizaka and T. Ishizaka, 'Immunoglobulin E: Biosynthesis and Immunological Mechanisms of IgE-mediated Hypersensitivity' in S. Gupta and R.A. Good (eds.), *Comprehensive Immunology* (Plenum, New York and London, 1979), vol. 6, pp. 153–178.

12. G. Moller (ed.), 'Immunoglobulin E', *Immunological Reviews, 41* (1978) (Munksgaard, Copenhagen).

13. M. Joseph, A.B. Tonnel, A. Capron and C. Voisin, 'Enzyme Release and Superoxide Anion Production by Human Alveolar Macrophages, Stimulated with Immunoglobulin E', *Clin. Exp. Immunol., 40* (1980), pp. 416–422.

14. P. Brandtzaeg, 'Transport Models for Secretory IgA and Secretory IgM', *Clin. Exp. Immunol., 44* (1981), pp. 221–232.

15. T.B. Tomasi, L. Larson, S. Challacombe and P. McNabb, 'Mucosal Immunity: the Origin and Migration Patterns of Cells in the Secretory System', *J. Allergy Clin. Immunol., 65* (1980), pp. 12–19.

16. J. Bienenstock and A.D. Befus, 'Mucosal Immunology', *Immunology 41* (1980), pp. 249–270.

17. D.B.L. McClelland, 'Bacterial and Viral Infections of the Gastrointestinal Tract' in P. Asquith (ed.), *Immunology of the Gastrointestinal Tract* (Churchill Livingstone, Edinburgh, 1979), pp. 214–245.

18. C. Andre, H. Bazin and J.F. Heremans, 'Influence of Repeated Administration of Antigen by the Oral Route on Specific Antibody-producing Cells in the Mouse Spleen', *Digestion, 9* (1973), pp. 166–175.

19. R.M. Rothberg, S.C. Kraft and S.M. Michalek, 'Systemic Immunity After Local Antigenic Stimulation of the Lymphoid Tissue of the Gastrointestinal Tract', *J. Immunol., 111* (1973), pp. 1906–1913.

20. C. Griscelli, P. Vassalli and R.T. McCluskey, 'The Distribution of Large Dividing Lymph Node Cells in Syngeneic Recipient Rats after Intravenous Injection', *J. Exp. Med., 130* (1969), pp. 1427–1451.

21. G.L. Asherson, M. Zembala, M.A.C.C. Pevera, B. Mayhew and W.R. Thomas, 'Production of Immunity and Unresponsiveness in the Mouse by Feeding Contact Sensitizing Agents and the Role of Suppressor Cells in the Peyer's Patches, Mesenteri Lymph Nodes and Other Lymphoid Tissues', *Cell Immunol., 33* (1977), pp. 145–155.

22. L.K. Richman, J.M. Chiller, W.R. Brown, D.G. Hanson and N.M. Vaz, 'Enterically Induced Immunologic Tolerance. I Induction of Suppressor T Lymphocytes by Intragastric Administration of Soluble Proteins', *J. Immunol., 121* (1978), pp. 2429–2434.

23. R.R.A. Coombs and P.G.H. Gell, 'Classification of Allergic Reactions Responsible for Clinical Hypersensitivity and Disease' in P.G.H. Gell, R.R.A. Coombs and P.J. Lachman (eds.), *Clinical Aspects of Immunology,* 3rd edn. (Blackwell Scientific Publications, Oxford, 1975), pp. 761–781.

24. S.I. Wasserman, 'The Mast Cell and the Inflammatory Response', in J. Pepys and A.M. Edwards (eds.), *The Mast Cell; its Role in Health and Disease* (Pitman Medical Press, Tunbridge Wells, UK, 1979), pp. 9–20.

25. W.E. Parish, 'The Clinical Relevance of Heat-stable Short-term Sensitizing

Anaphylactic IgG Antibodies (IgG S-TS), and of Related Activities of IgG_4 and IgG_2', *Brit. J. Derm., 105* (1981), pp. 223–231.

26. W.E. Parish, 'Evidence for Human IgG Antibodies Anaphylactically Sensitizing Man' in M.K. Bach (ed.), *Immediate Hypersensitivity: Modern Concepts and Developments* (Marcel Dekker, New York & Basel, 1978), pp. 277–299.

27. M.H. Lessof, D.G. Wraith, T.G. Merrett, J. Merrett and P.D. Baisseret, 'Food Allergy and Intolerance in 100 Patients. Local and Systemic Effects', *Quart. J. Med., New Series, 49* (1980), pp. 259–271.

28. A. Danneus, S.G.O. Johansson, T. Foucard and S. Ohman, 'Clinical and Immunological Aspects of Food Allergy in Childhood. I Estimation of IgG, IgA and IgE Antibodies to Food Antigens in Children with Food Allergy and Atopic Dermatitis', *Acta paed. Scand., 66* (1977), pp. 31.

29. W.E. Parish, 'Detection of Reaginic and Short-term Sensitizing Anaphylactic or Anaphylactoid Antibodies to Milk in Sera of Allergic and Normal Persons', *Clinical Allergy, 1* (1971), pp. 369–380.

30. B.F.J. Goodwin, 'IgE Antibody Levels to Ingested Soya Protein Determined in a Normal Adult Population', *Clinical Allergy, 12* (1982), pp. 55–62.

31. S. Freier, 'Paediatric Gastrointestinal Allergy', *Clinical Allergy, 3* (1973) supplement, pp. 597–618.

32. S.L. Bahna and D.C. Heiner, *Allergies to Milk* (Grune & Stratton, New York & London, 1980).

33. E. Bleumink, 'Food Allergy and the Gastrointestinal Tract' in P. Asquith (ed.), *Immunology of the Gastrointestinal Tract* (Churchill Livingstone, Edinburgh, 1979), pp. 195–213.

34. C.D. May, 'Food Hypersensitivity' in S. Gupta and R.A. Good (eds.), *Comprehensive Immunology* (Plenum Medical Book Co., New York & London, 1979), vol. 6, pp. 321–354.

35. J.B. Kirsner, 'Experimental 'Colitis' with Particular Reference to Hypersensitivity Reactions of the Colon', *Gastroenterology, 40* (1967), pp. 307–311.

36. H.J.F. Hodgson, B.J. Potter, J. Skinner and D.P. Jewell, 'Immune-complex Mediated Colitis in Rabbits. An Experimental Model', *Gut, 19* (1978), pp. 225–232.

37. M. Gunther, B. Aschaffenburg, R.H. Matthews, W.E. Parish and R.R.A. Coombs, 'The Level of Antibodies to the Proteins of Cow's Milk in the Serum of Normal Human Infants', *Immunology, 3* (1960), pp. 296–306.

38. K.B. Taylor, D.M. Sidney, C. Truelove and R. Wright, 'Serologic Reactions to Gluten and Cow's Milk Proteins in Gastrointestinal Disease', *Gastroenterology, 46* (1964), pp. 99–108.

39. B. Kletter, J. Gery, S. Freier and A.M. Davies, 'Immune Responses of Normal Infants to Cow Milk', *Int. Arch. Allergy, 40* (1971), pp. 656–666.

40. A. Ferguson and F. Carswell, 'Precipitins to Dietary Proteins in Serum and Upper Intestinal Secretions of Coeliac Children', *Brit. Med. J., i* (1972), pp. 75–77.

41. H. Valdimarsson, 'Effector Mechanisms in Cellular Immunity' in M.J. Hobart and I. McConnell (eds.), *The Immune System* (Blackwell Scientific Publications, Oxford, 1975), pp. 179–196.

42. K. Bendtzen, 'Biological Properties of Lymphokines', *Allergy, 33* (1978), pp. 105–119.

43. J. Morley, R.A. Wolstercroft and D.C. Dumonde, 'Measurement of Lymphokines' in D.M. Weir (ed.), *Handbook of Experimental Immunology*, vol. 2. *Cellular Immunology*, 3rd edn. (Blackwell Scientific Publications, Oxford, 1978), pp. 27.1–27.28.

44. L. Hudson and F.C. Hay, *Practical Immunology*, 2nd edn. (Blackwell

Scientific Publications, Oxford, 1980).

45. D.M. Weir (ed.), *Handbook of Experimental Immunology. Cellular Immunology*, 3rd edn. (Blackwell Scientific Publications, Oxford, 1978).

46. B.B. Mishell and S.M. Shiigi (eds.), *Selected Methods in Cellular Immunology* (W.H. Freeman, San Francisco, 1980).

47. H.F. Dvorak and M.E. Hammond, 'Cutaneous Basophil Hypersensitivity' in M.K. Bach (ed.), *Immediate Hypersensitivity: Modern Concepts and Developments* (Marcel Dekker, New York and Basel, 1978), pp. 659–692.

48. P.W. Askenase and J.E. Atwood, 'Basophils in Tuberculin and "Jones-Mote" Delayed Reactions of Humans', *J. Clin. Invest., 58* (1976), pp. 1145–1154.

49. P. Asquith (ed.), *Immunology of the Gastrointestinal Tract* (Churchill Livingstone, Edinburgh, 1979).

50. S.C. Truelove and D.P. Jewell, 'The Intestine in Allergic Diseases' in P.G.H. Gell, R.R.A. Coombs and P.J. Lachmann (eds.), *Clinical Aspects of Immunology*, 3rd edn. (Blackwell Scientific Publications, Oxford, 1975), pp. 1441–1465.

51. G.K.T. Holmes, P. Asquith and W.T. Cooke, 'Cell-mediated Immunity to Gluten Fraction III in Adult Coeliac Disease', *Clin. exp. Immunol., 24* (1976), pp. 259–265.

52. A. Ferguson, 'Delayed-hypersensitivity Reactions in the Small Intestine' in W.A. Hemmings (ed.), *Antigen Absorption by the Gut* (MTP Press, Lancaster, UK, 1978), pp. 145–154.

53. L. Juhlin, 'Incidence of Intolerance to Food Additives', *Int. J. Dermatol., 19* (1980), pp. 548–551.

54. L. Rosenhall, 'Hypersensitivity to Analgesics, Preservatives and Food Colorants in Patients with Asthma or Rhinitis', *Acta Univ. Upsaliens, 269* (1977), pp. 1–117.

55. G. Michaelsson and L. Juhlin, 'Urticaria Induced by Preservatives and Dye Additives in Food and Drugs', *Brit. J. Derm., 88* (1973), pp. 525–532.

56. L. Juhlin, 'Recurrent Urticaria: Clinical Investigation of 330 Patients', *Brit. J. Derm., 104* (1981), pp. 369–381.

57. R.P. Warin and R.H. Champion, *Urticaria* (W.B. Saunders Co. London and Philadelphia, 1974), pp. 39–66.

58. B.S.M. Stenius and M. Lemola, 'Hypersensitivity to Acetylsalicylic Acid (ASA) and Tartrazine in Patients with Asthma', *Clinical Allergy, 6* (1976), pp. 119–129.

59. J.K. Weltman, R.P. Szaro and G.A. Settipane, 'An Analysis of the Role of IgE in Intolerance to Aspirin and Tartrazine', *Allergy, 33* (1978), pp. 273–281.

60. N. Weliky and D.C. Heiner, 'Hypersensitivity to Chemicals, Correlation of Tartrazine Hypersensitivity with Characteristic Serum IgD and IgE Immune Response Patterns', *Clinical Allergy, 10* (1980), pp. 375–394.

61. E. Valverde, J.M. Vich, J.V. Garcia-Calderon and P.A. Garcia-Calderon, '*In vitro* Stimulation of Lymphocytes in Patients with Chronic Urticaria Induced by Additives and Food', *Clinical Allergy, 10* (1980), pp. 691–698.

62. A. Szozerlik, R.J. Gryglewski and G. Czerniawska-Mysik, 'Relationship of Inhibition of Prostaglandin Biosynthesis by Analgesics to Asthma Attacks in Aspirin-Sensitive Patients', *Br. Med. J., i* (1975), pp. 67–69.

63. J.G. Gerber, N.A. Payne, O. Okbz, A.S. Nies and J.A. Oates, 'Tartrazine and the Prostaglandin System', *J. Allergy Clin. Immunol., 63* (1979), pp. 289–294.

64. B.B. Vargaftig, J.C. Bessot and G. Pauli, 'Is Tartrazine-induced Asthma Related to Inhibition of Prostaglandin Biosynthesis', *Respiration, 39* (1980), pp. 276–282.

65. J.J. Kreindler, J. Slutsky and Z.H. Haddad, 'The Effect of Food Colours and Sodium Benzoate on Rat Peritoneal Mast Cells', *Ann. Allergy, 44* (1980), pp. 76–81.

66. W.E. Parish, 'Immunological Tests to Predict Toxicological Hazards to Man' in J.W. Gorrod (ed.), *Testing for Toxicity* (Taylor & Francis Ltd, London, 1981), pp. 297–315.

67. D.L. Archer, 'Immunotoxicology of Foodborne Substances: an Overview', *J. Food Protection, 41* (1978), pp. 983–988.

68. D.L. Archer, J.A. Bukovic-Wes and B.G. Smith, 'Inhibitory Effect of an Anti-oxidant, Butylated Hydroxyanisole, on the Primary *in vitro* Immune Response', *Proc. Soc. Exp. Biol. Med., 154* (1977), pp. 289–294.

69. C.M. Hampton and R.L. Myers, 'Effects of Cyclamate Calcium on the Immune Response in Rabbits', *Arch. Envir. Hlth,* (1976), pp. 51–53.

70. J.G. Vos and J.A. Moore, 'Immune Suppression as Related to Toxicology', *C.R.C. Critical Reviews in Toxiology* (1977), pp. 67–101.

71. J.B. Silkworth and L.D. Loose, 'Assessment of Environmental Contaminant-induced Lymphocyte Dysfunction', *Envir. Hlth Perspectives, 39* (1981), pp. 105–128.

72. J. Mertin and D. Hughes, 'Specific Inhibiting Action of Polyunsaturated Fatty Acids on Lymphocyte Transformation Induced by PHA and PPD', *Int. Archs. Allergy appl. Immunol., 48* (1975), pp. 203–210.

73. J. Mertin, 'Effect of Polyunsaturated Fatty Acids on Skin Allograft Survival and Primary and Secondary Cytotoxic Response in Mice', *Transplantation, 21* (1976), pp. 1–4.

74. J.V. Friend, S.O. Lock, M.J. Gurr and W.E. Parish, 'Effect of Different Dietary Lipids on the Immune Responses of Hartley Strain Guinea-pigs', *Int. Archs. Allergy appl. Immunol., 62* (1980), pp. 292–301.

75. K.J. Anderson, P. McLaughlan, M.E. Devey and R.R.A. Coombs, 'Anaphylactic Sensitivity of Guinea-pigs Drinking Different Preparations of Cows' Milk and Infant Formulae', *Clin. Exp. Immunol., 35* (1979), pp. 454–461.

76. P. McLaughlan, K.J. Anderson and R.R.A. Coombs, 'An Oral Procedure to Determine the Sensitizing Capacity of Infant Feeding Formulae', *Clinical Allergy, 11* (1981), pp. 311–318.

77. R.R.A. Coombs, M.E. Devey and K.J. Anderson, 'Refractoriness to Anaphylactic Shock after Continuous Feeding of Cow's Milk to Guinea-pigs', *Clin. exp. Immunol., 32* (1978), pp. 263–271.

78. L.M.J. Heppell and P.J. Kilshaw, 'Immune Responses of Guinea-pigs to Dietary Protein', *Arch. Allergy appl. Immunol., 68* (1982), pp. 54–59.

79. C. Andre, J.P. Vaerman and J.F. Heremans, 'Oral Immunization of Rats with Human Serum Albumin: Interference with Intestinal Absorption and Tolerogenic Effect' in W.A. Hemmings (ed.), *Antigen Absorption by the Gut*, (MTP Press, Lancaster, 1978), pp. 65–71.

80. E.E.E. Jarrett, D.M. Haig, W. McDougall and E. McNulty, 'Rat IgE Production II. Primary and Booster Reaginic Antibody Responses Following Intradermal or Oral Immunization', *Immunology, 30* (1976), pp. 671–677.

81. H. Bazin and E. Platteau, 'Production of Circulating Reaginic (IgE) Antibodies by Oral Administration of Ovalbumin to Rats', *Immunology, 30* (1976), pp. 679–684.

3 TERATOGENICITY AND REDUCED FERTILITY RESULTING FROM FACTORS PRESENT IN FOOD

A.B.G. Lansdown

Introduction

For many hundreds of years, man has been aware of the numerous structural and metabolic deformities that occur in the newborn of his own race and in other species. Many interesting cases have been described in the literature but until relatively recent times, few attempts were made to understand the causes of these abnormalities. The classical observations by Gregg[1] on the relationship between maternal rubella infection in pregnancy and the birth of children with heart disease and other deformities probably marked the beginning of teratology as we know it today. The subsequent demonstration of the teratogenic potential of the hypnotic drug, thalidomide, in the early 1960s gave a tremendous impetus to epidemiological and experimental studies designed to seek out other drugs, environmental chemicals and conditions which are potentially harmful to the developing fetus.

The word teratology, from its Greek derivation, implies a study of gross malformations and monstrosities arising from abnormalities in prenatal development. Nowadays, teratology is taken in a wider sense to include all aspects of abnormal prenatal development leading to structural, functional and behavioural defects with or without reduced post-natal viability. As a consequence of the numerous investigations conducted mainly in the last 20 years, a wide range of drugs, infectious agents and environmental contaminants have been identified as having teratogenic properties in one or more species. Although the main interest in this period has been on the teratogenicity of drugs and therapeutic agents, more recently the possibility of teratogens being present in food and drink has attracted attention.

It will be readily appreciated that for normal fetal development to occur, a mother should have a diet which contains adequate amounts of all essential constituents, and that abnormal fetal development may result from deficiencies in proteins, amino acids, lipids, minerals and vitamins. Whereas a chronic food shortage as occurs in many underprivileged societies is a well-known cause of fetal mortality and intrauterine growth retardation, it is now appreciated that structural and functional deformities are additional complications of starvation. Some nutrients which are essential for normal fetal development in small quantities may be teratogenic if present in the maternal diet at supra-optimal levels. Hypervitaminosis A is a well-known example of this situation.

In addition to the many substances present in food with an acknowledged nutritional role, a wide variety of other substances may be ingested with the diet either by intention or accidentally and which have no nutritional benefit. A number of these have been identified as teratogens in humans or in laboratory animals. Thus one identifies three groups of teratogenic agents in food:

1. Food constituents that have a nutritional value
2. Food additives included to enhance the appearance or acceptability of the food product
3. Food contaminants

According to Nishimura and Tanimura,[2] none of the current food additives have been claimed to be responsible for hazards to human offspring with epidemiological evidence. Up until recently, few publications have appeared in the literature on this subject. This may indicate either that teratogenicity studies have not been undertaken on food additives or that the results were negative in which case they would be unlikely to be accepted for publication. Alternatively, it may be that fetal safety evaluation studies have been conducted on certain food additives but the reports have been submitted in special submissions for governmental purposes, and as such would not be openly available.

By far the greatest teratogenic risk to humans from food concerns the contaminants. This is a large and heterogeneous group of substances and includes minerals, agricultural preparations (pesticides and herbicides), chemicals used in the packaging and preparation of the food, and plant constituents (alkaloids, fungal toxins, etc.).

In the present review, the problems of teratogenicity and reduced fertility arising from the ingestion of substances present in food is discussed with particular attention to agents in the last two categories. Because available information suggests that some of these teratogenic substances may act by impairing the availability of nutrients essential for normal fetal growth, the initial part of the review will relate to the influence of deficiencies and excess levels of essential food constituents in fetal growth in humans or other species. Much of the available information on the potential teratogenicity of environmental agents has been provided by experimental studies in suitable laboratory animals, and for this reason the subject is discussed, as far as possible, from a comparative standpoint.

Epidemiological and Experimental Aspects

A wide variety of substances which may be ingested in the food can be detrimental to a developing human fetus if ingested by a pregnant woman at a vulnerable stage in gestation. The onus is upon food manufacturers, agricultural scientists and regulatory authorities to identify offending agents and to take appropriate steps to remove them from the food as eaten or to restrict their use in the environment. Strongest evidence that a particular substance is teratogenic to man is provided by large-scale population studies such as those carried out in Japan and Iraq relating to accidental mercury poisoning. Chronic maternal alcoholism is a further illustration of teratogenicity arising from food where irrefutable clinical evidence exists.

In the main, teratological studies now tend to be prospective in outlook rather than retrospective. That is, by carefully conducted laboratory trials in animals, one hopes to identify most fetotoxic and teratogenic agents amongst new drugs, food additives and likely food contaminants before a human risk is encountered. Thus the experimental studies set up in the laboratory are anthropocentric in design, but, because no non-human species will reliably predict the human teratogenic risk for a wide range of agents, at least two different species of animals are recommended in pre-clinical studies.[3]

A well-defined series of pre-clinical laboratory tests has been in existence for some time for use in assessing the teratogenicity of drugs, but they are equally relevant in the safety evaluation of other substances such as those found in food. The basic scientific principles are:

1. Susceptibility to teratogenesis depends upon the genotype of the conceptus and the manner in which this interacts with environmental factors.
2. The susceptibility of a conceptus to teratogenic influence varies with its developmental stage at the time of exposure.
3. Teratogenic agents act by specific mechanisms on developing cells and tissues to initiate abnormal embryogenesis. The manifestations

of this include prenatal mortality (abortion or stillbirth), structural malformation, retarded intrauterine growth or post-natal development, and functional or behavioural disorders.

4. Manifestations of deviant development increase in degree as dosage increases from a no-effect level to a totally lethal level.[4]

Pre-clinical studies in laboratory animals are designed to show whether a particular chemical is accompanied by fetotoxic or teratogenic risks and the conditions under which these occur. Important in the safety evaluation of food substances, additives and contaminants is the route by which they are administered. Normally, this will be in the diet or by intragastric dosage and requires that in any non-human species used, the test substance will be absorbed from the gastrointestinal tract in a way similar to that seen in humans.

The ability of different species to absorb foreign substances from the intestine varies widely, however, probably on account of the wide variations in gastric acidity, digestive enzyme systems, specialisation to a particular form of diet, bacterial flora and morphological characteristics of the mucosal surfaces. For most known teratogens, there is a close correlation between the amount absorbed into the maternal circulation and the concentration present in the fetus at sensitive stages in development.[5] Thus variations in intestinal absorption contribute significantly to the interspecies differences seen in response to teratogens administered orally.

Several examples may be given to illustrate this point. Studies in Japan and elsewhere have shown that when pregnant women consumed seafood contaminated with methyl mercury, their offspring exhibited an increased risk of neurological damage.[6,7] In humans, mercury is absorbed intestinally in either inorganic or organic form but in other species variations exist. Rats do not readily absorb metallic mercury or inorganic salts from the gut but, by action of the bacterial flora present, are able to methylate these agents to organic forms which are absorbed.[8] Destruction of the bacteria in intestinal extracts by autoclaving or with antibiotics prevented this methylation process.

Although interspecies variations in susceptibility to orally administered teratogens are undoubtedly attributable in part to differences in the rate of absorption, species sensitivity and metabolic patterns of the offending agent are also important factors. Gillette[5] discussed at length some variations which exist in a way in which humans and other species metabolise teratogenic drugs. However, most interspecies variations in response to teratogens are attributable to differences in the rate of metabolism and not type. Hexobarbital, for example, is metabolised within minutes in the mouse and rabbit but has a half-life of several hours in humans. Imipramine is variously teratogenic in rats, mice and rabbits[9] yet has a questionable effect on the human fetus at recommended therapeutic levels, probably due to variations in the metabolism of imipramine and its excretion pattern.[10]

For the greater part of pregnancy, the capacity of the fetal liver to metabolise transplacentally acquired compounds is strictly limited by the low levels of metabolising enzymes present. Some metabolism of foreign compounds occurs in the placenta but the largest part takes place in the maternal liver.[11,12] The relative importance of enzyme systems in the maternal liver for metabolism of foreign compounds varies greatly from one species to another and where a particular pathway may predominate in one species it may be absent altogether in a second.[13]

Alcohol is teratogenic in humans and several non-human species. Premature delivery and neonatal mortality are complications of chronic alcoholism in pregnant women, with surviving offspring exhibiting the 'fetal alcohol syndrome' which is discussed below. In contrast, adult rats have several times the ability of an adult woman to metabolise alcohol and they void it more rapidly from the tissues.[14] In this species, alcohol produced minimal teratogenicity with some placental damage and changes in maternal tissues. Caffeine also is metabolised at different rates in animals and humans.[15] Metabolic rates expressed in terms of half-life, were considerably longer in dogs than in humans, rhesus monkeys or rats, probably due to increased plasma protein binding, tissue distribution and delayed excretion patterns.

Many studies to evaluate the teratogenic properties of food chemicals and contaminants have used dosage levels far in excess of those reasonably likely to be consumed by a pregnant woman even in the most extreme conditions. For example, sodium chloride is an essential dietary constituent and would not under normal conditions give rise to ill effect. Yet when pregnant mice were dosed with 1900–2500 mg/kg at about mid-gestation, fetal stunting and increased mortality were recorded.[16,17] This level is at least ten times greater than the average amount consumed each day by a 60 kg person. No structural abnormalities were seen in this mouse study. Clearly such a high concentration of sodium chloride would be expected seriously to disturb the osmotic balance in the body with a net metabolic change which is reflected as a non-specific effect on the fetus.

Coffee is widely consumed throughout the world and the teratogenicity of caffeine has been subject to a long controversy. The epidemiological studies of Lechat and his colleagues suggest that there may be a teratogenic risk associated with heavy coffee drinking in pregnancy, i.e. at least eight cups a day,[18] but this issue awaits further clarification. Various types of abnormalities have been produced in experimental laboratory animals, but at repeated oral doses equivalent to 38-76 cups of coffee each day in humans.[19] Such a massive dose absorbed into the maternal circulation at one time can hardly reflect the human risk where up to eight cups of coffee would normally be consumed in several hours.

In summary, differences in the rates at which humans and experimental animals metabolise known teratogenic compounds are clearly related to the teratogenic responses seen. The variations are not consistent for a wide range of teratogens and this makes selection of a suitable test species difficult. Experience with several drugs has shown that although the old world monkeys are phylogenetically close to humans, there is no reason to suppose that they will be more suitable than other laboratory animals in teratological studies. They tend to metabolise many compounds at different rates to man and exhibit marked variations in sensitivity and response. Currently, rabbits, rats and mice are the most frequently used species in teratological studies, but on occasions other species including monkeys, dogs, cats, guinea pigs and hamsters are selected. Choice of these species for use in predictive teratogenicity has been largely based on the knowledge of their reproductive behaviour, the developmental patterns of the offspring, proven susceptibility to known teratogens administered orally, and regular availability of the animals. Experience gained in experimental teratogenicity of drugs has largely served as a guideline in most recent studies for the prediction of hazards with food chemicals and additives.

Food Constituents

Natural Constituents

A maternal diet adequate in a range of essential nutrients is important for normal fetal development, but variations exist between different mammalian species in their relative requirements. Widespread studies in America and elsewhere have shown that diets deficient in protein in pregnancy cause retarded fetal growth, neonatal death and occasionally congenital deformity.[20,21] In mild forms, nutritional deprivation in pregnancy may not be injurious to the fetus since the pregnant mother is usually able to adapt her metabolism to compensate for the deficiency. In a general dietary deficiency, protein is a limiting factor.[22]

Protein. Protein deficiency in human pregnancy has been associated with a wide spectrum of developmental and behavioural anomalies in the fetus and new-born including cardiovascular, skeletal, mental and immunological deformities. Experimental studies in rats and other species indicate that many of these defects are not necessarily caused by defects in developmental processes but are due to retarded differentiation resulting from mitotic inhibition, that is, cell numbers in various tissues are lower than in normal.[23]

Although protein deficiency is an established cause of retarded and abnormal fetal development with reduced viability, shortages of certain amino acids may also be detrimental for fetal development and in some instances these may be damaging as a total protein deficiency. Pregnant rats, for example, require adequate amounts of tryptophan in pregnancy and deficiencies result in congenital cataract in the offspring.[24] Omission of lysine, methionine, valine or isoleucine from the diet leads to fetal death or to the birth of young showing impaired growth and mental development.[25]

Higher than normal levels of certain amino acids in the maternal circulation may also be a cause of fetal distress. Hyperphenylalaninaemia, in humans caused by a genetically determined inability to metabolise the amino acid, is a well-known cause of abnormal fetal development.[26] A similar situation has been demonstrated in rats injected with high doses of phenylalanine in pregnancy.[27] Excessive levels of dietary adenine, lysine and leucine are known to cause fetal mortality and occasionally structural malformation in rats but the mechanisms for these are not known.[28,29]

Carbohydrates. Carbohydrates are important in maintaining pregnancy but the role of sugar deficiency in causing fetal distress has not been well studied.[22] However, excessive levels of blood galactose as occurs in some women with a heterozygous genetical condition, is a cause of retarded growth with mental and optic deformities.

Lipids. Lipids are essential for fetal survival and prenatal mortality is a common consequence of maternal dietary deficiencies in pregnancy. Deformities are rarely seen. Dietary deficiencies in rats of certain essential lipids, such as linoleic acid, lead to retarded fetal growth and defects in the synthesis of brain cerebrosides.[30] Post-natal growth retardation was depressed if the lipid deficiency was prolonged into the lactation period. A lipid-sufficient diet, given to mothers after delivery, resulted in a 'catch-up' growth in the young. Prenatal lipid deprivation in rats led to abnormal behaviour post-natally, the brain weight was less than normal yet these offspring showed an increased learning capacity.[31]

The influence of excessive dietary lipid in pregnancy on fetal development seems to depend on the type of lipid administered and the duration of the period of dosage. High cholesterol levels in rabbits led to retarded fetal growth and increased mortality[32] but the offspring of pregnant rats fed a diet containing 20 per cent safflower oil were similar to the controls.[33] Reduced exploratory activity and an improved learning capacity were seen in the offspring of rats fed a 20 per cent cocoa butter diet in pregnancy.

Vitamins

Vitamin A (retinol). Vitamin A is derived from foods of animal origin. It is essential for normal fetal growth and development but is harmful if present in excess. Hypovitaminosis A has been shown experimentally in rats and rabbits to be a cause of ovum degeneration, failure of the fertilised ovum to implant and a range of congenital abnormalities involving the eyes, lip and palate, brain, urinogenital and cardiopulmonary tracts.[34,35] The type of effect seen was clearly related to the time in gestation at which hypovitaminosis was imposed and its severity.

Vitamin A is freely soluble in fats and fat solvents and is stable at normal cooking temperatures. However, prolonged exposure of foods to bright sunlight or the fats turning rancid results in an oxidation of vitamin A and its consequent destruction. Thus a prolonged feeding of such diets in pregnancy poses a potential teratogenic risk to the offspring.

Although normal eye development depends upon the availability of adequate amounts of vitamin A, the role of the vitamin in epithelial cell metabolism is uncertain. Recent views are that vitamin A is required for protein synthesis, changes in extracellular collagen and mucopolysaccharides. Defects in myelination may also accompany vitamin A deficiency.[36]

Particularly high levels of vitamin A have been recorded in fish liver and have been associated with human illness. Occurring in pregnancy, hypervitaminosis A is highly teratogenic in many species resulting in neurological, optic and skeletal abnormalities. Currently it is generally held that these fetal defects arise from abnormalities in growth patterns, probably from a loss of synchrony in mitosis in susceptible organ systems.[37] Impaired DNA synthesis in neuroepithelial tissues and mitotic inhibition, and defective mucopolysaccharide synthesis have also been suggested.[38]

Vitamin D. Calcium absorption from the intestinal tract and its regulation in body tissues is dependent on adequate levels of dietary vitamin D. Thus vitamin D has an important function in osteogenesis. Maternal vitamin deprivation leads to a congenital rickets-like condition and dental caries. In view of the importance of vitamin D in calcium metabolism, the principal teratogenic effects of maternal dietary deficiency involve the skeleton and dentition.

Excessive consumption of vitamin D in human pregnancy does not appear to cause a maternal hypercalcaemia or to be harmful for fetal development. Calcium uptake is regulated by parathyroid hormone and except where disease occurs, the fetus is protected from the adverse effects of excessively high levels. This differs from the situation in pregnant rats and rabbits where high dietary levels of vitamin D lead to low birthweights and reduced concentrations of calcium and phosphate in the offspring.[39] Abnormalities involving the skeletal and cardiovascular systems have been recorded.[40] Young rats subjected to excessive vitamin D prenatally failed to maintain a normal growth rate and exhibited abnormal osteogenesis. Excessive dietary vitamin D in rats before mating led to changes in the oestrus cycle and impeded fertilisation.[41]

Vitamin E. The association between dietary fat and vitamin E deficiency was known to Mattill[42] who in 1927 showed that an increase in the rancidity of the diet led to sterility in rats. Pregnant females subject to vitamin E deficiency aborted their litters, possibly due to a specific requirement for this vitamin.[43] Vitamin E acts as antioxidant and the deprived rats showed an excessively high metabolic rate with fatty degeneration in their muscles resembling a muscular dystrophy.

When vitamin E-deficient rats were given 2 mg of DL-α-tocopherol acetate intragastrically on the tenth day of gestation, a high resorption rate was observed and 37 per cent of the survivors exhibited abnormalities.[44] The type and severity of the abnormalities varied according to the amount of vitamin E present in the diet. Interestingly, the level of vitamin present in the fetuses did not correlate closely with the maternal vitamin intake.[45] In another study, vitamin dosing in rats subject to chronic vitamin E deficiency resulted in high maternal mortality and fetal death but without evidence of structural malformations.[46]

Vitamin K. Vitamin K is necessary for prothrombin synthesis and the normal coagulation of the blood. Thus patients subject to vitamin K deficiency are prone to excessive haemorrhages. Minimum requirements for this vitamin are low in humans and since most foods contain small quantities the effects of hypovitaminosis K are not well known. In pregnancy, the fetal vitamin K requirements are derived largely from the maternal circulation. In cases of deficiency, the offspring become liable to prolonged haemorrhage through defects in prothrombin synthesis. Infantile haemorrhagic disease in underprivileged societies is a serious clinical problem associated with vitamin K deficiency in some parts of the world.

In cattle, vitamin K availability is impaired when animals eat sweet clover. This plant contains the vitamin K analogue, dicoumarol, which inhibits vitamin K metabolism and causes symptoms resembling vitamin K deficiency.[47] The calves frequently die of haemorrhage. In rabbits, hypovitaminosis K has been shown to cause abortion with the survivors showing an increased tendency to haemorrhage. Fetal congenital abnormalities of blindness, mental retardation, hypoplastic nasal structures and neonatal mortality have been reported in a few cases of vitamin K treatment in human pregnancies.[48]

Water Soluble Vitamins of Groups B and C. Many studies have been conducted on the teratogenic action in rats on dietary deficiencies involving vitamins of groups B and C as well as the influence of anti-vitamins such as 6-aminonicotinamide, desoxypyridoxine and folate antagonists (e.g. aminopterin, pteroylglutamic acid (PGA)). Vitamin C appears to be necessary for normal health and fetal development in only a few species including humans, monkeys and guinea pigs. Fetal death and abortion in mid-gestation is a common consequence of vitamin C deficiency.[49] Histological examination of the tissues of these fetuses provided evidence for defects in collagen, elastin and mucopolysaccharide synthesis.

In humans, scurvy is a well-known complication of vitamin C deficiency, but the effects of vitamin C shortage in pregnancy are less well understood.[22] A comparison between vitamin C-deficient and normal pregnancies did not reveal an increased rate of structural abnormalities but the risk of premature delivery was greater.

Excessive vitamin C during pregnancy also seems to cause intrauterine mortality with stillbirths and abortions occurring in humans, guinea pigs, rats and mice.[50] Male guinea pigs given high doses of vitamin C showed infertility.[51]

The influence of vitamin B deficiencies in pregnancy vary according to the species involved and the particular vitamin concerned. The responses range from abortion and stillbirth to congenital deformity; but some deficiencies are not accompanied by obvious teratogenic effect.[22] Much of the available knowledge on vitamin B deficiencies in pregnancy relates to experimental studies in rats, but in some instances these observations differ from the results of vitamin B deficiencies in human pregnancies. Deficiencies of thiamine, niacin, pyridoxine, vitamin B_{12} and pantothenic acid in pregnant rats lead to intrauterine mortality or retarded growth. A lack of riboflavine in the diet of rats from 60 days before mating resulted in no young being born, while a deficiency from 40 days prior to mating resulted in a 50 per cent incidence of structural deformities involving the skeleton, brain and cardiovascular system.[52] A similar range of structural damage followed maternal niacin deficiencies (or niacin deficiency caused by feeding rats a diet containing the niacin antagonist, 6-aminonicotinamide). Deficiencies in dietary folic acid or prenatal treatment with folate antagonists are also well-known causes of teratogenesis in rats.

Dietary deficiencies of pantothenic acid and choline appear to be associated with a more specific pattern of fetal abnormalities in rats. For example, pregnant rats deprived of pantothenic acid produced offspring with abnormalities which chiefly involved the brain, but other systems were affected occasionally.[53] Choline deficiency in pregnancy was found to be a cause of post-natal hypertension and muscular incoordination in rats.[54] In contrast, prenatal choline deprivation in pigs led to high neonatal mortality and structural changes consisting of fatty liver, haemorrhage and urinogenital deformities.[55] This observation is an indication of the interspecies variation in the requirements in pregnancy for various vitamins. (Prenatal death was a more frequent result of nutritional deprivation of riboflavine, thiamine and pantothenic acid in pregnant pigs.)

Published studies on the effects of excessive levels of the B group vitamins in pregnancy are not numerous,[22] possibly due to a lack of obvious effects seen in the few studies conducted. High levels of dietary pyridoxine for example, did not cause changes in litter size, fetal viability or increased requirements for the vitamin post-natally.[56] Alternatively, an apparent lack of teratogenicity due to high levels of vitamin B in pregnancy may be due to the ability of pregnant mothers to regulate their blood levels and thus the amount passing transplacentally. It may well be that these vitamins of the B complex are not conducive to abnormal fetal development. However, in the light of species variations in response to other teratogenic influences, one is not justified in concluding that a no-effect response in rats means that human babies and the offspring of other species will be similarly unaffected.

Minerals

Certain minerals are essential in pregnancy for normal maternal health and fetal development. The absolute requirements vary according to species and the consequences of dietary deficiencies differ accordingly. The subject has received considerable attention over the last 50 years and several excellent reviews are available and should be consulted for more detailed information.[56–58]

When considering the problem of teratogenesis and reduced fertility associated with the minerals present in the diet, one recognises two main groups, namely, those minerals required for normal structural development and physiological or biochemical functions, and those which have no presumed nutritional role in mammals, and which are occasionally highly toxic. In the former group are included calcium, phosphate, iron, potassium, sodium, magnesium, copper and a number of trace metals. Lead, cadmium and mercury are the best known examples of minerals in the second category.

Minerals with a Nutritional Value

Calcium. Calcium is a major structural component of bone and thus a dietary deficiency or metabolic disturbance in pregnancy leading to a decreased availability of metal to the fetus could be expected to lead to skeletal deformity. However, this situation rarely arises since calcium levels in the maternal circulation and in the fetus are largely regulated by parathyroid function. In dietary calcium deficiency in pregnancy, blood calcium levels are maintained by mobilisation of reserves in the maternal skeleton. Extreme calcium deprivation in cattle is detrimental in pregnancy and often the calves are born weak and and non-viable.[59] In other species, lower than normal levels of calcium have been recorded in fetal bone-ash but without obvious structural abnormalities.[60,61] Smaller than normal litters may be an indication of intrauterine death or reduced fertility.

In assessing the importance of calcium in relation to normal fetal development, one has to take account of the availability of phosphate also; optimum skeletal ossification depends on an adequate supply of both calcium and phosphate.[62] This implies that any dietary constituent or contaminant which interferes with the availability of these moieties to the developing fetus, is a potential cause of abnormal bone formation. Strontium compounds which are widely distributed in nature are known to be calcium antagonists through clinical and experimental studies. Even though strontium ions may be necessary in some species for early ossification, an early teratological study in rabbits showed this element to be teratogenic,[63] presumably due to its ability to impair calcium metabolism. In a later study in pregnant rats, high doses of strontium nitrate at susceptible stages of gestation, did not cause gross fetal deformities but led to a reduced calcium content in the fetal bone-ash.[64] The strontium levels in the fetuses reached a 'saturation' point above which further metal did not accumulate. This suggests that some regulatory system exists in the fetus or placenta which limits the uptake of this metal irrespective of maternal blood levels. Experience has shown that although strontium is a known bone-seeking mineral and theoretically can impair ossification if present at sufficiently high levels, the risk of this occurring in human pregnancies seems to be low.[65]

A more serious risk associated with the presence of strontium compounds in food arises from the radioactive compounds which occur in the atmosphere in the vicinity of nuclear reactors or atomic bomb fallout. These products of the degeneration of uranium have been demonstrated in milk and dairy produce, and they have been associated with fetal teratogenesis. Pregnant rats and mice exposed to ^{90}Sr-compounds have been shown to produce small fetuses with a lowered viability and skeletal malformations.[66,67]

Although an excessively high maternal blood calcium level as occurs in certain diseases may be harmful to the fetus, excessive calcium absorption is not usually a problem in adults (including pregnant patients) since blood levels are regulated by normal homeostatic mechanisms. However, in rats, subject to hyercalcaemia due to high dietary calcium levels the young were smaller than normal, their hair growth patterns were abnormal and they exhibited a reduced postnatal growth rate.[68]

Magnesium. Magnesium is a ubiquitous element in the human body and small amounts occur in all tissues. The metal appears to be concentrated in mitochondria and is an essential constituent of many enzyme systems. Because of the widespread occurrence of magnesium in most foods, deficiencies are rarely a problem in human pregnancies but in experimental situations, magnesium deficiency is teratogenic and fetotoxic.[69] Minimal magnesium deficiency in pregnancy was associated with congenital oedema and haematological deformities. In humans, dietary magnesium deficiency is known to lead to behavioural and muscular abnormalities.[70]

Zinc. Zinc is an essential trace element and is involved in several enzyme systems including carbonic anhydrase and carboxypeptidase. Nausea, diarrhoea and skin disorders are well-documented symptoms of zinc deficiency. Maternal zinc deficiency in rats throughout gestation led to retarded fetal growth, increased intrauterine mortality and a 50 per cent incidence of teratogenesis.[71] Malformations in prenatal zinc deficiency affect every organ system but disorders of the nervous system were of high incidence. At shorter periods of zinc deficiency, the effects seen varied according to the duration of the deficiency period and the stage in pregnancy when it was given. Maternal zinc deprivation in the pre-implantation stage of rats led to impaired cleavage patterns and blastocyte formation *in vitro*,[72] but if mothers were given a normal zinc-sufficient diet from day six to the end of gestation, then fetal wastage was not increased significantly and fetal malformations were only slightly more numerous. However, if the zinc deficiency occurred at a period of high organogenesis, then the effects were more profound. In the period from six to 14 days gestation, zinc deficiency led to a rapid fall in maternal plasma zinc levels and a high rate of fetal growth abnormality and mortality.[22]

Current views are that zinc deficiency leads to fetal malformation and retarded growth through an inability to synthesise proteins, lipids and nucleic acids resulting in mitotic inhibition and possibly chromosomal damage.

Dietary zinc deprivation in pregnant rats leads to a rapid decline in plasma zinc levels and symptoms of zinc deficiency. Although zinc is stored in the maternal skeleton, this is not readily mobilised except in cases of calcium deficiency. Although the maternal and fetal blood zinc levels show no correlation, a clear relationship has been shown between the fetal serum zinc level and the severity of the growth impairment.[73]

As in the case of calcium, abnormal fetal development may result from the action of other substances present in the food or in the serum which impair the availability of the zinc ions. Zinc deficiency has been shown in rats exposed to normal diets to which ethylenediaminetetra-acetic acid (EDTA) had been added.[74] The effects were prevented if additional zinc was given in the diet at 1000 p.p.m. EDTA is used therapeutically in the treatment of heavy-metal poisoning, and also in the softening of water. Thus EDTA may be ingested in the diet by contamination and through its ability to chelate zinc, may cause zinc-deficiency symptoms.

Other Trace Minerals. In addition to calcium, magnesium and zinc, at least eight other minerals are presently known to be essential for life even though for some, only minute quantities are required. These additional elements include iron, copper, iodine, cobalt, manganese, molybdenum, selenium, chromium and tin. The elements nickel, arsenic, barium, fluoride, bromine and vanadium may be essential trace minerals in some species but much remains unclear as to the function that they perform. Clinical and experimental studies have shown that deficiencies of some of these minerals in pregnancy may lead to fetal distress and mortality.[19] Conversely, supra-optimal levels of other trace elements may be detrimental for fetal development.[75]

Dietary copper deficiency is a well-known cause of sway-back in lambs. This condition is characterised by a limb incoordination resulting from neurological damage. Demyelination, a lack of Nissl substance, defects in the Golgi apparatus and neurofibril abnormalities have been identified with dietary copper deficiency in both sheep and guinea pigs.[76] Fetal rats responded to copper deficiency by showing a high rate of intrauterine death, with survivors exhibiting anaemia, oedema and reduced viability after birth. Skeletal and dermatological abnormalities have also been recorded.[22,77]

Iron is an important requirement in pregnancy and maternal iron deficiency is a major cause of congenital anaemia. In rats, the young are weak at birth and frequently die early.[78] In humans, iron deficiency in pregnancy is characterised by hypochromic microcytic anaemia in the mother and this correlates well with the level of maternal dietary deficiency.[79] Anaemia in the new-born may persist until later in life.[80]

Dietary supply of iodine is essential for normal thyroid function and a maternal deficiency in pregnancy leads to hypothyroidism and fetal cretinism.[81] The condition has been recognised for many years in human infants, but has also been demonstrated in other mammalian species. Reduced neonatal viability may result and on occasions be accompanied by overt damage.

Manganese is important for the survival and structural development of many fetal mammals and in cases of maternal dietary deficiency the young die early in the post-natal period. The main congenital defect noted is that of ataxia and muscular incoordination resulting from impaired development of the otoliths in the inner ear. This defect has been related to a failure in mucopolysaccharide synthesis for which manganese is obligatory.[22,82] Defects in the preosseous chrondrogenesis are probably responsible for skeletal defects as seen in fetal rats from mothers deprived of manganese in pregnancy.

Studies of fetal defects resulting from deficiency or excessive maternal levels of other trace minerals are not numerous and those that are available were mostly conducted using the rat as a test species.

The teratogenicity of several trace metals has been reviewed by Ferm[75] but in many cases the levels administered to test animals were far in excess of expected levels that a person would consume in his diet. For example, arsenic (sodium arsenate), administered at 45 mg/kg on days six to eleven of pregnancy in the hamster, produced cleft palate, exencephaly, agnathia and anophthalmia.[83] In humans, the few reports of acute arsenic poisoning in pregnancy suggest that damage to the fetal nervous system and neonatal death are likely effects depending on the time of exposure in pregnancy. Interestingly, rabbits do not seem to be susceptible to arsenic teratogenicity.[84]

In summary, it is clear that maternal dietary deficiencies of any of the trace minerals may lead to abnormal fetal development and a reduced chance of fetal survival depending on the species involved, the severity of the deficiency in pregnancy and the gestational age of the conceptus. Correction of the dietary intake levels may avert the teratogenic effects for some metals but this will vary according to the nature of the deficiency and the severity of the damage resulting from it. For the most part this aspect of teratology has not been widely examined clinically or experimentally.

An important aspect of abnormal fetal development resulting from a lack of essential trace metals relates to metal 'deficiency' caused by chelating agents present in the diet. EDTA and nitrilotriacetic acid (NTA) are examples of chelating agents. Wilson[4] drew attention to this problem stating that some drugs and environmental chemicals have the ability to sequester scarce metals to create a dietary deficiency. In the case of metals such as lead, cadmium and mercury this obviously has a beneficial effect both for the mother and fetus but a deprivation of magnesium, zinc, calcium or manganese can only be detrimental.

Toxic Metals

Lead. Lead occurs widely in the environment and is a well-known contaminant of drinking water. Further, certain foods cooked in water containing lead have been shown to concentrate the metal and hence be a potential hazard to the consumer.[85] Nutritionists believe that lead in the diet has no beneficial value and that the risks of its ingestion may be serious. Lead poisoning has been recognised for well over 100 years particularly among workers exposed to lead in the course of their work. The principal hazards reported include anaemia, resulting from induced defects in haem formation in the erythroid cells of the bone marrow, renal damage and mental disturbances. Children are particularly at risk from lead poisoning since lead ion is absorbed more readily from the intestinal tract at this age compared to that seen in older individuals. Much has been published in the last century on the role of lead contamination in causing mental retardation in children.[86,87]

The influence of lead ingestion by pregnant women has been studied for many years also, and early reports in which increased abortions and stillbirths have been described refer to people exposed to very much higher concentrations of lead than would be expected in contaminated food or water nowadays. In 1860, Paul[88] studied 123 pregnancies and reported that 73 terminated in abortions and only 50 infants survived. The incidences of abortions and stillbirth reported subsequently have varied, presumably on account of variations in the level of lead consumed and the concentration of lead in the maternal circulation at critical periods in fetal development. Interestingly, stillbirths and neonatal mortalities were high in pregnancies where the mother *or* father were exposed to high lead exposure in their occupation.[89] Estimates indicate that the risk of an infant dying *in utero* or neonatally are increased by up to two or three times following exposure to lead in gestation, but of course the risk will depend on the maternal blood lead levels.

Attempts to correlate maternal blood concentrations of lead with fetal distress have been few and even now, the minimum lead level in the maternal or cord blood that is consistent with fetal damage is not known accurately. Nogaki,[90] studying 104 pregnancies in Japanese women, reported a spontaneous abortion rate of 84.2 per 1000 pregnancies in lead-exposed cases compared to 59.1 per 1000 in non-exposed women. Blood lead in the first group ranged from 110 to 317 μg lead/100 ml whole blood. Later studies suggest that blood lead concentrations of greater than 100 μg/100 ml may or may not be associated with severe symptoms and that at levels in the range 60 to

100 μg/100 ml some mild and non-specific clinical symptoms may occur. Although the blood lead 'threshold' at which there is an increased risk to the fetus is probably less than 60 μg/100 ml; it would seem that the high rates of abortion and mortality seen in the early studies were accompanied by maternal blood levels very much higher than are seen today.[91]

The critical level of lead in at-risk fetuses is not known but at concentrations of 5 to 25 μg/100 ml in whole blood, there was no relationship between cord blood lead and birthweight.[92]

In view of the clear propensity of lead to cause mental retardation in exposed children, some attempts have been made to examine the possibility that such a change may accompany prenatal exposure. Although suggestions that this may occur were given by two studies conducted in the Glasgow area recently, the conclusions drawn are controversial.[93,94] The statistical analysis used and the possible additional complications of maternal alcoholism and excessive smoking tend to make these observations unreliable. Thus, there is still an urgent need for further studies to be carried out into this problem.

Several experiments have been reported where laboratory animals have been used to examine the teratogenic effects of prenatal exposure to lead. The observations have been inconsistent and in many instances are at variance with the human situation. For example, the offspring of pregnant hamsters injected with 25–50 mg/kg lead acetate on day eight of gestation showed a high incidence of tail abnormalities possibly attributable to local haemorrhage and blister formation.[95] Rats injected intravenously with similar doses of inorganic lead compounds showed a wider range of structural abnormalities than seen in the hamster.[96] In contrast, when inorganic lead was administered in the drinking water to rats, few malformations were seen,[97] possibly due to the low absorption of lead by the intestinal mucosa and a sub-teratogenic level of lead being present in the maternal and fetal blood. Rabbits were also refractory to the teratogenic effects of lead compounds administered in the diet.

When powdered metallic lead was administered orally to pregnant sheep throughout gestation 27 per cent of the young aborted.[98] The levels of lead achieved in the fetal and maternal circulation in that study were not given but it is likely that they were higher than those reported in an earlier study where the offspring exhibited behavioural changes.[99] In that work, ewes were fed 4.5 mg/kg of powdered lead in their diets from preconception through gestation, the maternal blood level reached was 34 μg/100 ml. The lambs born showed 24 μg/100 ml of lead in their blood and a reduced learning capacity at 2–4 weeks. Behavioural changes were not significant at blood lead levels of 17 μg/100 ml. This is interesting work, showing that in sheep at least, prenatal exposure to lead can lead to postnatal neurological damage.

The results of the several animal experiments strongly suggest that inorganic lead compounds are potentially harmful to the human fetus, but as yet this view lacks conclusive proof. It may be that the animal species used are more susceptible to lead-induced damage, or that the levels of lead attained in the plasma due to environmental or dietary exposure in the human cases examined were insufficient to impair fetal development.

Lead in the diet is believed to impair fertility and reproductive function in males. Hilderbrand and his colleagues[100] suggested that the decline of the Roman Empire was in part attributable to infertility in the ruling classes due to excessive consumption of wine to which lead compounds had been added as flavouring. This seemingly wild speculation has been supported by some recent clinical and experimental evidence. For example, lead workers in the battery industry were reported to have lower than normal sperm counts and they produced some abnormal spermatozoa.[101] Blood lead levels recorded were between 40 and 75 μg/100 ml. In animal experiments, male rats exposed to 5 to 100 μg/day of lead compound for 30 days exhibited a loss of libido, reduced testicular weight and decreased sperm motility.[100] Female rats were less severely affected but showed abnormalities in oestrous cycles and in ovarian function. In other experiments, reduced litter size, lower fetal weight and poor survival rates were reported in multigeneration studies using rats, mice, rabbits and guinea pigs.[91]

Lead is clearly a cause for concern in respect of its potential risk in reproductive performances and fetal development, even though more epidemiological evidence is required to show that it can cause abnormal fetal growth following prenatal exposure in man, and to show blood lead threshold levels consistent with such action. Current views suggest that if lead is teratogenic as supposed, then the resulting defects are probably attributable to the action of lead ions in inhibiting the metabolism or enzyme systems which involve trace metals such as zinc, copper and iron. In biochemical studies, lead has been shown to inhibit the biosynthesis of haem by inhibiting the zinc-dependent enzyme γ-aminolaevulinic acid dehydratase and also ferrochelatase. These effects of lead are well documented in the erythroid cells of the bone marrow but may well be evident in other tissues known to be sensitive to lead, i.e. brain, kidney, liver and heart.

Cadmium. Cadmium like lead does not appear to have a nutritional value in man yet may occur as a food contaminant and exert a toxic effect. Cadmium poisoning in humans was highlighted by the Japanese 'itai-itai' disease which affected people living in areas where the soil was rich in cadmium. The poisoning arose from the high cadmium content of vegetables grown there and used as human and animal foodstuffs. In other areas, food contamination with cadmium has been noted in the vicinity of lead and zinc mines.

Although cadmium is normally absent from the body at birth, if it is ingested with the food and absorbed into the circulation, then visceral damage may result. The kidney is affected principally and pathological changes involving the tubules and glomeruli have been described.[102] Cadmium uptake from various forms of food leads to renal concentrations several times greater than those seen in other organs.[103] Age is an important factor in determining the rate at which cadmium is absorbed from food; neonatal and young animals, and presumably humans also, tend to absorb appreciably more cadmium intestinally than adults.[104]

Although Wilson considered that mercury was the only metal positively shown to be a teratogenic risk to the human baby,[105] cadmium like lead has been shown to cause abnormal fetal development with reduced survival in several sub-human species. However, in most cases the doses administered (usually intraperitoneally) were considerably in excess of those normally encountered in the environment. For example, when hamsters were injected with 2 mg/kg of cadmium sulphate on day eight of gestation, the young exhibited deformities of the head, mesoderm and skeletal structures.[106] The regular observation of cleft palate in hamsters suggested to workers that maybe cadmium is 'a site-specific teratogen'. This teratogenic dose contrasts with the maximum daily intake allowed in the United States for human safety which was given as 71.4 μg or 72 μg/day for an average adult.[107]

Limited human studies have shown that only trace levels of cadmium are absorbed by the fetus.[108] Whereas in rodents cadmium uptake tends to be greater early in the post-implantation period than later. Dencker using radiolabelled cadmium showed that this was primarily due to the type of absorption involved, that is by the yolk sac endothelium.[109] From this site, the cadmium passed to the primitive gut and other endodermal structures. When the role of the yolk sac in fetal nutrition was ended, only negligible amounts of cadmium were absorbed. At this later stage small quantities of cadmium were detected in the antimesometrial portion of the placenta.

Thus in mice and hamsters at least, cadmium absorbed or injected into the maternal circulation at an appropriate stage in gestation is teratogenic and this is probably mediated by an impairment in early fetal nutrition via the yolk sac and embryonic gut. Cadmium may also result in fetal death in rodents by causing placental damage and other changes which resemble human toxaemia of pregnancy.[110,111]

Finally, biochemical and analytical studies have shown that cadmium as taken up by the maternal ovarian tissue leads to a disturbance in oestrogen secretion. Such an action would be a cause for interruption of the pregnancy and intrauterine death as has been reported in the mouse.[112]

Extrapolation of these results of animal tests to humans is difficult but one may expect that if maternal plasma cadmium levels in pregnancy exceed the threshold limit value, then fetal development may be impaired.

Experimental studies in the rat have shown that cadmium is particularly toxic to testicular cells. The ion is preferentially absorbed from the blood by late-elongated spermatids which become non-functional, an effect which may be reversed by zinc chloride pre-treatment.[113] Plasma cadmium decreases rapidly as metal is taken up by the rete testis fluid, zinc ion inhibits this build up. The ratio of cadmium in the rete to the unbound plasma concentration approaches 40, indicating an accumulation in the seminiferous tubules. Thus if cadmium is absorbed into the blood stream following absorption in the gut it is likely to be a cause of male infertility, although it is not clear to what extent human tissues are susceptible to damage by cadmium.

Mercury. For many years, mercury compounds have found widespread use in the agricultural industry as seed dressings and fungicides. In the 1950s, mercury-containing compounds were recognised as a severe threat to public health. Current views are that apart from its ability to cause renal and neurological damage in adults through eating contaminated food, mercury presents a clear teratogenic risk in pregnant women. Occupational exposure to metallic mercury has been associated with ovarian dysfunction[114] and the use of mercury-containing drugs by pregnant women shown to cause abortion.[115] Good epidemiological evidence has demonstrated that organic mercury compounds pose a more significant risk to the unborn child even though the mother may not exhibit overt signs of poisoning. Mercury had been suspected of being teratogenic for several years, but it was not until the publication of the surveys carried out in the Minamata Bay area of Japan that the magnitude of the risk was appreciated.[6,116] Organomercurial fungicide residues washed with rain water into the rivers and estuaries became concentrated in sea food which was eaten by the local population. Ingestion of contaminated food by the local population led to a 6 per cent incidence of cerebral palsy and mental retardation in the children born.[117] Later incidents of congenital neurological damage were reported in the Nigata region of Japan,[6] and in Iraq and Pakistan;[7] these last two incidents were associated with the consumption of bread made from mercury-contaminated grain. In these studies, the extent of the maternal and fetal risk was clearly related to the amount of mercury consumed and the levels attained in the maternal and fetal circulation. In the most severe cases, both mother and fetus died as a result of the poisoning.

Epidemiological studies indicate that mercury poisoning through food contamination may arise in several ways.[7] They are:

1. In bread made from fungicide-treated wheat or barley.
2. In meat or other animal products from livestock fed on contaminated barley.
3. Vegetables grown or stored in conditions contaminated with mercury.
4. Game birds that have fed on contaminated grain.
5. Fish or shellfish caught in rivers, lakes or estuaries in which grain or mercury-containing residues have accumulated or been dumped by farmers.
6. Milk from mothers who have ingested mercury in their diet.

In humans, the correlation coefficient between the amount of mercury ingested from bread and the resulting blood concentration was 0.85 for people over the age of 16 and 0.89 for those in the six to fifteen age group.[7] Mercury may pass to the offspring transplacentally or post-natally in the milk. The mercury level in the milk is proportional to the concentration in the maternal blood up to 2500 ng/ml. In Iraq, the average milk levels of mercury were 3 per cent of the mean concentration in blood but in only two out of 20 lactating mothers were levels exceeding 2200 ng/ml exhibited in the blood. In 11 infants exposed to mercury in milk only, the blood concentrations reached were either equal to or lower than those seen in the mother. Although no clear evidence of toxicity was seen in eight children who exhibited blood mercury concentrations above 500 ng/ml, these levels must be regarded as hazardous in the light of previous studies which showed that paralysis, ataxia and dysarthria occurred in up to 10 per cent of children aged nine and over. As a result of prenatal mercury poisoning, blood levels in the neonates varied and in one infant the level seen was three times greater than that seen in the mother. Severe brain damage was seen in three infants who exhibited serum mercury levels of 2500 ng/ml or greater. Depending upon the age of the fetus at exposure and its genetical susceptibility to drugs, the teratogenic blood level of mercury might be considerably lower than this.

Available data on methylmercury poisoning show clearly that transplacental passage does occur and that the fetal concentrations reached may exceed those seen in the maternal circulation. Neonatal mercury poisoning via the maternal milk did not result in blood levels greater than those seen in the mother. Clearly, the offspring is at much greater risk through intrauterine exposure than during lactation.

Numerous experimental studies on the teratogenicity of organomercurial compounds have been reported but the results are conflicting. In only one study where cats were used as the test species, were the macroscopic and histological observations consistent with those of the above-mentioned human studies.[118] Abortion, skeletal anomalies and neurological damage was reported in fetal cats following maternal dosage at 0.083 mg/kg or more on days 10 to 58 of pregnancy. In nervous tissue, the mercury level of 2.07 p.p.m. was consistent with neuronal degeneration and proliferation of glial cells in the cerebrum and cerebellum resembling that seen with mercury poisoning in human babies. In rats, mice and hamsters, the teratogenicity of organomercury compounds varied according to the dose, the period of pregnancy and the salt administered. Fetal hamsters did not exhibit brain lesions following maternal dosage with methylmercuric chloride (2–4 mg/kg) through most of gestation but died *in utero* or exhibited skeletal malformation.[119] Fetal rats showed brain abnormalities at about 1.4 mg/kg and mice given a single oral dose of 30 mg/kg showed congenital brain lesions, cleft palate, micrognathia and microglossa.[2] Intrauterine death and retarded fetal growth were usual observations in rats, mice and hamsters following maternal dosage with the various methylmercuric salts at 5.0 mg/kg.

Although observations of experimental teratology studies with organomercuric compounds differ in several respects from those of the clinical surveys, a serious teratogenic risk for man is clearly indicated both from prenatal and neonatal exposure to organomercurial compounds through eating contaminated food. Such is the current awareness of the human risks from mercury poisoning that recently the Japanese government discouraged the use of sperm whale meat for human food since levels of mercury contained in it averaged 2.3 p.p.m., that is six times the level considered acceptable in the diet for safety.[120]

Food Additives

These form a heterogeneous group of chemicals which have minimal nutritional value but which are included in the food to enhance its appearance, to flavour it and preserve it. The nature of their use indicates that these food additives should exhibit negligible side effects in the consumer and are unlikely to be teratogenic or a potential cause of infertility. In most fertility and teratogenicity tests conducted in experimental animals so far, using low to moderate doses to simulate human levels of consumption, this view has been supported and minimal evidence of toxicity seen.

However, as a result of some recent experimental studies on food additives, concern has arisen over the possibility that reproductive hazards may exist with certain agents. The artificial sweetening agents, saccharin, sodium glutumate and cyclamates in particular, have been subject to considerable screening. A Soviet report which suggested that the widely used food colouring amaranth might be embryotoxic and teratogenic created a stir in Western Countries and as a consequence, this food additive has been toxicologically examined in greater detail than most. In this work the levels fed to test animals were greatly in excess of those ever likely to be ingested by humans, but nevertheless the published findings were disturbing at the time.

Sweetening Agents

Sugar was clearly a luxury in the eighteenth century, but since then the level of consumption in Great Britain has increased to as much as 120 lb per head each year.[121] Apart from its energy value, sugar as ingested has little nutritional value and has been implicated as a potential cause of dental decay, diabetes and ischaemic heart disease. This has led in recent years to a search for artificial sweeteners to reduce sugar consumption. Thus, agents such as saccharin and cyclamates have achieved ready acceptance in Western Society.

Toxic effects have been reported in population studies on the use of artificial sweeteners. In a questionnaire completed by women known to have consumed saccharin in pregnancy, the commonest reported effects seen in new-born children were increased activity and irritability, and an extreme nervousness.[122] This will need further substantiation, but so far there seems to be no evidence for any lasting adverse effects in human babies resulting from high maternal ingestion of saccharin in pregnancy.

So far, no convincing experimental evidence of fetal risk has been produced in rats, mice or monkeys with saccharin, cyclamate or mixtures of the two.[123] In rats, saccharin passes transplacentally to the fetuses and is voided from the mother in the milk. Although ocular deformities were reported in the offspring of pregnant rats fed sodium cyclamate or saccharin in one study,[124] these observations have not been verified in subsequent work where doses of 5 per cent sodium cyclamate or 0.4 per cent sodium saccharin were given in the diet throughout most of the gestational period.[125] Rat and rabbit fetal development was not abnormal at maternal doses of 500 to 2500 mg/kg of a saccharin:cyclamate mixture (1:10 parts).[126] Teratogenesis was not seen in rhesus monkeys exposed to doses of up to 2000 mg/kg at various stages in pregnancy.[127]

Although experimental studies indicate that neither saccharin nor cyclamates cause overt structural abnormalities as a result of prenatal exposure, there is some evidence to suggest that at high doses they may be responsible for reduced viability and behavioural defects neonatally. For example, 5 per cent saccharin given to pregnant and lactating mice caused a high neonatal mortality and retarded post-natal growth in survivors.[122]

More concern has been created by the experimental observation that the metabolite of cyclamate, cyclohexylamine, caused retarded fetal growth and fetal death in mice at doses of 61 to 120 mg/kg[128] and also in rhesus monkeys at 25 to 75 mg/kg.[2] Structural abnormalities were not reported in surviving offspring.

In short-term toxicity studies, cyclohexylamine caused testicular damage in rats.[129] At 2000 p.p.m. in the diet, cyclohexylamine caused a significant reduction in testicular weight with lower spermatogenesis and atrophy of seminiferous tubules. At 6000 p.p.m. a complete arrest of spermatogenesis and loss of germinal epithelium was seen in some rats but in others reduced fertility was not shown in reproductivity tests. In other studies, chromosomal damage in spermatogonial cells has also been reported in rats but not mice following intraperitoneal injection of cyclohexylamine at 1 to 50 mg/kg.[130] Spermatozoal abnormalities were not seen in mice given calcium cyclamate at 500 mg/kg for up to ten weeks.[131] At present the mechanism for testicular damage is not known.

Colouring Agents

Evidence that food colouring agents are teratogenic is very limited. However, the controversies that were aroused by the report from workers in the USSR that the red dyestuff amaranth was fetotoxic and teratogenic in rats are of interest. Amaranth (red dye no. 2) is a widely used colouring material in the food industry and like other food colours was assumed to be non-toxic.

Shtenberg and Gavrilenko dosed rats orally with 1.5 or 15 mg/kg for seven or eight months[132] and reported a reduction in fertility in multi-generation studies and an increased number of stillborn and abnormal offspring. Lactation was less than normal. Subsequent studies using mice and rats dosed orally with amaranth or its metabolites at up to 1000 mg/kg throughout gestation have failed to confirm the terato-genicity although embryotoxicity was seen with some metabolites (a few offspring in the mouse study exhibited exencephaly, hydrocephaly and minor changes at 100 mg/kg).[133] Amaranth was fetotoxic in rats at doses of 30 mg/kg and above. The number of implantations in mice and rats was not significantly influenced by amaranth dosing. Negligible evidence for teratogenicity was seen in rabbits dosed orally with 1.5 to 15 mg/kg amaranth from days 6 to 18 of pregnancy. Studies in rats or mice with the chemically related dye Ponceau 4R and metabolites of amaranth have also yielded no evidence of reduced fertility or teratogenicity.[133]

Monosodium Glutamate (MSG)

Monosodium glutamate is a natural constituent of many foods but about 12 years ago was highlighted as the cause of the so-called 'Chinese Restaurant Syndrome'. Affected patients showed neuro-logical and behavioural changes. As a result of several clinical and experimental studies, it is now accepted that new-born infants are at greater risk from MSG than older individuals.[134] The lesions charac-teristic of MSG poisoning include retinal damage, degeneration in the arcuate nucleus of the hypothalamus, and neuroendocrine abnormali-ties, but these were seen usually after particularly high doses.[135,136] Considerable interspecies variations have been reported in the responses of young animals to MSG. Rats and mice exhibited severe brain damage after injection with 500 mg/kg MSG but young dogs or monkeys were not affected, possibly on account of their more advanced developmental state.[137]

The ability of MSG to cause fetal brain damage is still controversial and experimental results so far are contradictory. Hypothalamic lesions were identified in the offspring of mice dosed with 5 g/kg MSG in late pregnancy and also in a single fetal monkey following maternal dosage with 2 g/kg MSG.[138] However, similar changes have not been present in the brains of 59 other fetal monkeys where the mothers were dosed with 1-4 g/kg MSG orally or by subcutaneous injection. Further, no damage was seen in the brains of six other monkeys injected *in utero* with 6.8 g/kg MSG.[139] In these the mean plasma level for MSG was 400 times greater than normal values.

If the fetal human exhibits a similar order of sensitivity as the fetal monkey, then the predicted risk of acquiring neurological damage would seem to be very low. Available evidence shows that MSG induces prenatal neurological damage only in mice and then at a dose of 5 g/kg which is many times higher than the maximum human intake from food at any age (1.5 mg/kg/day).[140,141]

Antioxidants

Butylated hydroxyanisole (BHA) and butylated hydroxytoluene (BHT) are used as antioxidants in the food industry and as such have a high risk of being consumed in the diet. Although an early study indicated that BHT may be a cause of anophthalmia in the offspring of rats fed 0.5 per cent (250 mg/kg/day) in the diet for five months, this risk has not been substantiated in subsequent work on mice or rats at doses of up to 1000 mg/kg at periods of maximum teratogenic susceptibility.[142,143]

Mice fed 0.5 per cent BHA or BHT in their diet through mating and gestation produced young which were overtly normal at birth but which developed abnormal growth and behavioural patterns when the antioxidants were fed for three weeks during the weaning period.[144] It is unclear from this work whether these changes were attributable to neurological damage and if so whether this damage occurred prenatally or through *post-partum* exposure. Behavioural teratology is a comparatively new field and may be useful in further experimental studies with these food additives.

Alcohol

Alcohol occurs in many beverages and the problems associated with chronic alcoholism among men and women have been appreciated since early Greek and Roman times. The influence of chronic alcoholism in pregnant women and the fetal complications associated with it have been well documented in the literature. Giroud and Tuchmann-Duplessis[145] believed that chronic alcoholism was not a cause of true embryopathies but may lead to premature birth or congenital mental deformity. In later studies, the 'fetal alcohol syndrome' became recognised with affected infants exhibiting retarded prenatal growth, microencephaly, reduced palpebral fissures, cardiovascular abnormalities and borderline mental deficiency due to cerebral damage.[146] In an evaluation of 23 alcoholic pregnancies, four babies died in the first week, one was stillborn and of the survivors, 44 per cent were mentally retarded.[147] Six of these children showed structural and developmental abnormalities. The authors realistically recommended that pregnancies complicated by chronic alcoholism be terminated in the light of these observations. Variations in the type and severity of growth impairment are closely related to the volume of alcohol consumed, the general level of nutrition and the age and susceptibility of the fetus to deformity.

Although alcohol passes transplacentally in humans and may distort fetal development by a direct toxic action, the precise mechanism of teratogenesis is not known.[148] It may be that metabolites of alcohol or other constituents of the alcoholic beverages, such as lead, are in part responsible for the brain damage seen. Experimental studies in pregnant sheep have shown that alcohol leads to an initial metabolic acidosis in the fetus with a mixed type of acidosis occurring later.[149] The fetal cerebral blood flow was increased but oxygenation of the brain tissue was not changed.[150] A depression of the central nervous system was considered to be responsible for these and other cardiovascular changes seen.

Alternative explanations suggested for alcohol-induced fetal damage from human observations and animal models relate to the ability of alcohol to disturb the nutrition. Alcohol is known to suppress the ingestion of essential nutrients such as proteins, amino acids and vitamins.[121] Stillbirth and retarded intrauterine growth seen in chronic alcoholism are consistent with the effects of these nutrient deficiencies as shown in experimental animals.

When alcohol has been administered in the drinking water to

experimental animals at various stages in pregnancy the observations made included retarded intrauterine growth, reduced litter sizes and occasionally structural abnormalities. Variations seen may be attributed to differences in the dosage schedule, nutritional status and the time in pregnancy when the alcohol was administered.[151] Interestingly, when rabbits were given alcohol shortly before a successful mating, they failed to become pregnant,[152] possibly due to the action of the alcohol in inhibiting ovulation or the transport of spermatozoa in the fallopian tubes. Alcohol is known to block the release of luteinising hormone in the rat.[153] Rabbits dosed with 2 ml/kg/day after implantation, produced young with hydroencephaly, cystic kidney and optic deformity.[2] When mice were dosed with alcohol at a concentration similar to that seen in many alcoholic beverages (42 per cent) at 24 mg/kg many young died *in utero* and survivors exhibited paw defects. The significance of increased levels of alcohol dehydrogenase in the livers of neonatal animals subject to prenatal alcohol exposure is not clear. The young have frequently been shown to have poor post-natal growth and occasionally exhibit behavioural abnormalities.[2]

Brain damage is a normal feature of the fetal alcohol syndrome in humans but this has not frequently been reported in experimental studies. Retarded brain growth was noted in studies with rats and guinea pigs but this was probably only one manifestation of a general fetal growth retardation. Histological changes were noted in the brains of fetal rats when mothers were dosed with 20 mg/kg of 40 per cent alcohol from 11 to 20 days of gestation.[154]

In summary, available information on chronic alcoholism in pregnancy points to a clear risk for fetal survival and post-natal growth, in the human baby. This is well supported by mechanistic and predictive studies in experimental animals. Several of these studies show that the fetal brain is particularly susceptible to alcoholic damage. The mechanism by which the fetal damage arises is not clear but the speculation that nutritional defects may be involved is interesting. Clearly, there is still a large amount of clinical and experimental work necessary to understand the complications of chronic alcoholism in pregnancy.

Impotence and reduced fertility in men are known consequences of chronic alcoholism and testicular biopsies have shown a reduced gonadal mass. This is due largely to a degeneration in the germinal epithelium and a reduction in the diameter of the seminiferous tubules. Leydig cells are not affected. At the same time, plasma testosterone levels may be depressed by up to 50 per cent or more.[155] The mechanism for these changes is not known but explanations suggested include a direct effect of alcohol or its metabolites on susceptible tissues *per se*, impairment in vitamin A metabolism, and interference in the hypothalamic-pituitary axis resulting in a reduction in gonadotropin secretion.[156] The last possibility now seems unlikely in view of recent clinical evidence that the hypothalamic-pituitary axis is less sensitive to provocative stimulation than normal in alcoholic men and that usually their plasma gonadotropin levels are normal. The suggestion that alcohol disturbs gonadal function through vitamin A antagonism is based on the concept that alcohol inhibits alcohol dehydrogenase, an enzyme involved in the conversion of retinol into its biologically active form retinal, which is necessary for normal spermatogenesis. The observation that chronic vitamin A deficiency occasionally predisposes to reduced plasma testosterone levels supports this view.

Experimental studies where alcohol has been administered to male dogs, rats and mice, have provided similar evidence of testicular damage to that described above in biopsies from human patients, without significantly adding to the knowledge of the mechanisms involved. Rats, for example, fed ethanol in their diets equivalent to 36 per cent of the total calorific value, developed testicular atrophy and other changes consistent with a reduced plasma testosterone level.[156] Unlike the human conditions, Leydig cell function was disturbed in these rats.

Since alcohol clearly leads to impaired testicular function and spermatogenesis, it is conceivable that a paternal effect may be responsible for abnormal fetal development or other reproductive disorders. This was demonstrated experimentally in a study where mice were given 0.1 ml of 40–60 per cent ethanol and then mated. The resulting litter sizes were appreciably reduced and intrauterine mortality increased possibly as a consequence of dominant lethal mutations in the spermatozoa.[157] When male rats were given 6–10 per cent alcohol in their diets, they showed a loss of libido and resulting litters were smaller and fetal mortality increased.[158] The mutagenicity of alcohol in males as a cause of abnormal fetal development is an interesting possibility but there is obviously much still to be learned as to the precise mechanism concerned.

Food Contaminants

Plant Substances

Potato Blight. Anencephaly and spina bifida share many epidemiological features and clearly are the most serious deformities seen in human babies. In 1972 Renwick hypothesised that consumption of blighted potatoes in early pregnancy was a major cause of these abnormalities.[159] However, discrepancies were noticed in his interpretation of the available information, such that the original hypothesis is now disputed.[160] Renwick suggested that the risk of neural tube defects could be reduced by up to 95 per cent if pregnant women totally avoided eating potatoes. Subsequent studies have failed to support this view.

If blighted potatoes are a cause of neural tube deformities in human infants, it is surprising that in the ten years following the publication of the hypothesis, the teratogenic moiety is still unidentified. Renwick correlated human fetal deformities with the potato blight fungus *Phytophthora infestans* and speculated that the teratogenicity may be attributable to toxic metabolites of the fungus, products of the infected and deteriorating potato tuber, or natural constituents of the potato synthesised in response to light. Of the last group, solanines are known to form in greening and sprouting potatoes, they are bitter in taste and highly toxic but apparently not teratogenic. Solanine fed to pregnant rats caused intrauterine mortality but no structural abnormalities were seen. In man, only about 1 per cent of these alkaloids is absorbed intestinally and most is hydrolysed to be voided in the faeces and urine. Numerous other substances falling in the former two categories were listed but like solanine and related alkaloids, the evidence available for their adverse influence in human pregnancy is negligible.

Experimental studies conducted in an attempt to evaluate the teratogenic risks associated with blighted potatoes have provided equivocal information. In one study using pregnant marmosets, osseous defects were seen in the skulls of four animals and presumed to be attributable to the potato extracts.[161] However this association was not substantiated in a later study where female rhesus monkeys and marmosets were fed 10 g/kg/day of blighted potato in their diet for two weeks prior to conception and then for six weeks.[162] Blighted potato extracts were not teratogenic in rats[163] but dried ground potato sprout preparations from seven varieties of potato produced cranial abnormalities in Simonsen S_1 hamsters.[164] Solanine and veratrum alkaloids produced similar effects. Interestingly neither the peel nor tuber material was teratogenic from one variety of potato with highly teratogenic sprouts. The need for further toxicological evaluation of the perplexing problem of potato teratogenicity was emphasised by another recent study where freeze-dried extracts of potatoes infected with *Phytophthora infestans* or *Alternaria solani* were fed to pregnant rabbits, hamsters of the Simonsen strain, and miniature swine.[165] Teratogenesis was absent in the hamster, but lesions of neurulation and anencephaly were noticed in rabbit and swine respectively.

The interpreation of the data presented is difficult. Although there may be some geographic and temporal relationship between outbreaks of potato blight in various parts of the world and increased neural tube abnormalities in children, the problem is far from clear. The absence of teratogenicity in some species of laboratory animals and different forms of structural abnormality in others do not clarify the situation.

Caffeine. Tea and coffee are normal in the diets of many communities throughout the world. They exert a stimulating effect due to the presence of caffeine and related xanthine derivatives. Analysis of many different blends of tea showed that the caffeine contained ranged from 50-80 mg in each cup. Cups of coffee contain about 100 mg of caffeine.[121] Caffeine is used therapeutically as a central nervous system stimulant but it has been shown to cause chromosomal damage in cultured cells and teratological effects in experimental animals. Although caffeine is possibly mutagenic, the estimated risk for humans is small.[2]

Caffeine is transferred transplacentally in the seven-week-old human fetus but does not accumulate in the preimplantation embryo and is not teratogenic. In rabbits [^{14}C]caffeine has been identified in the endometrium, uterine secretions and in preimplantation blastocyst.[166] No detrimental effects on the developing embryo were reported. In other studies, teratogenic effects have been caused by caffeine in mice, rats and rabbits but at doses far in excess of normal human ingestion.[2,19]

Prolonged caffeine dosage in mice at 100 or 250 mg/kg did not adversely influenced spermatozoal morphology or viablity suggesting that this compound is unlikely to be a cause of male infertility.[131]

Mycotoxins. The toxicity of mycotoxins in food has been appreciated since the Middle Ages and incidents such as St. Anthony's Fire and dry gangrene due to toxins present in rye contaminated with *Claviceps purpurea* are well documented. Hood[167] remarked recently that mycotoxins are among the world's most significant food contaminants, many are highly toxic to man and some like aflatoxin and rubratoxin are carcinogenic.

Mycotoxins have been examined for their teratogenicity mainly in mice at doses approaching the maternal toxic levels. When other test species have been used a wide interspecies variation in fetal responses was reported.[168] Aflatoxin caused a 90 per cent fetal mortality in mice dosed with 8 or 12 mg/kg on day nine of pregnancy, but was without effect at 2 or 4 mg/kg on days eight and thirteen respectively. A variety of structural malformations with increased mortality occurred in pregnant hamsters given 4 mg/kg intraperitoneally on day eight of pregnancy. Intrauterine growth retardation was reported when rats were dosed with 1 mg/kg on day 17.

Fetal mortality or intrauterine growth retardation was a usual observation with most of the mycotoxins examined experimentally, but structural malformations were occasionally seen.[167] They were significantly increased in mice dosed with cytochylasin, elymoclavin, ochratoxin A and rubratoxin. Although there was a range in species sensitivity, ochratoxin A induced teratological responses in rats, mice, hamsters, pigs and cattle including growth retardation, intrauterine mortality and structural abnormalities of the skeleton, brain, limbs and snout.[169,170] Ochratoxin A is highly toxic but abnormal fetal development is possibly due to impaired carbohydrate or protein metabolism or damage to the prechordal mesoderm. Although this toxin is particularly harmful to the smaller species, it is less toxic in ruminants on account of the low intestinal absorption and the ability of bacteria to convert it into innocuous compounds which are voided in the faeces.

Ergot is an alkaloid found in the fungus *Claviceps purpurea*. It has a marked vasoconstrictor action and in pregnancy may be a cause of fetal abnormality by its action in reducing placental blood flow.[171] Thus although ergot may adversely influence fetal development by a direct action, its action is primarily indirect, that is by causing placental insufficiency.

Extrapolation of these observations on the teratogenicity of fungal toxins in animals to humans is complex. Many of the alkaloids are very toxic and have a well-known pharmacological action which has been utilised medically. Man has learned by experience which foods overtly contaminated with fungi are likely to be harmful and avoids them accordingly. Much has been written on the subject and readers are referred to excellent monographs prepared by the Food Protection Committee[172] and the United States Food and Nutrition Board[173] for further reference.

Other Vegetable Toxins. The range of non-fungal toxins present in plants used in various parts of the world for food is large and some of them have been shown to exert adverse effects on reproduction in man or experimental animals. One such compound is hypoglycin A (α-amino-β-methylene cyclopropanyl-propionic acid) which is found in the West Indian Ackee fruit (*Blighia sapida*). This fruit is eaten by the local population but is a well known cause of food poisoning or vomiting sickness.[121] Hypoglycins A and B are amino acid derivatives of leucine and may be extracted from the unripe fruit. Persaud and his colleagues have shown that hypoglycin A administered to pregnant rats caused a high rate of fetal malformations (encephalocele and syndactyly) which were increased when hypoglycin A was given with leucine or carnitine.[174] They considered that, in the rat at least, hypoglycin A acts by inhibiting acyl dehydrogenase flavin-dependent oxidation. Hypoglycin B teratogenicity was reduced in the rat by administration of supplementary leucine, suggesting that it causes fetal abnormality by impairing the amino acid availability.

In regions of India, *Lathyrus sativa* seeds are sown with wheat as a drought-resistant crop. Thus *Lathyrus* seeds may occur in food in small quantities. However, if eaten to excess the seeds give rise to the disease known as lathyrism which may cause crippling paralytic conditions. The chemical β-aminopropionic-nitrate (BAPN) extracted from the seeds is a powerful drug used in the treatment of skin disorders but is also a potent teratogen in experimental animals.[175] BAPN inhibits the cross-linking of collagens and can cause structural deformities especially in the skeletal system.

Bracken (*Pteridium aquilinum*) is used as a food in several countries but is now known to be carcinogenic in rats, mice, guinea pigs and cows, and mutagenic in mice. Although it caused mild fetal growth retardation in mice, it was not teratogenic.[2,176]

Carrageenin is a complex polysaccharide obtained from some red seaweeds, which has been employed as a stabiliser and emulsifying agent in the food industry. However since this compound was shown to produce intestinal ulceration in several species of laboratory animal,[177] its use has been restricted. Other studies have shown that carrageenin administered in the diet to pregnant animals is both fetotoxic and teratogenic.[178] Increased fetal mortality was observed in pregnant rats fed 5 per cent carrageenin in their diet. Surviving offspring exhibited pulmonary damage when examined six months post-natally. These observations indicate that carrageenin should be used in food with caution as safer stabilisers are available nowadays.

Agricultural Chemicals

A large and diverse range of agricultural chemicals is used in various countries for controlling insect pests, fungal infections of crops, and for undesired vegetation. The extent to which these products occur in the environment varies according to local government legislation. Many of these products or their residues gain ëntry to the human diet in vegetables grown in treated conditions or through the meat of animals feeding on treated pastures and where the residues are not removed in subsequent processing or cooking. The ability of the various agricultural products to impair reproductive function or fetal development in humans, farm animals or laboratory animals has received considerable attention in recent years, and has been a fertile area of research.

Insecticides. From the few epidemiological studies reviewed, the risks of reduced fertility or fetal malformation in humans arising from accidental consumption of pesticides before or in pregnancy appears to be low.[105] A possible association was suggested between intrauterine mortality and fetal limb deformity from a retrospective study of 18 pregnancies in Japan where mothers had consumed insecticides in their food in a particular season.[179] In a later study, only four infants in 240 pregnancies studied prospectively exhibited structural abnormalities attributable to insecticide exposure.[180] When tissues of these children were examined for chlorinated hydrocarbon pesticide residues, the levels seen were within the ranges seen in the general adult population. DDT or DDE blood levels in patients during or following abortion were not significantly different from those seen in normal pregnancies.[181]

Animal studies on the teratogenicity of pesticides have been numerous and have shown considerable interspecies differences in response to oral or in diet administration. In most cases, the doses used to induce teratogenesis were greatly in excess of those likely to be ingested by humans through food contamination. The variations in fetal responses relate to the type of pesticide tested, the period of dosage during pregnancy, the length of the gestational period of the test species and its suceptibility to teratogenesis.

In a comparative survey of 19 pesticides for which animal data exist, the larger species, dogs, monkeys and pigs, tend to be more susceptible to prenatal risks than rodent species.[105] For example, offspring of these three species showed an increased rate of structural abnormality, intrauterine mortality or growth retardation in response to maternal doses of carbaryl of less than 50 mg/kg throughout pregnancy. Rat fetuses and neonates were not reported to be structurally abnormal at 300 mg/kg but reduced fertility was evident in multigeneration studies at 2000-3000 p.p.m. in the diet. Teratogenicity was seen in pregnant guinea pigs or rabbits dosed at 300 mg/kg and 50-200 mg/kg respectively. In the case of the insecticide diazinon, pig and dog fetal development was impaired at doses less than 10 mg/kg throughout pregnancy, whereas doses of up to 95 mg/kg were required to cause fetal mortality and possibly malformations in rat fetuses. No effect was reported in rabbits dosed at 7-30 mg/kg.

In multigeneration studies, the rat seems to be the choice species and in studies conducted over three generations, aldrin, carbaryl, dieldrin and malathion have been shown to cause a reduced rate of fertility.[105] Recent information obtained from a study of 101 men involved in the production of carbaryl showed no significant evidence to relate carbaryl exposure to infertility or an inability of these men to father children, thus contradicting predictions made by laboratory animal studies.[182]

Despite the numerous observations on the influence of DDT on avian reproduction, this insecticide is not embryotoxic and did not adversely influence spermatogenesis in mice.[131] It has been shown to delay oestrus in dogs and reduce libido, but this may be due to hormonal effects or other metabolic changes. Because of the wide variation in species responses, doses administered and the duration of treatment, meaningful comparisons of the teratogenic potentiation of the different pesticides are difficult. Wilson[105] considered that most of the organochlorine pesticides (other than DDT) are accompanied by some teratogenic risk in experimental animals and as a class may be more active than the organophosphorus compounds.

Polychlorinated Biphenols (PCB). The PCBs occur widely in nature and are present in such commodities as plastics, insulating materials, cutting oils, synergists for pesticides and many other products. They may be produced by irradiation of DDT and thus can be produced in the sterilisation of food containing pesticide residues.

Transplacental passage of PCBs has been demonstrated in experimental animals and since PCBs have been detected in human fetuses. the possibility of a teratogenic risk should be considered.[183] Studies in Japan have indicated that PCBs may cause fetal skin defects following maternal ingestion and that the residue may be located in the fetal brain, heart, liver and kidney but at generally lower levels than seen in adult organs.[184] These infants had dark skins, acne and eye disease (Yushio oil disease) associated with PCB levels of 65-839 μg/kg which is far higher than normal levels of exposure. From the available human toxicity data, it is difficult to make a prediction as to the fetal risk associated with PCBs or the threshold risk level. Although available information indicates that they are readily absorbed into the body from food, are deposited in fat and muscle, and pass across the placenta, they are not well metabolised. Further studies are obviously necessary to determine whether a long-term risk is associated with continued build-up of these compounds in the body from the prenatal period into post-natal life.

The teratogenicity of PCBs has been studied in several species of animals. In pregnant rats fed up to 5000 p.p.m. PCB in their diets major fetal deformity was not seen although some intrauterine growth retardation was seen at higher levels which were toxic to the mothers.[184] The lowest fetotoxic dose in the rat of kanechlor-500 was 28.6 mg/kg each day[2] which is considerably higher than the estimated human consumption in Japan (0.96 μg/kg each day).

Mixed PCBs fed to sows at 2 p.p.m. through pregnancy resulted in reduced litter size but not deformity.[185]

Herbicides. At least 100 different herbicides have been used to kill unwanted plants in the last 30 years but until recently none were suspected of being harmful to the human fetus. However, Agent Orange (2,4,5-trichlorophenoxyacetic acid, i.e. 2,4,5-T) which was introduced into the United States as long ago as 1944 was believed to be responsible for an increased miscarriage rate following its use as a defoliating agent in the Vietnam war. As a consequence of this the use of Agent Orange was banned by the Environmental Protection Agency (EPA).[186] Since then 2,4,5-T has been extensively examined toxicologically with a view to assessing its teratogenic and carcinogenic properties.

Population studies have been initiated in Vietnam in the local population exposed to 2,4,5-T in their food, with a view to identifying the actual risks involved. The study is still in progress and several lawsuits are pending, but at a recent conference sponsored by the American Farm Bureau Federation, the so-called Dispute Resolution Conference, the general conclusion reached was that the available information does *not* indicate that a human risk exists from exposure to 2,4,5-T. This committee was obviously aware that at high doses (15–50 mg/kg) 2,4,5-T produced various structural abnormalities, growth retardation and prenatal mortality in rat, mouse and hamster fetuses following maternal oral dosing at susceptible stages of pregnancy but that the offspring of rabbits, sheep or monkeys were not affected.[187,188]

Arising from the common concern surrounding the use of 2,4,5-T came the awareness that the principal risk probably lay with the highly toxic contaminant 2,3,7,8-tetrachlorodibenzo-*p*-dioxin (TCDD or Dioxin). TCDD is clearly the most toxic of the chlorinated dioxin series and it occurs in the synthesis of 2,4,5-T by the alkaline hydrolysis of 1,2,4,5-tetrachlorobenzene. The lethal dose for guinea pigs is about 1 μg whereas in the rat given 1-10 μg/kg/day no obviously lethal lesion is produced. The animal usually 'fades away' and dies in several weeks.

In Vietnam, the Agent Orange used was probably contaminated with about 30 μg of TCDD per gram but in the United Kingdom, 2,4,5-T as sold as a general herbicide and brushwood killer contains less than 0.1 μg/g.[188] This means that the contamination level in forests at the time of spraying is less than 0.05 μg per square metre. This will be subject to degradation by bacterial action and due to sunlight.

In teratological studies in animals, TCDD dosage led to cleft palate in some strains of mice but not rats.[189] When TCDD was added to 2,4,5-T and administered to pregnant rats and mice, it appears that the effects of the TCDD are evident at levels of 10 p.p.m.[190] The teratogenicity of TCDD alone has been examined in other species including the monkey, hamster, rabbit, sheep and even reindeer and in most cases no skeletal deformities were reported.[191–193]

In the light of available clinical and experimental information reviewed at the Dispute Resolution Conference, the committee concluded that TCDD is not likely to be injurious to pregnant women or their offspring. It is likely, however, that concern over the safety of this compound will continue for some time.[194] Close surveillance of the long-term effects of the massive chemical explosion in Seveso (Italy) in 1976, where dioxin was dispersed over a wide area, will hopefully provide further evidence of the toxic influence of this compound on reproductive function, and the post-natal development of children exposed to dioxin prenatally.[195]

Fungicides. Apart from the mercury compounds which are discussed above, a number of fungicides used in agriculture and in the home have been examined for their teratogenicity in experimental animals. Most of these would not normally be used as preservatives in the food industry but may enter the body of a pregnant woman in the same way as mercury, DDT and 2,4,5-T. Wilson[105] reviewed some of the many predictive investigations on such compounds in captan, difolatan, disulphiram, folpet and griseofulvin. Although the experimental techniques vary considerably among workers, the general conclusion reached was that fungicides as a group are not highly embryotoxic even when administered at doses approaching maternal toxicity. The levels of fungicides dosed to pregnant animals were in most cases far higher than were likely to be consumed by a pregnant woman under normal circumstances.

Discussion

Prenatal growth and development in mammals may be impaired by a wide variety of drugs, environmental agents and substances present in food. The presumed mechanisms for abnormal fetal development due to different agents vary but those conditions which lead to alterations in the pre-implanation embryo, the patterns of cell division and differentiation in the fetus, or the nature of the intrauterine environment are likely to cause fetal abnormality. This may be manifest either in terms of a generalised growth retardation, malformation in one or more organ systems, and in extreme cases pre- or post-natal mortality. The type and magnitude of the response relate directly to the stage in gestation of the exposure and the amount of the offending substance consumed by the pregnant mother.

Much of the available information that substances occurring in human diet may be teratogenic under certain circumstances is derived from experiments conducted in laboratory animals. Frequently, the doses administered were considerably higher than would be consumed by humans under normal circumstances, but the work was intended to establish whether or not a teratogenic risk was associated with the compound and which tissues or organ systems are susceptible. Unfortunately, for a number of food chemicals, a teratogenic response has been demonstrated in some non-human species but not others, and in some cases different issues have been affected. This renders the extrapolation of the results of predictive tests in animals to humans difficult. Bearing in mind the wide interspecies differences in response and sensitivity to the various agents, it cannot be assumed that a positive response in rats or monkeys is indicative of teratogenicity in the human baby following inadvertent ingestion of an offending chemical by a pregnant mother.

In conclusion, substances such as alcohol and mercury, and possibly the food contaminants lead, cadmium and agricultural chemical residues are likely to be harmful to the human baby through maternal ingestion at susceptible stages in pregnancy, but for the most part, chemicals added to food to enhance its appearance, flavour and preservation seem to be without serious risks of teratogenesis. Infertility is an additional problem associated with alcohol, cadmium and some other chemicals which may occur in food.

The problem of teratogenicity and infertility arising from substances present in the human diet is a difficult one to evaluate. The human diet varies enormously from one part of the world to another and such factors as social class, religion and local customs are important. When unexplained increases in fetal abnormalities or reproduction failure have occurred, epidemiological surveys have usually led to a range of experimental studies being conducted in animals in an attempt to identify the cause. In the case of food, the studies precipitated by the 2,4,5-T and mercury-contaminated grain and seafood are notable examples. Whereas on some occasions misleading results have been given by predictive studies, in the main suitably designed laboratory studies have provided much useful information as to the nature and mechanism of teratogenic action of substances found in the diet.

Acknowledgements

I am grateful to Mr F.M. Sullivan, Reader in Pharmacology at Guy's Hospital Medical School for his helpful advice in preparing this review. The helpful discussions with Miss Margaret Cook, teratologist at Wyeth Laboratories, are also appreciated.

References

1. N.M. Gregg, 'Congenital Cataract Following German Measles in the Mother', *Trans. Ophthal. Soc. Aust., 3* (1941), p. 80.

2. H. Nishimura and T. Tanimura, *Clinical Aspects of the Teratogenicity of Drugs* (Excerpta Medica, Amsterdam, 1976), p. 284.

3. T.V.N. Persaud, 'Aspects of Teratology', *W.I. Med. J., 17* (1968), p. 74.

4. J.G. Wilson, 'Current Status of Teratology' in J.G. Wilson and F.C. Fraser (eds.), *Handbook of Teratology* (Plenum, New York, 1977), vol. 1, pp. 47–74.

5. J.R. Gillette, 'Factors that Affect Drug Concentration in Maternal Plasma' in J.G. Wilson and F.C. Fraser (eds.), *Handbook of Teratology* (Plenum Press, New York, 1977), vol. 3, pp. 35–78.

6. T. Takeuchi and H. Matsumoto, 'Minamata Disease of Human Foetuses' in H. Nishimura and J.R. Miller (eds.), *Methods for Teratological Studies in Experimental Animals and Man* (Igaku Shoin, Tokyo, 1969), p. 280.

7. F. Bakir, S.R. Damluji, L. Amin-Zaki, M. Murtaiha, A. Khaladi, N.Y. Al-Rawi, S. Tikriti, H.I. Dhahir, T.W. Clarkson, J.C. Smith and R.A. Doherty, 'Methylmercury Poisoning in Iraq', *Science, 181* (1973), p. 230.

8. I.R. Rowland, M.J. Davies and P. Grasso, 'Methylation of Methylmercury by Gastrointestinal Contents in the Rat', *Biochem. Soc. Trans., 3* (1975), p. 502.

9. K.H. Harper, A.K. Palmer and R.E. Davies, 'The Effect of Imipramine upon the Pregnancy of Laboratory Animals', *Arzneim-Forsch., 15* (1965), p. 1218.

10. J.V. Dingell, F. Sulser and J.R. Gillette, 'Species Differences in the Metabolism of Imipramine and Desmethylimipramine (DMI)', *J. Pharmacol. Exp. Ther., 143* (1964), p. 14.

11. G. Feuer, 'The Link Between the Mother and Foetus in Drug Metabolism', *Rev. Can. Biol., 32*, suppl. automne (1973), p. 113.

12. M.R. Juchau and M.G. Pedersen, 'Drug Biotransformation Reactions in the Human Foetal Adrenal Gland', *Life Sci., 12* (1973), p. 193.

13. R.T. Williams, 'Species Variation in Drug Biotransformations' in B.N. La Du, H.G. Mandel and E.L. Way (eds.), *Fundamentals of Drug Metabolism* (Williams and Wilkins, Baltimore, 1971), pp. 187–205.

14. L.S. Kennedy and T.V.N. Persaud, 'Acute Alcohol Intoxication in the Pregnant Rat' in T.V.N. Persaud (ed.), *Advances in the Study of Birth Defects* (MTP Press, Lancaster, 1979), vol. 2, pp. 223–238.

15. R.L. Smith and J. Caldwell, 'Drug Metabolism in Non-human Primates' in D.V. Parke and R.L. Smith (eds.), *Drug Metabolism from Microbe to Man* (Taylor and Francis Ltd, London, 1977), pp. 331–356.

16. J. Dancis and D. Springer, 'Fetal Homeostasis in Maternal Malnutrition: Potassium and Sodium Deficiency in Rats', *Pediat. Res., 4* (1970), p. 345.

17. H. Nishimura and S. Miyamoto, 'Teratogenic Effects of Sodium Chloride in Mice', *Acta Anat., 74* (1969), p. 121.

18. M.F. Lechat, I. Borlee, A. Bouckaert and C. Misson, 'Caffeine Study', *Science, 207* (1980), p. 1296.

19. P.E. Palm, E.P. Arnold, R.C. Rachwell, J.C. Leyczek, K.W. Teague and C.J. Kensler, 'Evaluation of the Teratogenic Potential of Fresh-brewed Coffee and Caffeine in the Rat', *Toxicol. appl. Pharmacol., 44* (1978), p. 1.

20. A. Lechtig, H. Delghado, R.E. Lasky, R.E. Klein, R.L. Engle, C. Yarburgh and J.P. Habicht, 'Maternal Nutrition and Foetal Growth in Developing Societies', *Am. J. Dis. Child., 129* (1975), p. 434.

21. B.Y. Iba, J.D. Niswander and L. Woodville, 'Relation of Prenatal Care to Birthweights, Major Malformations and Newborn Deaths of American Indians', *Health Services Report, 88* (1973), p. 697.

22. L.S. Hurley, 'Nutritional Deficiencies and Excesses' in J.G. Wilson and F.C. Fraser (eds.), *Handbook of Teratology* (Plenum Press, New York, 1977), vol. 1, pp. 261–308.

23. R.L. Naeye, W. Blanc and C. Paul, 'Effects of Maternal Nutrition on the Human Foetus', *Pediatrics, 52* (1973), p. 494.

24. R.L. Pike, 'Congenital Cataract in Albino Rats Fed Different Amounts of Tryptophane and Niacin', *J. Nutr., 44* (1951), p. 191.

25. S. Zamenhof, S.M. Hall, L. Grauel, E. van Marthens and M.J. Donahue, 'Deprivation of Amino Acids and Prenatal Brain Development in Rats', *J. Nutr., 104* (1974), p. 1002.

26. R.O. Fisch, D. Doeden, L.L. Lansky and J.A. Anderson, 'Maternal Phenylketonuria', *Am. J. Dis. Child., 118* (1969), p. 847.

27. S.A. Luse, A. Rhys and R. Lessey, 'Effects of Maternal Phenylketonuria on the Rat Foetus', *Am. J. Obstet. Gynecol., 108* (1970), p. 387.

28. T. Fujii and H. Nishimura, 'Side Preponderant Forelimb Defects of Mouse Foetuses Induced by Maternal Treatment with Adenine', *Okajimas Fol. Anat. Jap., 49* (1972), p. 75.

29. Y. Niiyama, K. Kishi, S. Endo and G. Inoue, 'Effects of Diets Devoid of One Essential Amino Acid on Pregnancy in Rats Maintained by Ovarian Steroids',

J. Nutr., 103 (1973), p. 207.

30. C. Alling, A. Bruce, I. Karlsson, O. Sapia and L. Svennerholm, 'Effect of Maternal Essential Fatty Acid Supply on Fatty Acid Composition of Brain, Liver, Muscle and Serum in 21 Day Old Rats', *J. Nutr., 102* (1972), p. 773.

31. J.L. Smart, J. Dobbing, P.R. Adlard and A. Lynch, 'Vulnerability of Developing Brain: Relative Effects of Growth Restriction During Foetal and Suckling Periods on Behaviour and Brain Composition of Adult Rats', *J. Nutr., 103* (1973), p. 1327.

32. D.B. Zilversmit, M. Remington and L.B. Hughes, 'Foetal Growth and Placental Permeability in Rabbits for High Cholesterol', *J. Nutr., 102* (1972), p. 1681.

33. R.F. Borgman, R.G. Bursey and B.C. Caffrey, 'Influence of Maternal Dietary Fat upon Rat Pups', *Am. J. Vet. Res., 36* (1975), p. 799.

34. G.E. Lamming, G.W. Salisbury, R.L. Hays and K.A. Kendall, 'The Effect on Incipient Vitamin A Deficiency on Reproduction in the Rabbit. Decidua, Ova and Fertilisation', *J. Nutr., 52* (1954), p. 217.

35. J.G. Wilson and J. Warkany, 'Cardiac and Aortic Arch Anomalies in the Offspring of Vitamin A Deficient Rats Correlated with Similar Human Anomalies', *Pediatrics, 5* (1950), p. 708.

36. B. Palludan, *A-avitaminosis in Swine* (Munksgaard, Copenhagen, 1966).

37. G.M. Morriss, 'Morphogenesis of the Malformations Induced in Rat Embryos by Maternal Hypervitaminosis A', *J. Anat., 113* (1972), p. 241.

38. D.M. Kochhar, 'Studies of Vitamin A-induced Teratogenesis: Effects on Embryonic Mesenchyme and Epithelium and on Incorporation of ^{3}H-Thymidine', *Teratology, 1* (1968), p. 299.

39. A. Ornoy, L. Nebel and Y. Menczel, 'Impaired Osteogenesis of Foetal Long Bones Induced by Maternal Hypovitaminosis D_2', *Arch. Path., 87* (1969), p. 563.

40. W.F. Friedman and L.F. Mills, 'The Relationship Between Vitamin D and Craniofacial and Dental Anomalies of the Supravalvular Aortic Stenosis Syndrome', *Pediatrics, 43* (1969), p. 12.

41. L. Nebel and A. Ornoy, 'Effect of Hypervitaminosis D_2 on Fertility and Pregnancy in Rats', *Israel J. Med. Sci., 2* (1966), p. 14.

42. H.A. Mattill, 'The Relation of Vitamins B and E to Fertility and Pregnancy in Rats', *Am. J. Physiol., 79* (1926–27), p. 305.

43. J. Green and J. Bunyan, 'Vitamin E and the Biological Oxidant Theory', *Nutr. Abs. Revs, 39* (1969), p. 321.

44. D.W. Cheng, T.A. Bairnson, A.N. Rao and S. Subbammal, 'Effect of Variations of Rations on the Teratogenicity in Vitamin E Deficient Rats', *J. Nutr., 71* (1960), p. 54.

45. D.W. Cheng, K.G. Braun, B.J. Braun and K.G. Udani, 'Tocopherol Content of Maternal and Foetal Rat Tissues as Related to Vitamin E Intake During Gestation', *J. Nutr., 74* (1961), p. 111.

46. H.A. Kenney and C.E. Roderuck, 'Fatal Syndrome Associated with Vitamin E Status of Pregnant Rats', *Proc. Soc. Exp. Biol. Med., 114* (1965), p. 257.

47. L.M. Roderick, 'The Pathology of Sweet Clover Disease in Cattle', *J. Am. Vet. Med. Assoc., 75* (1929), p. 314.

48. A.P. Kraus, S. Perlow and K. Singer, 'Danger of Dicurmarol^R Treatment in Pregnancy', *J. Am. Med. Assoc., 139* (9149), p. 758.

49. M.P. Martin, E. Bridgeforth, W.J. McGanity and W.J. Darby, 'The Vanderbilt Cooperative Study of Maternal and Infant Nutrition. X. Ascorbic Acid', *J. Nutr., 62* (1957), p. 201.

50. E.P. Samborskaya and T.D. Ferdman, 'The Mechanism of Termination of

Pregnancy by Ascorbic Acid', *Bull. Exp. Biol. Med.*, *62* (1966), p. 934.
51. W. Nauweiler, 'Die hypervitaminose und ihre Beziehung zur Schwangerschaft', *Intern. Z. Vitaminforsch.*, *22* (1951), p. 392.
52. U. Bologna and V. Picconi, 'Effeti' Dell Ipovitaminosi B_2 Sulla Gravidanza e Sui Nati del Ratto Albino', *Soc. Ital. Biol. Speriment.*, *26* (1956), p. 219.
53. J. Lefebvres-Boisselot, 'Role Tératogèns de la Déficience en Acide Pathothénique Chez le Rat', *Ann. Med.*, *52* (1951), p. 225.
54. C.C. Kratzing and J.J. Perry, 'Hypertension in Young Rats Following Choline Deficiency in Maternal Diets', *J. Nutr.*, *101* (1971), p. 1657.
55. M.E. Ensminger, J.P. Bowland and T.J. Cunha, 'Observations on the Thiamine, Riboflavine and Choline Needs for Sows for Reproduction', *J. Anim. Sci.*, *6* (1947), p. 409.
56. M.F. Schumacher, M.A. Williams and R.L. Lyman, 'Effect of High Intakes of Thiamine, Riboflavine and Pyridoxine on Reproduction in Rats and Vitamin Requirements of the Offspring', *J. Nutr.*, *86* (1965), p. 343.
57. E.M. Widdowson, 'Trace Elements in Foetal and Early Post-natal Development', *Proc. Nutr. Soc.*, *33* (1974), p. 275.
58. A. Giroud, 'Nutrition and the Embryo', *Fed. Proc.*, *27* (1968), p. 163.
59. E.B. Hart, E.V. McCollom, H. Steenbock and G.C. Humphrey, 'Physiological Effect on Growth and Reproduction of Rations Balanced from Restricted Resources', *Wis. Exp. Stn. Res. Bull.*, *17* (1911), p. 160.
60. J.W. Bawden and J.W. Osborne, 'Tracer Study on the Effect of Dietary Calcium Deficiency during Pregnancy in Rats', *J. Dent. Res.*, *41* (1962), p. 1349.
61. H. Itoh, S.L. Hansard, J.C. Glenn, F.H. Hoskins and D.M. Thrasher, 'Placental Transfer of Calcium in Pregnant Sows on Normal and Limited Calcium Rations', *J. Anim. Sci.*, *26* (1967), p. 335.
62. W.M. Cox and M. Imboden, 'The Role of Calcium and Phosphorus in Determining Reproductive Success', *J. Nutr.*, *11* (1936), p. 147.
63. F. Lehnerdt, 'Zur Frage der Substitution das Calciums im Knochensytem durch Strontium', *Ziegler's Beitr.*, *46* (1909), p. 468.
64. A.B.G. Lansdown, R.C. Longland and P. Grasso, 'Reduced Foetal Calcium without Skeletal Malformations in Rats Following High Maternal Doses of a Strontium Salt', *Experientia (Basel)*, *28* (1972), p. 558.
65. E.M. Widdowson, R.A. McCance, G.E. Harrison and A. Sutton, 'Effects of Giving Phosphate Supplement to Breast-fed Babies on Absorption and Excretion of Calcium, Strontium, Magnesium and Phosphate', *Lancet*, *ii* (1963), p. 1250.
66. B.J. Hopkins and B.W. Casarett, 'Gross Developmental Anomalies Induced by Strontium-90 in the Rat Embryos', *University of Rochester Atomic Energy Report*, UR.-643.
67. S. Hiraoka, 'The Transplacental Effects of the Radio-Strontium-90 upon the Mouse Embryos', *Acta. Anat. Nippon*, *36* (1961), p. 161.
68. A. Fairney and A.A. Weir, 'The Effect of Abnormal Calcium Levels on the Offspring of Rats', *J. Endocrinol.*, *48* (1970), p. 337.
69. L.S. Hurley and G. Cosens, 'Congenital Malformations and Foetal Anaemia Resulting from Magnesium Deficiency in Rats', *Fed. Proc.*, *30* (1971), p. 516.
70. N.S. Scrimshaw and V.R. Young, 'The Requirements of Human Nutrition', *Scientific American*, (1976), September, p. 51.
71. L.S. Hurley and H. Swenerton, 'Congenital Abnormality Resulting from Zinc Deficiency in Rats', *Proc. Soc. Exp. Biol. Med.*, *123* (1966), p. 692.
72. L.S. Hurley and R.E. Schrader, 'Abnormal Development of Pre-implantation Rat Eggs after Three Days of Maternal Dietary Zinc Deficiency', *Nature (Lond.)*, *254* (1975), p. 427.
73. L.S. Hurley and G. Cosens, 'Reproduction and Prenatal Development in Relation to Dietary Zinc Levels' in W.G. Hoekstra, J.W. Suttie, H.E. Ganther and

W. Mertz (eds.), *Trace Element Metabolism in Animals* (University Park Press, Baltimore, 1974), vol. 2, pp. 516–518.

74. H. Swenerton and L.S. Hurley, 'Teratogenic Effects of a Chelating Agent and Their Prevention by Zinc', *Science, 173* (1971), p. 62.

75. V.M. Ferm, 'The Teratogenic Effects of Metals on Mammalian Embryos' in D.H.M. Woolham (ed.), *Advances in Teratology* (Academic Press, New York, 1972), vol. 5, pp. 51–75.

76. P.A. Cancilla and R.M. Barlow, 'Structural Changes in the Nervous System in Swayback (Enzootic Ataxia) in Lambs', *Acta Neuropathol., 6* (1966), p. 260.

77. G.A. Hall and J.M. Howell, 'The Effect of Copper Deficiency on Reproduction in the Female Rat', *Br. J. Nutr., 23* (1969), p. 41.

78. B.L. O'Dell, B.C. Hardwick and G. Reynolds, 'Mineral Deficiencies in Milk and Congenital Malformation in the Rat', *J. Nutr., 73* (1961), p. 151.

79. F.E. Hytten, 'Nutrition in Pregnancy', *Practitioner, 212* (1974), p. 459.

80. T.R.C. Sisson and C.J. Lund, 'The Influence of Maternal Iron Deficiency on the Newborn', *Am. J. Clin. Nutr., 6* (1958), p. 376.

81. A.H. Parmelee, E. Allen, I.F. Stein and H. Buxbaum, 'Three Cases of Congenital Goitre', *Am. J. Obstet. Gynecol., 40* (1973), p. 257.

82. R.E. Shrader, L. Erway and L.S. Hurley, 'Mucopolysaccharide Synthesis in the Developing Inner Ear of the Manganese-deficient and Pallid Mutant Mice', *Teratology, 8* (1973), p. 257.

83. R.D. Hood and S.L. Bishop, 'Teratogenic Effect of Sodium Arsenate in Mice', *Arch. Envir. Hlth, 24* (1972), p. 62.

84. F.F. Snyder and H. Speert, 'The Placental Transmission of Neoarsphenamine in Relation to the Stage of Pregnancy', *Am. J. Obstet. Gynecol., 36* (1938), p. 579.

85. M.R. Moore, M.A. Hughes and D.J. Goldberg, 'Lead Absorption in Man from Dietary Sources', *Int. Arch. Occup. Envir. Hlth, 44* (1979), p. 81.

86. Lead, *Airborne Lead in Perspective* (National Academy of Sciences, Washington, 1972).

87. J.L. McNeil, J.A. Ptasnik and D.B. Croft, 'Evaluation of Long-term Effects of Elevated Blood Lead Concentrations in Asymptomatic Children', *Arch. Ind. Hyg. Toxicol., 26*, suppl. (1975), p. 97.

88. C. Paul, 'Memoires Originaux', *Arch. Gen. Med.* (1860), p. 513.

89. C. Lund, 'The Effect of Chronic Lead Poisoning on Reproductive Capacity', *Nord. Hyg. Tidskr., 18* (1936), p. 12.

90. K. Nogaki, 'On Action of Lead on Body of Lead Refinery Workers: Particularly on Conception, Pregnancy and Parturition in Care of Females and on the Vitality of Their Newborn', *Acta. Med. (Fukuoka), 27* (1957), p. 1314.

91. Swedish Council of Environmental Information, *Workshop on The Utility of an Evaluated Data Bank in Teratology and its use for a State-of-the-Art-Report on Adverse Effects of Lead on Reproduction and Development* (Nynashamn. Sweden, 1977), pp. 22–31.

92. J.J. Gershanik, G.T. Brooks and J.A. Little, 'Blood Lead Levels in Pregnant Women and Their Offspring', *Am. J. Obstet. Gynecol., 119* (1975), p. 508.

93. A.D. Beattie, M.R. Moore, A. Goldberg, M. Finlayson, J. Graham, E. Mackie, J. Main, D.A. McLaren, M.A. Murdoch and G.T. Stewart, 'Role of Chronic Low Lead Level Exposure in the Aetiology of Mental Retardation', *Lancet, i* (1975), p. 589.

94. M.R. Moore, P.A. Meredith and A. Goldberg, 'A Retrospective Analysis of Blood Lead in Mentally Retarded Children', *Lancet, i* (1977), p. 717.

95. V.H. Ferm and D.W. Ferm, 'The Specificity of the Teratogenic Effect of Lead in the Golden Hamster', *Life Sci., 10* (1971), p. 35.

96. R.M. McClain and B.A. Becker, 'Teratogenicity, Foetal Toxicity and

Placental Transfer of Lead Nitrate in Rats', *Toxicol. Appl. Pharmacol., 31* (1975), p. 72.

97. C.A. Kimmel, L.D. Grant and C.S. Sloan, 'Chronic Lead Exposure Assessment of Development Toxicity', *Teratology, 13* (1976), p. 274.

98. T.L. Carson, G.A. Van Gelder, F.C. Karas and W.B. Buck, 'Slowed Learning in Lambs Prenatally Exposed to Lead', *Arch. Envir. Hlth, 29* (1974), p. 154.

99. R.M. Sharma and W.B. Buck, 'Effect of Chronic Lead Exposure on Pregnant Sheep and Their Progeny', *Vet. Toxicol., 18* (1976), p. 186.

100. D.C. Hilderbrand, R. Der, W.T. Griffin and M.S. Fahim, 'Effect of Lead Acetates on Reproduction', *Am. J. Obstet. Gynecol., 115* (1973), p. 1058.

101. I. Lancranjan and H.J. Popesco, 'Reproductive Ability of Workmen Occupationally Exposed to Lead', *Arch. Envir. Hlth, 30* (1975), p. 396.

102. P. Grasso, 'Renal Responses to Cadmium', *Fd Cosmet. Toxicol., 13* (1975), p. 470.

103. R.P. Sharma, J.C. Street, P.L. Verma and J.L. Shupe, 'Cadmium Uptake from Feed and its Distribution to Food Products of Livestock', *Envir. Hlth Perspect., 28* (1979), p. 59.

104. B. Engstrom and G.F. Nordberg, 'Factors Influencing Absorption and Retention of Oral ^{109}Cd in Mice: Age Pre-treatment and Subsequent Treatment with Non-Radioactive Cadmium', *Acta Pharmacol. Toxicol., 45* (1979), p. 315.

105. J.G. Wilson, 'Environmental Chemicals' in J.G. Wilson and F.C. Fraser (eds.), *Handbook of Teratology* (Plenum Press, New York, 1977), vol. 1, pp. 357–385.

106. J.E. Mulvihill, S.H. Gamm and V.H. Ferm, 'Facial Formation in Normal and Cadmium-treated Golden Hamsters', *J. Embryol. Exp. Morphol., 24* (1970), p. 393.

107. *Joint FAO/WHO Expert Committee on Food Additives*, 16th Report, Geneva 1972.

108. S. Chaube, H. Nishimura and C.A. Swinyard, 'Zinc and Cadmium in Normal Human Embryos and Foetuses', *Arch. Envir. Hlth, 20* (1973), p. 237.

109. L. Dencker, 'Cadmium: Accumulation in the Embryos, Placentae and Ovaries Related to the Effects on Reproduction', *Acta. Pharmacol. Toxicol., 39* suppl. 1, (1976), p. 9.

110. J. Parizek, 'Vascular Changes at Sites of Oestrogen Biosynthesis Produced by Parenteral Injection of Cadmium Salts: the Destruction of the Placenta by Cadmium Salts', *J. Reprod. Fertil., 7* (1964), p. 263.

111. J. Parizek, 'The Peculiar Toxicity of Cadmium During Pregnancy – an Experimental "Toxaemia of Pregnancy" Induced by Cadmium Salts', *J. Reprod. Fertil., 9* (1965), p. 111.

112. A.B. Kar, R.P. Das and J.N. Karkun, 'Ovarian Changes in Prepuberal Rats after Treatment with Cadmium Chloride', *Acta Biol. Med. Gen., 3* (1959), p. 372.

113. R.L. Dixon, I.P. Lee and R.J. Sherins, 'Methods to Assess Reproductive Effects of Environmental Chemicals: Studies of Cadmium and Boron', *Envir. Hlth Perspect., 13* (1976), p. 59.

114. S. Panova and S. Ivanova, 'Changes in the Ovarian Function and Some Functional Indices of Liver of Women in Professional Contact with Metallic Mercury', *Akush. Ginekol., 15* (1976), p. 136.

115. J. Afonso and R. De Alvarez, 'Effects of Mercury on Human Gestation', *Am. J. Obstet. Gynecol., 80* (1960), p. 145.

116. Report of an expert committee, Methylmercury in Fish. *Nord. Hyg. Tidskr.,* suppl. 4 (1971).

117. R.W. Miller, 'How Environmental Effects on Child Health are Recognised'. *Pediatrics, 53* (1974), p. 792.

118. K.S. Khera, 'Teratogenic Effects of Methylmercury in the Cat: Note on the use of this Species as a Model for Teratogenicity Studies', *Teratology, 6* (1972), p. 139.

119. S.B. Harris, J.G. Wilson and R.H. Prinz, 'Embryotoxicity of Methylmercuric Chloride in Golden Hamsters', *Teratology, 6* (1972), p. 139.

120. J.F. Beary, 'Mercury in Sperm Whale Meat', *Science, 206* (1979)

121. S. Davidson, R. Passmore and J.F. Brock, *Human Nutrition and Dietetics*, 5th edn (Churchill Livingstone, Edinburgh, 1973).

122. D. Stone, E. Matulka and B. Palaski, 'Do Artificial Sweeteners Ingested in Pregnancy Affect the Offspring', *Nature (Lond.), 231* (1971), p. 53.

123. R.W. West, 'The Exposure of Foetal and Suckling Rats to Saccharin from Dosed Maternal Animals', *Toxicol. Lett., 4* (1979), p. 127.

124. J. Lederer and A.M. Pottier-Arnould, 'Influence de la Saccharine sur le Dévelopement de l'Embryon Chez la Rate Gestante', *Diabète, 21* (1973), p. 13.

125. G. Luckhaus and L. Machemer, 'Histological Examination of Perinatal Eye Development in the Rat after Ingestion of Sodium Cyclamate and Sodium Saccharin During Pregnancy', *Fd Cosmet. Toxicol., 16* (1978), p. 7.

126. E.E. Vogin and B.L. Oser, 'Effects of Cyclamate:Saccharin Mixture on Reproduction and Organogenesis in Rats and Rabbits', *Fed. Proc., 28* (1969), p. 743.

127. J.G. Wilson, 'Abnormalities of Intrauterine Development in Non-human Primates', *Acta Endocrinol., 166*, suppl. (1972), p. 261.

128. B.A. Becker and J.E. Gibson, 'Teratogenicity of Cyclohexylamine in Mammals', *Toxicol. Appl. Pharmacol., 17* (1970), p. 551.

129. I.F. Gaunt, M. Sharratt, P. Grasso, A.B.G. Lansdown and S.D. Gangolli, 'Short-term Toxicity of Cyclohexylamine in the Rat', *Fd Cosmet. Toxicol., 12* (1974), p. 609.

130. H.S. Legator, K.A. Palmer, S. Green and K.W. Petersen, 'Cytogenetic Studies in Rats of Cyclohexylamine, a Metabolite of Cyclamate', *Science, 165* (1969), p. 1139.

131. A.J. Wyrobeck and W.R. Bruce, 'Chemical Induction of Sperm Abnormalities in Mice', *Proc. Nat. Acad. Sci. USA, 72* (1975), p. 4425.

132. A.I. Shtenberg and E.V. Gavrilenko, 'Influence of the Food-dye Amaranth upon the Reproductive Function and Development of Progeny in Tests on Albino Rats', *Vopr. Pitan., 29* (1970), p. 66.

133. K.S. Larsson, 'A Teratologic Study with the Dyes Amaranth and Ponceau 4R in Mice', *Toxicology, 4* (1975), p. 75.

134. L.J. Filer, 'Monosodium Glutumate Metabolism in the Foetus and Newborn' in J. Dancis and J.C. Hwang (eds.), *Perinatal Pharmacology: Problems and Priorities* (Raven Press, New York, 1974), pp. 177–193.

135. A. Lamperti and G. Blaha, 'Ovarian Function in Hamsters Treated with Monosodium Glutumate', *Biol. Reprod., 21* (1979), p. 923.

136. B.L. Oser, K. Morgareidge and S. Carson, 'Monosodium Glutamate Studies in Four Species of Neonatal and Infant Animals', *Fd Cosmet. Toxicol., 4* (1975), p. 7.

137. R. Heywood and A.N. Worden, 'Glutamate Toxicity in Laboratory Animals' in L.J. Filer, S. Garrattini, M.R. Kare, W.A. Reynolds and R.J. Wurtman (eds.), *Glutamic Acid: Advances in Biochemistry and Physiology* (Raven Press, New York, 1979), pp. 203–215.

138. J.W. Olney, O.-L. Ho, V. Rhee and B. Schainker, 'Cysteine-induced Brain Damage in Infant and Foetal Rodents', *Brain Res., 45* (1972), p. 309.

139. J.W. Olney, 'Excitotoxic Amino Acids: Research Applications and Safety Implications' in L.J. Filer, S. Garrattini, M.R. Kare, W.A. Reynolds and R.J.

Wurtman (eds.), *Biochemistry and Physiology* (Raven Press, New York, 1979), pp. 287–319.

140. W.A. Reynolds, N. Lemkey-Johnston and L.D. Stegink, 'Morphology of the Foetal Monkey Hypothalamus After *in utero* Exposure to Monosodium Glutumate' in L.J. Filer, S. Garrattini, M.R. Kare, W.A. Reynolds and R.J. Wurtman (eds.), *Glutamic Acid: Advances in Biochemistry and Physiology* (Raven Press, New York, 1979), pp. 217–299.

141. S. Garrattini, 'Evaluation of the Neurotoxic Effects of Glutamic Acid' in R.J. Wurtman and J.J. Wurtman (eds.), *Nutrition and the Brain* (Raven Press, New York, 1979), vol. 4, pp. 79–124.

142. A.R. Johnson, 'A Re-examination of the Possible Teratogenic Effects of Butylated Hydroxytoluene (BH) and its Effect on the Reproductive Capacity of the Mouse', *Fd Cosmet. Toxicol., 3* (1965), p. 371.

143. D.J. Clegg, 'Absence of Teratogenic Effect of Butylated Hydroxyanisole (BHA) and Butylated Hydroxytoluene in Rats and Mice', *Fd Cosmet. Toxicol., 3* (1965), p. 387.

144. J.D. Stokes and C.L. Scudder, 'The Effects of Butylated Hydroxyanisole and Butylated Hydroxytoluene on Behavioural Development in Mice', *Dev. Psychobiol., 7* (1974), p. 343.

145. A. Giroud and H. Tuchmann-Duplessis, 'Congenital Abnormalities Role of Exogenous Factors', *Path. Biol., 10* (1962), p. 119.

146. K.L. Jones and D.W. Smith, 'Recognition of the Foetal Alcohol Syndrome in Early Infancy', *Lancet, ii* (1973), p. 999.

147. K.L. Jones, D.W. Smith, A.P. Streissguth and N.C. Mirianthopoulos, 'Outcome of Offspring of Chronic Alcoholic Women', *Lancet, i* (1973), p. 1076.

148. L.N. Cook, R.J. Shott and B.F. Andrews, 'Acute Transplacental Ethanol Intoxication', *Am. J. Dis. Child., 129* (1975), p. 1075.

149. L.I. Mann, A. Bhakthavathsalen, S. Liu and P. Makowski, 'Placental Transport of Alcohol and its Effect on Maternal and Foetal Acid Base Balance', *Am. J. Obstet. Gynecol., 122* (1975), p. 837.

150. L.I. Mann, A. Bhakthavathsalen, S. Liu and P. Makowski, 'Effect of Alcohol of Foetal Cerebral Function and Metabolism', *Am. J. Obstet. Gynecol., 122* (1975), p. 845.

151. L.A. Kennedy and T.V.N. Persaud, 'Acute Alcohol Intoxication in the Pregnant Rat' in T.V.N. Persaud (ed.), *Advances in The Study of Birth Defects* (MTP, Lancaster, 1979), pp. 223–238.

152. R.R. Chaudhury and M. Matthews, 'Effect of Alcohol on the Fertility of Female Rabbits', *J. Endocr., 34* (1966), p. 275.

153. J.D. Keiffler and M.M. Ketchel, 'Blockade of Ovulation in the Rat by Ethanol', *Acta Endocr., 65* (1970), p. 117.

154. A. Suzuki, C. Morita and K. Nakamura, 'A Study on Morphological Changes in Rat Brain Transplacentally Treated with Ethanol Prenatally', *Proc. Congen. Anom. Res. Assoc. Japan, 7* (1967), p. 65.

155. D.H. Van Thield, R. Lester and R.J. Sherins, 'Hypogonadism in Alcoholic Liver Disease: Evidence for a Double Effect', *Gastroenterology, 67* (1974), p. 1188.

156. D.H. Van Thield, J.S. Gavala, R. Lester and M.D. Goodman, 'Alcohol-induced Testicular Atrophy', *Gastroenterology, 69* (1975), p. 326.

157. F.M. Badr and S.R. Badr, 'Induction of Dominant Lethal Mutation in Male Mice by Ethyl Alcohol', *Nature (Lond.), 253* (1975), p. 134.

158. R.W. Klassen and T.V.N. Persaud, 'Experimental Studies on the Influence of Male Alcoholism on Testicular Function, Pregnancy and Progeny' in T.V.N. Persaud (ed.), *Advances in the Study of Birth Defects* (MTP, Lancaster,

1979), vol. 2, pp. 239–256.

159. J.H. Renwick, 'Hypothesis: Anencephaly and Spina Bifida are Usually Preventable by Avoidance of a Specific but Unidentified Substance Present in Certain Potato Tubers', *Br. J. Prev. Soc. Med., 26* (1972), p. 67.

160. R.W. Smithells, 'Environmental Teratogens of Man', *Br. Med. Bull., 32* (1976), p. 27.

161. D.E. Poswillo, D. Sopher and S. Mitchell, 'Experimental Induction of Fetal Malformation with Blighted Potato: a Preliminary Report', *Nature, 239* (1972), p. 462.

162. J.R. Allen, R.J. Marler, C.F. Chesney, J.P. Helgeson, A. Kelman, K.G. Weckel, E. Traisman and J.W. White, 'Teratogenicity Studies on Late Blighted Potatoes in Non-human Primates (*Macaca mulatta* and *Sanguinus labiatus*)', *Teratology, 15* (1977), p. 17.

163. S. Chaube, C.A. Swinyard and R.H. Daines, 'Failure to Induce Malformation in Foetal Rats by Feeding Blighted Potatoes to Their Mothers', *Lancet, i* (1973), p. 329.

164. R.F. Keeler, S. Young, D. Brown, G.F. Stallknecht and D. Douglas, 'Congenital Deformities Produced in Hamsters by Potato Sprouts', *Teratology, 17* (1978), p. 327.

165. R.P. Sharma, C.C. Willhite, M.T. Wu and D.K. Salunkhe, 'Teratogenic Potential of Blighted Potato Concentrate in Rabbits, Hamsters and Minature Swine', *Teratology, 18* (1978), p. 55.

166. S. Fabro and S.M. Sieber, 'Caffeine and Nicotine Penetrate the Preimplanation Blastocyst', *Nature, 223* (1969), p. 410.

167. R.D. Hood, 'Effects of Mycotoxins on Development' in T.V.N. Persaud (ed.), *Advances in the Study of Birth Defects* (MTP, Lancaster, 1979), vol. 2, pp. 191–210.

168. J.A. Dipaulo, J. Elis and H. Erwin, 'Teratogenic Response by Hamsters, Rats and Mice to Aflatoxin B_1', *Nature, 215* (1967), p. 638.

169. P.E. Still, A.W. Maclin, W.E. Ribelin and E.B. Smalley, 'Relationship of Ochratoxin A to Foetal Death in Laboratory Animals', *Nature, 234* (1971), p. 563.

170. A.W. Hayes, R.D. Hood and H.L. Lee, 'Teratogenic Effect of Ochratoxin A in Mice', *Teratology, 9* (1974), p. 93.

171. J. Grauwiler and H. Schon, 'Teratological Experiments with Ergotamine in Mice, Rats and Rabbits', *Teratology, 7* (1973), p. 227.

172. Food Protection Committee, *Toxicants Occurring Naturally in Food* (National Academy of Sciences/National Research Council, Washington, 1966).

173. I.E. Liener, *Toxic Constituents of Plant Foodstuffs* (Academic Press, New York, 1969).

174. T.V.N. Persaud and S. Kaplan, 'The Effects of Hypoglycin A, a Leucine Analogue on the Development of Rat and Chick Embryos', *Life Sci., 9* (1970), p. 1305.

175. A. Abramovich and F.C.H. Devoto, 'Anomalous Maxillofacial Patterns Produced by Maternal Lathyrism in Rat Foetuses', *Arch. Oral Biol., 13* (1968), p. 823.

176. Y. Yasuda, T. Kihara and H. Nishimura, 'Embryotoxic Effects of Feeding Bracken Fern (*Pteridium aquilinum*) to Pregnant Mice', *Toxicol. appl. Pharmacol., 28* (1974), p. 264.

177. R. Marcus and J. Watt, 'Seaweeds and Ulcerative Colitis in Laboratory Animals', *Lancet, ii* (1969), p. 489.

178. R. Marcus and J. Watt, 'Potential Hazards of Carrageenin', *Lancet, i* (1980), p. 602.

179. D. Ogi and Y. Hamada, 'Pesticide Exposed Human Foetuses: Case Reports

of Intrauterine Deaths and Congenitally Defect Babies', *Jap. Obstet. Gynecol. Soc., 17* (1965), p. 569.

180. J.J. Nora, A.H. Nora, R.J. Sommerville, R.M. Hill and D.G. Macnamara, 'Maternal Exposure to Potential Teratogens', *J. Am. Med. Assoc., 202* (1967), p. 1065.

181. J.A. O'Leary, J.E. Davies and W.F. Edmundson, 'Transplacental Passage of Pesticides', *Am. J. Obstet. Gynecol., 107* (1970), p. 65.

182. M.D. Whorton, T.H. Milby, H.A. Stubbs, B.H. Avishia and E.Q. Hull, 'Testicular Function Among Carbaryl-exposed Employees', *J. Toxicol. Envir. Hlth, 5* (1979), p. 929.

183. A. Curley, V.W. Burse and M.E. Grim, 'Polychlorinated Biphenyls: Evidence of Transplacental Passage in the Sherman Rat', *Fd Cosmet. Toxicol., 11* (1973), p. 471.

184. K. Shiota, T. Tanimura and H. Nishimura, 'Effects of Polychlorinated Biphenols on Pre- and Post-natal Development in Rats', *Teratology, 10* (1974), p. 97.

185. L.G. Hanson, C.S. Byerly, R.L. Metcalf and R.F. Bevill, 'Effect of Polychlorinated Biphenyl Mixture on Swine Reproduction and Tissue Residues', *Am. J. Vet. Res., 36* (1975), p. 23.

186. P. Gunby, 'Plenty of Fuel for Agent Orange Dispute', *J. Am. Med. Assoc., 242* (1979), p. 593.

187. K.D. Courtney, D.W. Gaylor, M.D. Hogan, H.L. Falk, R.R. Bates and I. Mitchell, 'Teratogenic Action of 2,4,5-T', *Science, 168* (1970), p. 864.

188. Editorial, '2,4,5-T where Next?', *Lancet, ii* (1979), p. 1114.

189. T.F.X. Collins and C.H. Williams, 'Teratogenic Studies with 2,4,5-T and 2,4-D in the Hamster', *Bull. Envir. Contam. Toxicol., 6* (1971), p. 559.

190. J.G. Vos, J.A. Moore and J.F. Zinkl, 'Toxicity of 2,3,7,8-tetrachlorodibenzo-*p*-dioxin (TCDD) in C57BL/6 Mice', *Toxicol. Appl. Pharmacol., 29* (1974), p. 229.

191. K.D. Courtney and J.A. Moore, 'Teratology Studies with 2,4,5-trichlorophenoxyacetic Acid and 2,3,7,8-tetrachlorodibenzo-*p*-dioxin', *Toxicol. Appl. Pharmacol., 20* (1971), p. 396.

192. D. Neubert and I. Dillman, 'Embryotoxic Effects in Mice Treated with 2,4,5-trichlorophenoxyacetic Acid and 2,3,7,8-tetrachlorodibenzo-*p*-dioxin', *Naunyn.-Schmiedeberg's Arch. Pharmacol., 272* (1972), p. 243.

193. R.J. Kociba, P.A. Keeler, C.N. Park and P.J. Gehring, '2,3,7,8-tetrachlorodibenzo-*p*-dioxin (TCDD): Results of a 13-Week Oral Toxicity in Rats', *Toxicol. Appl. Pharmacol., 35* (1976), p. 563.

194. J.B. Scrimgour and F. Cockburn, 'Congenital Abnormalities', *Lancet, ii* (1979), p. 1349.

195. G. Reggani, 'Medical Problems Raised by the TCDD Contamination in Seveso, Italy', *Arch. Toxicol., 40* (1978), p. 161.

4 CARCINOGENS IN FOOD

P. Grasso

The presence of toxic substances in food has been known to man from the earliest years and over the centuries man has learned to process such foods making them fit for human consumption or to avoid them altogether. By contrast, it is only comparatively recently, in fact within the last half-century, that the existence of carcinogenic substances in food has been recognised. In the western world, the quantities of such compounds present in food is very small indeed, but even these small quantities have caused considerable concern since it is thought in some quarters that they might be responsible for some tumours in man hitherto thought to be of spontaneous origin. It is probable that the hazard from the very small amounts of carcinogens in food is very low indeed but at the moment there are no reliable means for making a quantitative assessment of this hazard.

In this chapter, attention will be principally directed to carcinogens found in the food consumed in the western world but some important carcinogenic agents present in the food in other parts of the world will also be mentioned.

Carcinogens in Edible Plants

A typical example is the toxic glucoside cycasin. It occurs in the nuts, roots and leaves of cycad plants, which flourish in tropical and subtropical regions of the world. The flour made from the nut has been the staple food for the local inhabitants since antiquity while the leaves have been occasionally eaten by cattle and man. Neurological disorders are common in both cattle and local inhabitants and suspicion fell on this widely used plant. A toxic glycoside cycasin was isolated from this plant and it was shown to be neurotoxic and carcinogenic in experimental animals.[1] The native populations are aware of the toxic properties of these foods, and the toxin is removed by crushing and washing the nut during the preparation of the flour. Rats maintained on diets containing up to 10 per cent of home-made Guamanian cycad flour did not show any evidence of a carcinogenic response but the purified glycoside induced tumours in liver and various organs of treated rats.[1,2] The flour prepared from nuts of one particular cycad plant *Encephalartos hildebrandti* produced liver, kidney and lung tumours when fed to rats[3] probably because it contains a glycoside chemically similar to cycasin.

These glycosides are apparently carcinogenic to animals through the formation of the unstable aglycone, methylazoxymethanol, following attack by β-D-glucosidase enzymes derived from bacterial flora or from epithelial cells lining the intestinal wall. Both cycasin and methylazoxymethanol are readily transferred by way of the placenta to fetuses in pregnant rats, the offspring developing tumours late in adult life.[4] Despite the clear carcinogenic response in experimental animals, the only reported epidemiological survey to date indicates that the mortality rate from hepatomas (1961–1966) in natives of the Miyalo Islands (Okinawa), who had subsisted on cycads during and prior to 1959, showed no significant difference from that of a control population.[5]

Another carcinogen synthesised by edible plants is found in oil of sassafras. This oil is present in small amounts in cinnamon, nutmeg and mace and is used as a flavouring (up to 1 p.p.m.) in soft drinks. Sassafras oil contains up to 93 per cent safrole (*p*-allyl methylenedioxybenzene) which has been shown to be hepatocarcinogenic when fed to rats in the diet for 67 weeks at a level of 0.5 per cent.[6] Feeding studies in dogs have shown extensive liver damage at levels of 40 and 80 mg safrole/kg body weight, but no tumours.[7] Although no epidemiological data are available on the effects of human exposure to dietary safrole, there are no indications that populations consuming articles of diet containing sassafras suffer from liver or any other type of cancer to a greater extent than populations with limited exposure to this chemical.

Some ten years back, it was demonstrated that the ingestion of bracken (*Pteridium aquilinum*) induced intestinal tumours in rats,[8] a finding which caused much speculation as to the possible hazard associated with the use of brackens processed for human food. In Japan, for example, the plant *Osmunda japonica* is used as a vegetable and for seasoning purposes, and its similarity to the bracken fern has led to the suggestion that the high incidence of gastric carcinomas among the Japanese could be attributed to this plant.[9] More recent data[10] suggest that the carcinogenic activity of bracken is largely located in the rhizome, but no tumours were induced in rats fed for four months on bracken starch prepared from the rhizomes in the way it is commonly prepared for human consumption in Japan. This suggests that the tumorigenic agent is water-soluble and was washed out during starch preparation. Although bracken fern is widely ingested as a delicacy in the United States and New Zealand, there is again no suggestion of an association with an increased tumour incidence in these countries.

Fungal Carcinogens

It is fitting that aflatoxin should take priority in a discussion of the hazards associated with toxic fungal products, since it has been the subject of more than a half of the 2000 or so publications on mycotoxins over the last decade. To date, more than a dozen related compounds, designated aflatoxins, have been isolated and characterised, of which aflatoxin B_1 appears to be the most thoroughly investigated. In particular, this toxin was shown to be a potent carcinogen to rats, and useful data were generated on the effects of aflatoxin when ingested by these animals for prolonged periods. In one study, for example, 15 p.p.b. ($b = 10^9$) aflatoxin B_1 in the diet was carcinogenic to all rats treated for 68 weeks.[13] In monkeys, as much as 10 μg/kg was given in groundnut meal but it produced no apparent untoward effects over three years.[14]

From the epidemiological surveys completed to date, there is considerable circumstantial evidence that, in some parts of the world, aflatoxin is toxic and carcinogenic to man at levels currently present in the daily diet in those areas. A recent example comes from a survey in Thailand which revealed that in an area with a high incidence of liver cancer (6 new cases/100,000 people/year), the average consumption of aflatoxin (0.045-0.77 μg/kg/day) was 9-10 times greater than in an area with a low cancer incidence (2 new cases/100,000 people/year).[15,16] The standard deviation for these intake levels was, however, quite broad, and one woman in the former group consumed up to 3,225 μg/kg in a single day.

Similarly, a correlation was found between high hepatoma incidence (15 cases/100,000/year) and the amount of aflatoxin consumed in one district of Uganda, where almost half the food samples contained over 1 p.p.m.[13] The *per capita* consumption of aflatoxin in this district was estimated to be as high as 0.002–2 mg/day, equal to about 0.4–40 μg/kg. To put these figures into perspective, it is worth noting that a daily dose level of 40 μg/kg aflatoxin induced hepatic carcinomas in monkeys

Table 4.1: Frequency of Liver Cancer in Populations Consuming High Levels of Aflatoxin

	Thailand		Uganda		Kenya		
	Ratbur area	Songkhla area	Karamajo district	Other districts	High altitude	Middle altitude	Low altitude
Incidence of liver cancer cases/ 100,000 people/year	6	2	15	1.4–3	3.11 0	10.8 3.28	12.12 5.44
Average aflatoxin consumption (μg/kg body weight/day)	0.045–0.077	0.005–0.008	0.4–40	1	4.88 3.46	7.84 5.86	14.81 10.03

Source: From IARC Monographs on the *Evaluation of Carcinogenic Risk of Chemicals to Man*. Some naturally occurring substances. vol. 10, IARC, Lyon, 1976).

when administered over a period of 4.5 years.[17] A comparison with data obtained in other areas of Uganda is tabulated in Table 4.1.

The table also shows the results of a recent survey[18] of three areas of different altitudes in the Muranga's district of Kenya which clearly indicate an increased frequency of liver cancer in those populations consuming greater quantities of aflatoxin. In addition, this survey shows that in the human as well as in laboratory animals, the male appears to be more susceptible than the female to the carcinogenic effects of the toxin.

Unfortunately, no reliable estimate of the safe intake level of aflatoxin in humans can be deduced from the available data. Several countries have set tolerance levels in specific crops, in the range 5-30 p.p.b. aflatoxin, based on the limits of sensitivity of analytical methods (principally thin-layer chromatography) as much as on toxicological considerations. Current techniques now facilitate the detection of aflatoxin B_1 in the range 1-5 p.p.b., as well as total aflatoxin at the 5 p.p.b. level, and judging by recent epidemiological findings, every effort should be made to restrict the presence of the toxin to these levels.

The need for strict measures to limit the presence of fungal toxins in food is all the more urgent in the light of recent indications that materials which represent no carcinogenic hazard in their own right may conceivably enhance the tumorigenic effect of aflatoxins. Thus, cyclopropenoid fatty acids (CFA), which occur in cottonseed meal, were found to be non-carcinogenic when fed to trout, whereas a purified ration containing 200 p.p.m. CFA and 4 p.p.b. aflatoxin B_1 caused severe liver damage in trout fry within 6 months.[19]

Although no mycotoxins have yet been shown to possess a carcinogenic potency comparable to that of aflatoxin, *Aspergillus* is not the only ubiquitous fungal species which can produce mycotoxin with neoplastic potential. Much attention in recent years has been given to metabolites of *Penicillium* species – this being the commonest of the 15 or more kinds of fungi causing mouldy rice. Strains of one species mentioned, *P. islandicum* spp, produce two known hepatotoxic and carcinogenic metabolites – cyclochlorotine (a chlorine-containing peptide) (CC) and luteoskyrin (LS). These mycotoxins differ in their sub-acute effects on the liver, in that necrosis has been noted in the early stages of feeding 50-500 μg LS to mice for 2 years, whereas 40-60 μg CC administered according to the same experimental regimen evoked preferential development of fibrosis and cirrhosis. The tumorigenic effects of LS and CC were comparable, in that hepatomas of various types were induced by both compounds in a dose-dependent manner in mice surviving for 216 days – the minimum period for tumour development.[20]

An enormous number of undesirable fungi contaminate our food. Some idea of the challenge can be gleaned from one report that, of 943 cultures of fungi isolated from human and animal foodstuffs, some 65 per cent produced materials which are toxic to animals.[21] Nevertheless, the number of mycotoxins that present a hazard to man is still relatively small. Future research may tell us whether any of the fungi that have been recently isolated from food produce toxins which may be a potential carcinogenic hazard to man.

Carcinogens Introduced During Food Processing and Preservation

Most carcinogens present in food are derived from natural sources. A limited number of important carcinogens result from time-honoured methods of cooking or preserving food. None arise from food processes consistent with 'good manufacturing practice', and of course, no carcinogens are allowed as food additives.

Polycyclic Aromatic Hydrocarbons

The polycyclic aromatic hydrocarbons (PAH) from one of the more important group of carcinogens from the point of view of health hazard since they are potent carcinogens and are probably found in a much wider variety of food than any other group of carcinogenic compounds. They are found in fresh and decaying vegetables of various types and in grilled and roasted meat or fish. At least 13 PAH have been identified in food but of these only six are carcinogenic. In vegetables PAH are thought to be end products of one or more biosynthetic pathways. The sharp increase in the PAH content of vegetable matter during the process of decay has been put forward as evidence that they result from some catabolic process in the dying plant. The high content of PAH in decaying vegetable matter accounts, to a large extent, for the PAH in soil and water. This in turn may contribute to the PAH content in vegetables.[22]

PAH are, however, formed in considerable amounts as a result of industrial activity and vegetables grown close to factories which emit a considerable quantity of smoke have a higher PAH content than vegetables grown in rural areas.[23] The difference, however, is quite small, and it is doubtful whether it represents any increase in the hazard from these polynuclear hydrocarbons.

A more important source of PAH is the grilling or broiling of fish and meat. The amount of PAH in this type of food is greater than in food cooked in any other way and considerably greater than that found in fresh vegetables. Several epidemiological studies have been made in an attempt to correlate consumption of grilled or roasted fish and meat with an increased incidence of cancer of the stomach[24] but no firm evidence in this direction has been produced so far. The factors leading to the formation of high PAH in foods cooked on an open flame are imperfectly understood but it has been shown that when fat is allowed to drip on to the open flame it leads to the formation of considerable amounts of PAH.[25] High oven temperatures have the same effects.[25]

Smoking is one of the time-honoured methods of preservation of fish and meat. The use of fire-wood or charcoal for this purpose leaves a coating of PAH on the food. Despite the technological advances made in the preservation of food, smoking is still very much in vogue in many parts of the world because of the traditional flavour that this method of preservation imparts. The work of Lijinski and Shubik[26] has shown that the flavour in the smoke from burning wood resides in a fraction that is soluble in water so that by bubbling the smoke through water, the tar containing the PAH sediments out, whereas the smoke flavour is retained in the aqueous medium. The solution of this smoke fraction in water is known as 'liquid smoke' and has been used successfully to flavour meat and fish preserved by more conventional methods.

The intake of PAH carcinogens varies considerably according to the geographical area. One potent carcinogen which often forms a substantial proportion of PAH mixtures in food is benzo (a) pyrene (BP) (Table 4.2). The estimated daily intake of this carcinogen in food by any one person in the United Kingdom is estimated to be at most 3 μg but the real amount ingested is likely to be much less than this.[27] The level of BP intake may be significantly higher in other countries such as Iceland or Japan where broiled or roasted meat and fish may contribute substantially to the total intake, but claims of an aetiological association between the high intake of BP and the high incidence of stomach cancer in Japan have not been adequately substantiated. Such correlations are, in any case, questionable on the basis of experiments in animals. In one of the more revealing studies, mice

Table 4.2: Examples of Polycyclic Aromatic Hydrocarbon Content in Some Common Foodstuffs

Foodstuff	3,4-Benzopyrene (p.p.b.)	1,2-Benzanthracene (p.p.b.)
Fresh vegetables	2.85–24.5	0.3–43.6
Vegetable oils	0.4–1.4	0.8–1.1
Coconut oil	43.7	98.0
Margarine	0.4–0.5	1.4–3.0
Coffee	0.3–1.3	1.3–3.0
Tea	3.9	2.9–4.6
Grain	0.19–4.13	0.40–6.85
Oysters and mussels	1.5–9.0	–
Smoked ham	3.2	2.8
Smoked fish	0.83	1.9
Smoked bonito	37	189
Cooked sausage	12.5–18.8	17.5–26.2
Singed meat	35.99	28–79
Broiled meat	0.17–0.63	0.2–0.4
Charcoal broiled steak	8.0	4.5
Barbecued beef	3.3	13.2
Barbecued ribs	10.5	3.6

Source: From P. Grasso and C. O'Hare, 'Carcinogens in Food', in Charles E. Searle (ed.), *Chemical Carcinogens* (A.C.S. Monograph 173 American Chemical Society, Washington, 1976).

fed on BP for several days at levels of approximately 4 and 40 μg/mouse/day developed no tumours, but tumours of the squamous region of the stomach occurred in 1/23 animals fed 80 μg/day.[28] No tumours of the glandular portion of the stomach were observed. Since there is no squamous portion in the stomach of man, this single tumour is not relevant in assessing risk to man but the absence of tumours of the glandular portion is important, since, it is unlikely that the dietary BP concentration in any part of the human gastro-intestinal tract would exceed these experimental levels. The detoxifying effect of the intestinal aryl hydrocarbon hydroxylase complex may well be sufficient to cope with these sorts of levels of BP.[29] Nevertheless, it would be prudent to endeavour to reduce the PAH content in food as much as is practicable.

The possible synergistic interactions of BP with other carcinogens present in food – such as aflatoxins, nitrosamines, or other polycyclic hydrocarbons, would appear to be worthy of some consideration though relatively little research has, to date, been channelled in this direction.

Nitrosamines

The nitrosamines form another group of carcinogens whose presence in food is attracting considerable attention. These compounds are not present naturally in food but result from the interaction of nitrite and secondary and tertiary amines. The nitrite is added to food as a preservative or it may be derived from the reduction of nitrate which is often added in place of nitrite. Despite the addition of these and other preservatives, both fish and meat undergo a slow process of degradation with liberation of free amines. Interaction between nitrites and secondary or tertiary amines yields the well-known nitrosamine compounds. A vast array of such compounds has been synthesised and tested for carcinogenic activity in laboratory animals.[30,31] The majority yielded a high incidence of tumours in a wide variety of animal species. Only a few failed to induce tumours.

Fortunately, only a very few of the wide range of nitrosamines that have been synthesised are found in food. The intensive investigation by Colby and colleagues and of other workers[32] have revealed that only four volatile nitrosamines are present: dimethylnitrosamine (DMN), diethylnitrosamine (DEN), nitrosopiperidine and nitrosopyrrolidine. Although as yet no knowledge is available on the presence of non-volatile nitrosamines in food, the presence of only a limited number of volatile nitrosamines out of the very wide range that is known to exist gives some degree of reassurance that the non-volatile compounds might also be found in a limited range.

Of these four compounds, DMN is by far the commonest nitrosamine encountered in human food. A large number of investigations have been carried out in a number of centres in order to estimate the content of DMN in human food. Results of the more important of these investigations are given in Table 4.3. The highest level found in meat for human consumption is 4 p.p.b. while that in fish is 9 p.p.b.[32] The daily intake in man does not exceed 1 p.p.b. per day calculated on the basis of average consumption data. There is some evidence that storage at household refrigerator temperatures and cooking increases the content of DMN in nitrate/nitrite-treated fish.[33]

Despite the very small amount of nitrosamines present in food, concern has been expressed about their possible role in causing human cancer. This concern stems from the potent carcinogenic activity of the four nitrosamines mentioned when they were administered to experimental animals. Admittedly, most of the doses used in animal experiments were of the order of several p.p.m. daily and thus considerably

Table 4.3: Nitrosamines Found in Common Types of Food (in μg/kg food)

Type of food	Nitrosodiethylamine	Nitrosodimethylamine	Nitrosopiperidine	Nitrosopyrrolidine
Vegetables and vegetable oil				
Beans	0–0.2	None in 16 species of vegetables	–*	–
Soya bean oil (freshly expressed or refined)	0–4	0–20	–	–
Cheeses		0.1–68	2-11	–
Fish				
Fresh or frozen	1–2	3–18	–	–
Salted or pickled	50–108	1–35	–	0–37
Smoked, baked or processed	up to 147	6–177	1	–
Meat				
'Cured meats'	up to 40	1–80	up to 60 (sausages)	10–105 (sausages)
Fried bacon	1–4	0.1–28	1	3–44
Uncooked bacon	0–2	1–9.5	none found	up to 17
Water	0.01–1.83 μg/l	0.8–3.3 μg/l	0.03 μg/l	0.09–0.2 μg/l
Alcoholic Beverages	max. 0.1	max. 10	–	–

* No investigations cited.

Source: From IARC Monographs on the *Evaluation of the Carcinogenic Risk of Chemicals to Man*, 'Some nitroso compounds', vol. 17 (IARC, Lyon, 1978).

greater (by an order of 10^5–10^6) than the amount taken in by man. There are, however, results from a few studies which indicate that even at low doses, DMN and DEN are capable of inducing cancer in some of the limited number of animals employed in the conventional type of experiment. Thus DMN at 2 p.p.m. in the diet induced a 30 per cent incidence of hepatocellular carcinoma in rats whereas DEN induced twice this incidence of hepatic tumours when given in drinking water at 0.075 mg/kg (= 1.5 p.p.m.).[34,35] The high incidence of tumours induced in rats at these relatively low experimental dosages makes it difficult not to entertain the possibility that a real hazard exists from the nitrosamines present in food.

Unfortunately, there is another aspect which makes this problem even more complicated and more difficult to assess. This aspect relates to the possible synthesis of nitrosamines *in vivo* from an interaction between amines liberated in the course of protein degradation in the digestive tract and nitrate present in food. It has been shown experimentally that the pH of the stomach of most species allows the formation of both dimethylnitrosamine and nitrosomorpholine when nitrite and dimethylamine or morpholine are introduced separately in this organ.[36] It has also been shown that a high incidence of liver tumours can be induced in rats if nitrite and morpholine are administered together in the diet for long periods of time.[37] It has not been possible so far to induce tumours by the administration of dimethylamine and nitrite in the rats but this failure may be due to local physiological factors which do not permit the formation of dimethylnitrosamine in these animals.

The chemical interaction between nitrite and secondary and tertiary amines can be accelerated by bacterial action and by the presence of certain catalysts – mainly thiocyanates which are present in the saliva.[38] Bacterial action is not an important source of nitrosamines in the stomach in the average healthy human adult because of the low stomach pH, which most bacteria cannot tolerate, but may have to be seriously considered in achlorhydrics. There is also the possibility that nitrosamine formation could occur in the intestine, particularly in the large intestine, where most of the intestinal flora in man exists. It would seem unlikely, however, that the necessary conditions for synthesis are present since nitrites/nitrates are rapidly absorbed from the small intestine so that the amount reaching the large bowel is very small indeed.

In theory, nitrosamines in food present a real carcinogenic hazard to man. They are potently carcinogenic in animals even at low doses and are present in many types of protein food preserved by the addition of nitrate. In practice this hazard would appear to be very small indeed since the amounts taken in by man are several orders of magnitude lower than the levels studied in animals so far. Furthermore, it is likely that the nitrosamine is released from food over several hours, as protein digestion progresses, so that the small concentration reaching the cells would allow adequate measures to be taken to 'mop up' the active radicles or to repair whatever damage they may have caused to the cell genome.

Food Additives

The danger of unwittingly adding a carcinogenic chemical to food was recognised as soon as the food processing industry came into being and legislative requirements for screening food additives for carcinogenic activity, epitomised in the notorious Delaney amendment, reflect this early concern. Three types of food additives have in particular given rise to concern about possible carcinogenic hazards.

Food Colours

Food colours were one of the earliest groups to be suspect as possible carcinogens because of their supposedly close relationship to 'coal-tar' – in fact they were called in some quarters 'coal-tar' colours.[39] In those early days it was thought that one of the most sensitive methods of testing compounds supposedly derived from 'coal-tar' was by repeated subcutaneous injection and most food colours were, as a consequence, tested by this route. This method of testing was adopted on the basis of two fallacious assumptions, one was that the sulphonated triphenylmethane, xanthine and azo colourings were potentially carcinogenic because in theory they were related to polycyclic aromatic compounds of the type known to exist in coal tar[39] and the other was that subcutaneous sarcoma was a true index of carcinogenic activity.[40]

Because of advances in chemistry, food colours are no longer looked upon as 'coal-tar'-derived and the demonstration that hypertonic solutions of glucose, common salt and surface active solutions produced local sarcomas led to the realisation that this route of administration was unsuitable for testing the carcinogenic activity of chemicals.[40] In addition the current trend to test chemicals by the same route as that of human exposure has reduced further any remaining interest in this route for testing chemicals for carcinogenic activity.

A very large proportion of the food colours currently in use have been tested for carcinogenic activity by the oral route thanks mainly to the programme carried out by the Food and Drugs Administration in the USA and by the British Industrial Biological Research Association in the UK. Two colours, Ponceau 3R and Ponceau MX,[41,42] produced hepatic tumours in rats and mice when given by the oral route and were promptly withdrawn because they were suspected of having carcinogenic activity. Ponceau MX was subsequently employed in the study of the mechanism of induction of this tumour and it was found that the compound was hepatotoxic. When it was administered for a prolonged period it first led to the development of multiple hyperplastic nodules which in turn led to the development of carcinomas.[43,44] This mechanism was different from that by which potent carcinogens such as nitrosamines and aflatoxin produce tumours. These compounds do not seem to require prolonged injury to do so. In fact tumours are known to develop by very low doses which do not lead to any detectable injury.[44] It is likely that compounds of this sort, which are powerful mutagens as well as carcinogens, induce cancer as a consequence of the damage they produce to the nuclear DNA.[45] Ponceau MX and its metabolites are not mutagenic (I. Rowlands, 1977, unpublished observations), and in all probability induce cancer by a different (? non-genetic) mechanism. It is most unlikely that during the brief period of their use they presented a carcinogenic hazard to man.

According to an IARC review, one other food colour, Brilliant Blue FCF, is suspected of possessing carcinogenic activity because an increased incidence of kidney tumours was observed in mice treated with this colouring.[46] There is, however, considerable doubt about the validity of this view since there was no dose-response relationship and the historical controls indicated that the incidence of renal tumours in the test animals was well within the limits of the expected incidence of this type of tumour in the particular strain of mice used.

Exhaustive tests on other food colours by the oral route have failed to reveal any carcinogenic potential in any of the colourings tested. (See *Food and Cosmetics Toxicology* for results.)

Sweeteners

The need for artificial sweeteners was recognised several years ago when carbohydrate restriction was established as an important factor in the control of diabetes. Saccharin was developed to fill this need and was used for 80 years[47] by diabetics as a substitute for sugar without suspicion of carcinogenicity. Other chemicals were introduced to provide low-calorie sweeteners in response to public demand particularly from the soft drinks industry. One of the most commercially successful of these chemicals was cyclamate. It was not as powerful a sweetener as saccharin and had to be used in relatively larger amounts, but it had the advantage of not leaving a bitter after-taste as saccharin does.[48]

Cyclamate was tested on its own for carcinogenic activity with negative results[49] but a combination of cyclamate and saccharin in the proportion employed in a popular and widely used formulation produced a high incidence of bladder tumours in the treated rats.[50] For unknown reasons cyclamate was implicated in the production of these tumours but it was shown, in subsequent experiments, that saccharin, on its own, given in doses which were the equivalent of 700 to 1000 times the estimated human intake could produce bladder tumours in the rat.[51,52] This effect was not readily reproducible and a number of negative results at comparable dose levels were obtained in experiments on rats.[53] One particular aspect of the experimental work carried out on saccharin was the demonstration that when rats were exposed *in utero* as well as during the weaning period and during their adult lives, the likelihood of developing bladder tumours was increased.[53]

These experimental results led to a severe restriction of the use of saccharin in Canada and to a threat of its being banned in the USA. The ban by the USA authorities has not materialised as yet (i.e. 1983) because of the public reaction against it. There was also a demand for some epidemiological studies to be undertaken before the ban was put in force.

A number of such studies has been carried out. The first study from Canada indicated that there was an excess of bladder cancer in male users of artificial sweeteners compared with non-users.[54] Comparable studies in the USA and UK failed to confirm this first report while a large study involving 3010 patients with bladder cancer and 5783 controls revealed that the relative risk of developing bladder tumours was similar in users or non-users of artificial sweeteners.[55] Other studies conducted on similar lines yielded the same reassuring results.[56]

Some doubts have been expressed recently by a group of workers who examined a sub-group of a particular population – in this case the sub-group studied consisted of non-smoking women – and found an increased incidence of tumours of the bladder. They argue that this sub-group has a very low 'baseline risk' since the natural incidence of bladder tumours is very low and is therefore better suited to demonstrate an increased tumour incidence.[57] However, statisticians have reservations about the validity of this approach and it may well be that future studies on the same lines will not confirm these findings.[56]

Despite the largely reassuring epidemiological data, the experimental findings of bladder tumours leave a question mark on the safety of saccharin. Extensive studies on the metabolism of saccharin yielded the important findings that saccharin appears to be excreted almost unchanged by the adult rat and it is, therefore, difficult to see how it could act as a carcinogen. But the production of tumours after *in utero* exposure suggests that a different situation may occur in pregnancy or that the fetal tissues may handle saccharin differently from those of the adult. The other source of concern is the finding that saccharin has a 'promoting' effect on vascular epithelium in the rat following the intravesical instillation of an 'initiating' dose of methylnitrosourea, a proximate carcinogen.[58] It is difficult at the moment to evaluate these findings in terms of hazard to man. Future work may provide some useful clues to the solution of this problem.

Antioxidants

Antioxidants do not appear to have given rise to much concern about cancer hazard. Recent work on butylated hydroxytoluene an antioxidant extensively employed as an additive to butter, cream and similar products seems to indicate that it might cause an increase in pulmonary tumours in mice when given in doses about 1000 or 2000 times the average human intake.[59] The validity of this tumour as an index of carcinogenic hazard to man is, however, in serious dispute.[60]

Pesticides

This group of chemicals is employed widely to protect seeds, growing plants and crops from devastation by insect pests. Some pesticides like the organophosphates and thiocarbamates break down spontaneously or are destroyed by bacterial action, so that little if any remains in the food by the time it is prepared for human consumption. Others as, for example, the organochlorine pesticides persist, and despite washing and cooking, leave measurable residues, although these never amount to more than a few μg/kg of food.

Two principal groups of pesticides have given rise to concern that they might present a carcinogenic hazard to man: the organochlorine pesticides and the thiocarbamates.

The organochlorine pesticides have been studied intensively toxicologically, and, with few exceptions, all produce liver enlargement and histological manifestation of hepatotoxicity, when administered to laboratory animals in doses close to the maximum tolerated dose level. In carcinogenicity tests it has been the practice to treat rats and mice with amounts that are close to the 'maximum tolerated dose'.[65] At these dose levels, hepatic neoplasia, consisting principally of benign lesions but sometimes also of malignant tumours, develops in the treated animals. Rats appear to be less liable than mice to develop hepatic neoplasia from organochlorine pesticides, in fact there are only a few publications reporting this effect in the rat. These results indicate that in this species neoplasia develops only after feeding rats at almost lethal levels.

Most regulatory authorities in Europe and the USA regard the induction of mouse hepatic neoplasia as indicative of a carcinogenic hazard to man. There are, however, indications that this type of tumour may be the consequence of prolonged, often severe hepatic damage[44,61] and some toxicologists consider that under these conditions the carcinogenic response has little or no relevance to man.[60] This view seems to be supported by the fact that organochlorine compounds are not mutagenic by any of the tests to which they have been subjected so far,[65] so that it is unlikely that they cause tumours by the same mechanism as the large group of potent carcinogens that are powerful mutagens.[45]

It must be kept in mind that man ingests very small amounts indeed of organochlorine pesticides. For example, McGill and Robinson,[62] reporting on organochlorine residues in complete, prepared meals in 1965 at the height of the period of contamination with these pesticides, estimated an intake of less than 1 μg/kg body weight of the DDT group of pesticides and 0.3 μg/kg body weight of the dieldrin group. Because of a voluntary ban on the use of these pesticides, the daily intake to-day must be very much smaller than these figures would indicate.

But even the 1965 figures compare very favourably with the amounts that have been shown to produce an unequivocal tumorigenic response in mice. In at least two carcinogenic studies in mice the minimal effective dose of DDT was 12.5 mg/kg body weight and that of dieldrin was 1 mg/kg body weight.[63,64]

Other classes of pesticides appear to present much less of a problem. Unlike the organochlorine pesticides, they degrade rapidly to relatively inactive products. An exception to this would appear to be the dithiocarbamates, a group of fungicides which form ethylenethiourea (ETU) on degradation. This compound persists long enough after spraying to find its way into food. Under experimental conditions it produces both thyroid and hepatic tumours in mice and thyroid tumours in rats when administered in very high doses.[75] The thyroid tumours have been shown to result from a hormonal imbalance (excessive secretion of thyroid-stimulating hormone) caused by the suppression of thyroxine, one of the principal pharmacological effects of ETU,[75] and have no significance in terms of the human exposure since the levels of ETU are well below the levels that are pharmacologically active in man. The liver tumours in mice are of the same type as those produced by phenobarbitone and organochlorine pesticides and therefore insufficient to be regarded as serious evidence of carcinogenic risk to man.

Food and Cancer

The preoccupation with the possible carcinogenic hazards from food additives and contaminants now seems to be complemented by concern about the safety of the major food components themselves. The epidemiological studies which indicated that the incidence of colonic cancer is very much lower in communities consuming a high-fibre diet than in Western Societies[66] may have contributed substantially to this concern. Other epidemiological studies appear to confirm the suspicion that diet may play a major and hitherto unsuspected role in the causation of cancer. Thus according to the report of the ARC/MRC committee convened to advise on research in the field of food and nutrition,[67] a survey based on 37 countries has shown that there was a strong correlation between colon cancer and the dietary intake of meat and fat. Although it is impossible to separate the effects of these two food constituents, because their consumption is very closely related, many believe that fats are now likely to be implicated in the causation of cancer.[67] The mechanism by which fats cause cancer is obscure but it is thought that they favour the multiplication of certain bacteria which may synthesise carcinogenic compounds from steroids or other chemicals found in food. Fats have also been implicated in the aetiology of mammary cancer in women but it is thought to play only a secondary role in this respect.[68] Intensive laboratory investigations have not succeeded in providing a scientific foundation for these hypotheses based on epidemiological observations but they have provided some data which indicate that the calorie intake of the diet as well as the proportion of protein and fat it contains may have a profound influence on the incidence of certain naturally occurring tumours.

Thus, reduction in the calorie intake by restricting food was shown by Tannenbaum[69] to lead to the reduction of the number of spontaneous hepatic tumours in mice compared with animals fed *ad libitum*. Recently Tucker[70] demonstrated that the incidence of liver tumours in mice and of pituitary, mammary and skin tumours in rats was significantly reduced by reducing the calorie intake by 20 per cent. The experiments indicate the importance of a high calorie intake in influencing the incidence of so called 'spontaneous' neoplasia in rodents. More attention should be given to these experimental results when thinking of the role of the diet in human cancer since from the dietary point of view Western Society is over-fed and simulates in this respect the rodent fed *ad libitum*.

Although no data are available on the influence of increased calorie intake on the natural incidence of tumours in rodents, there are some observations which indicate that an increase in some of the major nutritive components of the diet may increase tumour incidence in these species. For example, an increase in dietary protein increases the incidence of tumours of the bladder in rats,[71] while in mice, feeding a diet high in casein (20–40 per cent of protein intake) increased substantially the incidence of hepatic tumours.[72] The same result was observed when mice were maintained on a diet high in fats.[73] In rats, a high fat diet increased the sensitivity of the mammary gland to tumour formation by an oral dose of 7-12-dimethylbenz(a)anthracene.[74]

The results of animal studies clearly indicate that dietary factors have a profound influence on the incidence of certain spontaneous tumours. It also has been observed by several workers in recent years that a change in dietary constituents can alter the susceptibility of experimental animals to carcinogenic materials. In the case of safrole, for example, a significantly larger number of tumours were observed in male rats fed diets deficient in riboflavin and containing 12 per cent casein as opposed to normal diets.[11] In contrast, there appeared to be a significantly higher incidence of urinary bladder carcinomas in rats fed bracken fern and supplemented with large amounts of thiamine than in rats fed a bracken fern diet alone.[12] In both these experiments, dietary supplementation with the appropriate vitamin alone had no effect on the observed high tumour incidence.

The role of the diet in the causation of human cancer was reviewed by Doll and Peto.[76] They point out that although there is some evidence that diet may influence cancer incidence there is no clear evidence of any powerfully active substances naturally present in the American food. Furthermore, there is no evidence for the presence in food of carcinogens produced by microorganisms other than aflatoxin. This is present only in 'minute' amounts and it is questionable whether it is worthwhile spending a vast amount of funds to eliminate it altogether from food.

There seems to be some evidence that being overweight is associated with an increased cancer incidence in women but not in men. Cigarette smoking and lung cancer could possibly mask this association in men.

There is no clear evidence that synthetic food additives contribute to cancer in man. Saccharin and butylated hydroxytoluene appear to be the least likely to be implicated in this respect, but there is some uncertainty with regard to the nitrosamines formed from the addition of nitrates and nitrites although the risk is thought to be a very small one. No conclusions could be reached on any of the other additives because no epidemiological studies had been performed to investigate the cancer risk they might present.

Conclusion

The carcinogens that contaminate food originate from natural sources or from long-established practices of cooking or of preserving food. The amounts present in the food consumed by the developed countries are exceedingly small and although there is no established and acceptable method of quantifying the risk they present, it is most unlikely that they are responsible for any significant proportion of the deaths from cancer.

Despite intensive tests on food additives, often at levels very much higher than those present in food, there is no convincing evidence that any food additive presents any cancer hazard to man. The same comment most probably also applies to the minute doses of pesticides present in our food.

The major nutritive constituents of food, particularly proteins and fats, may have some role in influencing the natural incidence of tumours. The amount of food calories ingested is, however, likely to be more important in this respect than any single component.

References

1. G.L. Laqueur and M. Spatz, 'Toxicology of Cycasin', *Cancer Res., 28* (1968), pp. 2262–2267.

2. M.G. Young, O. Mickelsen, M.E. Campbell, G.L. Laqueur and J.C. Keresztesy, 'Cycad Flour Used by Guamanians: Effects Produced in Rats by Long-term Feeding', *J. Nutr., 90* (1966), pp. 153–156.

3. G.M. Mugers and P. Ndcrito, 'Tumours of the Liver, Kidneys and Lungs in Rats Fed *Encephalartos hildebrandtii*', *Br. J. Cancer, 22* (1968), pp. 563.

4. M. Spatz and G.L. Laqueur, 'Transplacental Induction of Tumours in Sprague-Dawley Rats with Crude Cycad Material', *J. Nat. Cancer Inst., 38* (1967), pp. 233–245.

5. T.S. Elizan, data presented to the 4th Conference on the *Toxicity of Cycads*, Bethesda, USA, April 1965, Marjorie G. Whiting (ed.), (Public Health Service, US Dept. of Health, Education and Welfare).

6. E.L. Long, 'Liver Tumours Produced in Rats by Feeding Safrole', *Archs Path., 75* (1963), pp. 595–604.

7. E.C. Hagan, W.H. Hansen, O.G. Fitzhugh, P.M. Jenner, W.I. Jones, Jean M. Taylor, Eleanor L. Long, A.A. Nelson and J.B. Brouwer, 'Food Flavourings and Compounds of Related Structure. II Subacute and Chronic Toxicity', *Fd Cosmet. Toxicol., 5* (1967), pp. 141–157.

8. I.A. Evans and J. Mason, 'Carcinogenic Activity of Bracken', *Nature (Lond.), 208* (1965), pp. 913–914.

9. H.F. Kraybill, 'Food Contaminants and Gastro-intestinal or Liver Neoplasia. Survey of Experimental observations', *Envir. Res., 2* (1969), pp. 231–246.

10. I. Hirono, 'Comparative Study of Carcinogenic Activity in Each Part of Bracken', *J. Nat. Cancer Inst., 50* (1973), pp. 1367–1371.

11. F. Homburger, T.F. Kelly, P.D. Bogdonoff, 'Effects of Diet on Toxicity of Safrole in Rats', *Toxicol. Appl. Pharmacol., 7* (1965), pp. 485.

12. A.M. Pamukcu, 'Effect of Coadministration of Thiamine on the Incidence of Urinary Bladder Carcinomas in Rats Fed Bracken Fern', *Cancer Res., 30* (1970), pp. 2671–2674.

13. M.E. Alpert, M.S.R. Huth. G.N. Wogan and C.S. Davidson, 'Association Between Aflatoxin Content of Food and Hepatoma Frequency in Uganda', *Cancer (Philad.), 28* (1971), pp. 253–260.

14. W.F.J. Cuthbertson, A.C. Laurensen and D.A.H. Pratt, 'Effect of Groundnut Meal Containing Aflatoxin on Cyanomolgus Monkeys', *Br. J. Nutr., 21* (1967), pp. 893–908.

15. R.C. Shank, J.E. Jordan, G.N. Wogan, A. Nondasuta and B. Subhamani, 'Dietary Aflatoxins and Human Liver Cancer. III. Field Survey of Rural Thai Families for Ingested Aflatoxins', *Fd Cosmet. Toxicol., 10* (1972), pp. 71–87.

16. R.C. Shank, N. Bhamarapravati, J.E. Gordon and G.N. Wogan, 'Dietary Aflatoxins and Human Liver Cancer. IV. Incidence of Primary Liver Cancer in Two Municipal Populations of Thailand', *Fd Cosmet. Toxicol., 10* (1972), pp. 171–179.

17. C. Gopalan, P.G. Tulpule and D. Krishnamurthi, 'Induction of Hepatic Carcinoma with Aflatoxin in the Rhesus Monkey', *Fd Cosmet. Toxicol., 10* (1972), pp. 519–521.

18. IARC Monographs on the *Evaluation of the Carcinogenic Risk of Chemicals to Man* (International Agency for Research on Cancer, Lyon, 1972), vol. 1, p. 145.

19. R.O. Sinnhuber, J.H. Wales and D.J. Lee, 'Cyclopropenoids, Carcinogens for Aflatoxin-induced Hepatoma in Trout', *Fedn Proc. Fedn Am. Socs exp. Biol., 25* (1966), p. 555.

20. K. Uraguchi, M. Saito, Y. Noguchi, K. Takahashi, M. Enomoto and T. Tatsuno, 'Chronic Toxicity and Carcinogenicity in Mice of the Purified Mycotoxin, Luteoskyrin and Cyclochlorotine', *Fd Cosmet. Toxicol., 10* (1972),

pp. 193–207.
21. C. Metel Christenen, 'Toxicity to Experimental Animals of 943 Isolates of Fungi', *Cancer Res., 28* (1968), p. 2293.
22. G. Grimmer and A. Hildebrandt, 'Content of Polycyclic Hydrocarbons in Different Vegetables. III. Hydrocarbons in the Human Surroundings', *Dt. Lebensmitt Rdsch., 61* (1965), pp. 237–239.
23. D.J. Tilgner, 'Food in a Carcinogenic Environment', *Fd. Mf., 45* (1970), pp. 47–50.
24. Y. Masuda, K. Mori and M. Kurastune, 'Polycyclic Aromatic Hydrocarbons in Common Japanese Food. I. Broiled Fish, Roasted Barley, Shoya and Caramel', *Gann, 57* (1966), pp. 133–142.
25. W. Lijinsky and A.E. Ross, 'Production of Carcinogenic Polynuclear Hydrocarbons in the Cooking of Food', *Fd. Cosmet. Toxicol., 3* (1967), pp. 343–347.
26. W. Lijinsky and P. Shubik, 'The Detection and Estimation of Polycyclic Aromatic Hydrocarbons in Liquid Smoke', *Fd Cosmet. Toxicol., 3* (1965), pp. 145–147.
27. Anonymous, 'Benzpyrene in the Dock', *Bull. Br. Ind. Biol. Res. Assoc., 10* (1971), pp. 64–65.
28. J. Neal and R.H. Rigdon, 'Gastric Tumours in Mice Fed Benzo ( ) pyrene. A Quantitative Study', *Tex. Rep. Biol. Med., 25* (1967), pp. 553–557.
29. L.W. Wattenberg, 'Studies of Polycyclic Hydrocarbon Hydroxylases of the Intestine Possibly Related to Cancer', *Cancer, 28* (1971), pp. 99–102.
30. H. Druckrey, R. Preussmann, S. Ivankovic and D. Schmähl, 'Organotropie Carcinogene Wirkungen Bei 65 Verschiedenen *n*-nitrosoverbindungen an BD-ratten', *Z. Krebsforsch, 69* (1967), pp. 103–201.
31. P.N. Magee, 'Toxicity of Nitrosamines: Their Possible Human Health Hazards', *Fd Cosmet. Toxicol., 9* (1971), pp. 207–218.
32. N.T. Crosby, J.K. Foreman, J.F. Palframan and R. Sayer, 'Estimation of Steam-volatile *N*-nitrosamines in Foods at the 1 μg/kg Level', *Nature (Lond.), 238* (1972), pp. 342–343.
33. F. Ender, G.N. Havre, R. Madsen, L. Ceh and A. Helgebostad, 'Studies on Conditions Under which *N*-nitrosodimethylamine is Formed in Herring Meal Produced from Nitrite-preserved Herring', *Z. Tierphysiol. Tierernahr. Futtermittelk., 22* (1967), pp. 133–140.
34. B. Terracini, P.N. Magee and J.M. Barnes, 'Hepatic Pathology in Rats on Low Dietary Levels of Dimethylnitrosamine', *Br. J. Cancer, 21* (1967), pp. 559–565.
35. H. Druckrey, A. Schildbach, D. Schmähl, R. Preussmann and S. Ivankovic, 'Quantitative Analyse der Carcinogen Wirkung von Diathylnitrosamin', *Arzneimittel-Forsch, 13* (1963), pp. 841–851.
36. P. Grasso and C. O'Hare, 'Carcinogens in Food' in Charles E. Searle (ed.), *Chemical Carcinogens* (ACS Monograph 173 American Chemical Society, Washington, D.C., 1976), pp. 701–728.
37. R.C. Shank and P.N. Newberne, 'Nitre-Morpholine-induced Hepatomas', *Fd Cosmet. Toxicol., 10* (1972), pp. 887–888.
38. E. Boyland, E. Nice and K. Williams, 'The Catalysis of Nitrosation by Thiocyanate from Saliva', *Fd Cosmet. Toxicol., 9* (1971), pp. 639–643.
39. J.W. Daniel, 'The Excretion and Metabolism of Edible Food Colours', *Toxicol. Appl. Pharmacol., 4* (1962), pp. 572–594.
40. P. Grasso and L. Golberg, 'Subcutaneous Sarcoma as an Index of Carcinogenic Potency', *Fd Cosmet. Toxicol., 4* (1966), pp. 297–320.
41. H.C. Grice, W.A. Mannell and M.G. Allmark, 'Liver Tumours in Rats Fed Ponceau 3R', *Toxicol. Appl. Pharmacol., 3* (1961), pp. 509–520.

42. Y. Ikeda, S. Horiuchi, K. Kobayashi, T. Furuja and K. Kohgo, 'Carcinogenicity of Ponceau MX in the Mouse', *Fd Cosmet. Toxicol., 6* (1968), pp. 591–598.

43. P. Grasso, A.B.G. Lansdown, Ida S. Kiss, I.F. Gaunt and S.D. Gangolli, 'Nodular Hyperplasia in the Rat Liver Following Prolonged Feeding of Ponceau MX', *Fd Cosmet. Toxicol., 7* (1969), pp. 425–442.

44. P. Grasso and T.J. B. Gray, 'Long Term Studies on Chemically-induced Liver Enlargement in the Rat. III. Structure and Behaviour of the Hepatic Nodular Lesions Induced by Ponceau MX', *Toxicology, 7* (1977), p. 327.

45. J. McCann, E. Choi, E. Yamasaki and B.N. Ames, 'Detection of Carcinogens as Mutagens in the *Salmonella*/microsome Test: Assay of 300 Chemicals'. *Proc. Natl. Acad. Sci. USA, 72* (1975), pp. 5135–5139.

46. IARC Monographs on the *Evaluation of the Carcinogenic Risk of Chemicals to Man*, 'Some Aromatic Amines and Related Nitro Compounds', (IARC, Lyon, 1978), vol. 16, pp. 171–197.

47. B. Crammer and R. Ikan, 'Properties and Syntheses of Sweetening Agents', *Chem. Soc. Rev., 6* (1977), pp. 431–453.

48. K.J. Parker, 'Alternatives to Sugar', *Nature (Lond.), 271* (1978), pp. 493–495.

49. P.G. Brantom, I.F. Gaunt and P. Grasso, 'Long-term Toxicity of Sodium Cyclamate in Mice', *Fd Cosmet. Toxicol., 11* (1973), pp. 735–746.

50. J.M. Price, C.F. Biava, B.L. Oser, E.E. Vogin, J. Steinfeld and H.L. Ley, 'Bladder Tumours in Rats Fed Cyclohexylamine or High Doses of a Mixture of Cyclamate and Saccharin', *Science, 167* (1970), pp. 1131–1132.

51. I.C. Munro, C.A. Moodie, D. Krewski and H.C. Grice, 'A Carcinogenicity Study of Commercial Saccharin in the Rat', *Toxicol. Appl. Pharmacol., 32* (1975), pp. 513–526.

52. J. Chowaniec and R.M. Hicks, 'Response of the Rat to Saccharin with Particular Reference to the Urinary Bladder', *Br. J. Cancer, 39* (1979), pp. 355–375.

53. IARC Monographs on the *Evaluation of the Carcinogenic Risk of Chemicals to Man* (IARC, Lyon, 1980), vol. 22, pp. 111–171.

54. G.R. Howe, J.D. Burch, A.B. Miller, B. Morrison, P. Gordon, L. Weldon, L.W. Chambers, G. Fodor and G.M. Winsor, 'Artificial Sweeteners and Human Bladder Cancer', *Lancet, ii* (1977), pp. 578–581.

55. R.N. Hoover and P. Strasser, 'Artificial Sweeteners and Human Bladder Cancer: Preliminary Results', *Lancet, i* (1980), pp. 837–841.

56. Editorial, 'Saccharin and Bladder Cancer', *Lancet, i* (1980), pp. 855–856.

57. A.S. Morrison and K.E. Buring, 'Artificial Sweeteners and Cancer of the Lower Urinary Tract', *N. Engl. J. Med., 302* (1980), pp. 537–541.

58. R.M. Hicks, J. Chowaniec and J. St. J. Wakefield, 'Experimental Induction of Bladder Tumours by a Two-stage System', in T.J. Slaga, T. Sivak and R.K. Boutwell (eds.), *Carcinogenesis*, vol. 2, *Mechanisms of Tumour Promotion and Cocarcinogenesis* (Raven Press, New York, 1978), pp. 475–489.

59. N.K. Clapp, R.L. Tyndall, R.B. Cumming and J.A. Otten, 'Effects of Butylatedhydroxytoluene Alone or with Diethylnitrosamine in Mice', *Fd Cosmet. Toxicol., 12* (1974), pp. 367–371.

60. P. Grasso and R.F. Crampton, 'The Value of the Mouse in Carcinogenicity Testing', *Fd Cosmet. Toxicol., 10* (1972), pp. 412–426.

61. F. Hutterer, F.M. Klion, A. Wengraf, F. Schaffner and H. Popper, Hepatocellular Adaptation and Injury: Structural and Biochemical Changes Following Dieldrin and Dimethylbutter Yellow', *Lab. Invest., 20* (1969), pp. 455–464.

62. A.E.J. McGill and J.A. Robinson, 'Organochlorine Insecticide Residues in Complete Prepared Meals. A 12-month Survey in S.E. England', *Fd Cosmet.*

Toxicol., 6 (1968), pp. 45–57.

63. B. Terracini, M.C. Testa, T.R. Cabral and N. Day, 'The Effect of Long-term Feeding of D.D.T. to Balb/c Mice', *Int. J. Cancer, 11* (1973), p. 747.

64. A.I.T. Walker, E. Thorpe and D.E. Stevenson, 'The Toxicology of Dieldrin (HEOD), I. Long-term Oral Toxicity Studies in Mice', *Fd Cosmet. Toxicol., 11* (1973), pp. 415–432.

65. IARC Monographs on the *Evaluation of the Carcinogenic Risk of Chemicals to Man*. 'Some Organochlorine Pesticides', (IARC, Lyon, 1974), vol. 5.

66. D.P. Burkitt, A.R.P. Walker and N.S. Painter, 'Effect of Dietary Fibre on Stools and Transit-times, and its Role in the Causation of Disease', *Lancet, ii* (1972), pp. 1408–1412.

67. 'Food and Nutrition Research', Report of the Agricultural Research Council and Medical Research Council Committee (1974) (HMSO, London, Elsevier Scientific Publishing Company, Amsterdam – New York), pp. 136–140.

68. F.J.C. Roe, 'Aetiology of Breast Cancer: a Brief Review', *Invest. Cell Pathol., 2* (1979), pp. 45–53.

69. A. Tannenbaum, Nutrition and Cancer' in F. Homburger (ed.), *The Physiopathology of Cancer*, Ind. Ed. (Hoeber – Harper – New York, 1959), pp. 517–562.

70. M.J. Tucker, 'The Effect of Long-term Food Restriction on Tumours in Rodents', *Int. J. Cancer, 23* (1979), pp. 803–807.

71. M.H. Ross and C. Bras, 'Influence of Protein Under- or Over-Nutrition on Spontaneous Tumour Prevalence in the Rat', *J. Nutr., 103* (1973), pp. 944–963.

72. A. Tannenbaum and H. Silverstone, 'The Genesis and Growth of Tumours. IV Effects of Varying the Protein (Casein) in the Diet', *Cancer, Res., 9* (1949), pp. 162–173.

73. J.B.M. Gellatly, 'The Natural History of Hepatic Parenchymal Nodule Formation in a Colony of C57BL Mice with Reference to the Effect of Diet', in W.H. Butler and P.M. Newberne (eds.), *Mouse Hepatic Neoplasia* (Elsevier Scientific Publishing Company, Amsterdam, 1975), pp. 77–110.

74. M.G. Cutler and R. Schneider, 'Sensitivity of Feeding Tests in Detecting Carcinogenic Properties in Chemicals: Examination of 7,12-Dimethylbenz (a) anthracene and Oxidised Linoleate', *Fd Cosmet. Toxicol., 11* (1973), pp. 443–457.

75. S.L. Graham, K.J. Davis, W.H. Hansen and C.H. Graham, 'Effects of Prolonged Ethylene Thiourea Ingestion on the Thyroid of the Rat', *Fd Cosmet. Toxicol., 13* (1975), pp. 493–499.

76. R. Doll and R. Peto, 'The Causes of Cancer. Quantitative Estimates of Avoidable Risks of Cancer in the United States Today', *J. Nat. Cancer Inst., 66* (1981), pp. 1191–1308.

5 THE MUTAGENICITY OF FOOD

D. Anderson and I.F.H. Purchase

Introduction

When viewed from a toxicological point of view, man's food is characterised by its great variety; obtained from a large number of sources and a large number of different species. Before eating it, he applies a variety of different methods of treatment in order to make the food palatable. This makes considerations of the toxicology of food extremely complex. Traditionally, even primitive societies have recognised that certain foodstuffs have acute toxic properties and these have resulted in taboos preventing the eating of these toxic plants or fishes. In more advanced societies the same toxic species are recognised as poisons and the public are educated in the knowledge of which are dangerous in order to avoid the type of incidents which result from acute poisoning.

In considering a problem such as mutagenesis, where the interval between the ingestion of a potentially mutagenic foodstuff and the observations of an effect is prolonged and in some cases may cover several generations, reliance cannot be placed on simple cause and effect relationships such as are seen in the acute toxic incidents. From this point of view foodstuffs have traditionally been considered safe. With the development of food processing and the use of chemicals to improve the quality of food, questions of safety have been raised and these have concentrated more and more on the potentially devastating but long-term effects of the ingestion of carcinogenic or mutagenic substances contained in food.

The emphasis on the environmental causes of chronic diseases in man derives from observations on variations and incidences of these diseases between different populations. An important, if not the major, component of the interaction of environment with man is via his food. Man consumes about 10 tons dry weight of food by the age of 50[1] and this may allow the environmental influences to have both beneficial and harmful effects on human homeostasis. Because of the large variety of human food, and the complexity of observing mutagenic events in man, it is difficult to carry out a systematic study linking food-borne mutagens with mutagenic events. Indeed, there appear to be no reports which attempt to study the relationship between ingestion of natural foods and heritable effects in man. A summary of the field must of necessity be fragmentary and based on the information generated in related subjects.

Principles of Testing Food for Mutagenicity

Foods, and even the individual components which go to make up foods, are exceedingly complex chemical mixtures. This presents particular problems to the testing of foodstuffs for mutagenic activity. The first of these problems is that of administration of the food component to the testing system, particularly with *in vitro* assay systems. Secondly, the food component can interfere with the end-point used for detecting genetic events. This would be particularly so when nutritional dependence on an amino acid, for example, is used as the detector end-point. The third problem is the identification of a suitable food to act as a negative control in laboratory experiments. The fourth problem is that administration of food components in sufficiently large concentrations to the assay system may produce nutritional imbalance which will confound the interpretation of the results. As a consequence, when whole components of food are to be tested for mutagenic activity, it would appear that *in vivo* systems would provide better results than *in vitro* systems, although limited work may be done in certain specialist circumstances using *in vitro* systems. The only alternative approach which is available for testing the mutagenicity of whole components of food is to provide extracts of the food which could then be used in either *in vitro* or *in vivo* systems.

According to Stich *et al.*,[275] food systems are complex colloidal mixtures composed of low-molecular-weight compounds, polymers such as proteins and polysaccharides and particulate matter consisting of aggregated protein, protein-lipid complexes and cell fragments. Mutagens in the food systems may be physically adsorbed to polymers or may be trapped in the particulate matter. Various extraction procedures can be carried out using different solvents, e.g. chloroform, methanol, ethyl acetate. The original foodstuff on each separated fraction can be 'spiked' with known amounts of mutagens and their recovery by the extraction and separation procedures measured. The 'spiking' mutagens used should actively simulate their normal state in the food product in order not to obtain misleading results. Since microbial test organisms can be coated with linoleic or oleic acid if they are present in the lipophilic extract such organisms may become impregnable to many mutagens. However, it is to be expected that each test system will produce a few negative results since no one bioassay can respond to all the various genotoxic effects a compound can produce.[276]

There are fewer problems associated with the testing of individual chemical constituents of food. These may be isolated and purified and used in a purified form in *in vivo* or *in vitro* testing systems. Typical examples of this would be the naturally occurring pyrolizidine alkaloids or the food contaminants produced by fungi (such as aflatoxin).

Background to Mutagenicity Testing

A mutation is a change in the genetic material of a cell which is permanent and is transmitted to successive generations. Mutagenicity testing, therefore, relies on test systems using a variety of organisms, which have known genetic characteristics. Changes in the genetic make-up of the organism result in observable end-points which can be scored to give an indication of mutagenic activity.

Although selectively advantageous mutations are an important contributing factor to evolutionary processes, the majority of mutational events appear to be deleterious to the organism. It has been estimated that there is a genetic element in at least 10 per cent of all human pathological conditions,[2] although the exact cause of these mutational events is not known. Considering the overall health of a population, it is clear that attempts should be made to limit the amount of genetic change in order to limit the amount of genetic disease.

Apart from the concern about diseases which are heritable as a consequence of germ cell mutations, there is growing evidence that somatic cell mutations may also be important. A large number, and possibly the majority, of chemical carcinogens are also mutagens and they are thought to produce their carcinogenic effect as a consequence of inducing mutations in somatic cells which lead to changes in cell behaviour and the production of tumours. Although this is a genuine concern in terms of the effects of mutation, carcinogenesis is only a secondary consideration here, but forms the subject of a separate chapter.

Damage to genetic material may occur at either the gene level or the chromosomal level. Gene mutations, or point mutations, are the consequence of the alteration in the type or sequence of base pairs within the DNA. If a base pair is replaced with a different base pair, the lesion produced in the DNA is known as a base pair substitution. If there is a change in the sequence of base pairs due to the addition or deletion of one or more base pairs, the subsequent mutation is known as a frame shift mutation. Both base pair substitution and frame shift mutation events lead to point mutation. Damage to the genetic material at the chromosomal level can lead either to changes in the structure of the chromosome or changes in the number of chromosomes.

The way in which genetic disease as observed in man can be related to gene and chromosomal abnormalities is shown in Table 5.1.

Mutagenicity Testing

Most systems used for testing foods or chemicals for mutagenicity are very specific with respect to the genetic end-point. As gene and chromosomal mutations are so different in their genetic end-points, testing of chemicals should take into account both types of mutation. However, there are some tests which detect generalised DNA damage but are non-specific with respect to the genetic end-point and can be used in the screening of chemicals for genetic activity. Reviews are available which describe in some detail tests available for mutagenicity testing.[3–5]

The techniques which are used in *in vitro* and *in vivo* genetic systems are very different in execution, although the underlying genetic principles may be similar.

In Vitro Mutagenicity Tests

These tests, which are based on microbial or cell culture techniques, have three important components: the chemical, the metabolic system and the genetic marker.

The Chemical. The mutagenic potency of chemicals varies over an enormous range of up to 10^9-fold. This emphasises the great importance of knowing precisely what the purity of chemicals used in mutagenicity testing systems is. There are several examples of impurities giving positive results when a chemical is tested, for example saccharin.[6,7] Another frequent problem encountered in the testing of chemicals is the provision of the correct solvent which does not

Table 5.1: Examples of Genetically Based Human Pathological Conditions

10% of Human Pathological Conditions Have a Genetic Element

Gene Abnormalities	Monogenic	Autosomal	Dominant	Huntingdon's chorea Achondroplasia
			Recessive	Phenylketonuria Cystic fibrosis
		Sex Linked		Haemophilia Muscular dystrophy
	Multigenic			Schizophrenia Epilepsy
Chromosomal Abnormalities	Non-disjunction	Autosomal		Mongolism Edward's syndrome
		Sex Linked		Klinefelter's syndrome (XX) Turner's syndrome (XO)
	Chromosome breakage			Ataxia Telangiectasia * Fanconi's anaemia *

* arise from autosomal recessive gene changes initially.

interfere with the genetic marker, but is suitable for dissolving the chemical.

The Metabolic System. Many chemical mutagens do not produce their mutagenic effect until they have been enzymatically activated to reactive species, which then react with DNA. Any *in vitro* testing system which does not provide the necessary metabolic activation will fail to detect these pro-mutagens, in spite of the fact that in *in vivo* situations the mutagenic activity may well be expressed following metabolism to the active mutagen. Prokaryotic organisms used as genetic markers do have some metabolic capacity. For example, it is known that *Salmonella typhimurium* has nitro-reductases capable of activating some nitrofuran compounds. However, in general the metabolic activity of prokaryotic organisms is low and is different from that in systems using liver homogenates where activation predominates as a result of reduction in detoxification enzymes. In the case of eukaryotic cells, such as yeasts, in culture, considerable metabolic activity is present during the log growth phase, but this may not reflect the full range of mammaliam metabolism. It might be expected that mammalian cells in culture would have metabolic systems which are similar to those of whole mammals. Whereas this may be true in the qualitative sense, the ability of mammalian cells *in vitro* to metabolise chemicals is limited by the low levels of enzyme activity. For these reasons auxiliary metabolic systems are normally included in any *in vitro* mutagenicity assay.

The most commonly used metabolic system is the so-called S9-fraction. In the standard protocol of Ames *et al.*[8] rats are injected intraperitoneally with Aroclor 1254 and killed on the fifth day when the livers are removed and homogenised. After centrifugation of the homogenate at 9,000 *g*, the supernate forms the so-called S9-fraction. This S9, with the necessary co-factors to provide energy, is added to the *in vitro* system and provides a remarkable degree of metabolic activation. Attempts have been made to use isolated hepatocytes to circumvent the imposed artificiality of the S9 system, but these testing systems also have their disadvantages, and have not been shown to be generally superior to S9.

The Genetic Marker. The frequency of genetic events tends to be extremely low, even when induced by potent chemical mutagens. In order to provide an easier detection of these low frequency events,

Table 5.2: Forward Mutation Assays for Detecting Point Mutations

Organism	Marker
Bacteria	
S. typhimurium	Resistance to 8-azaguanine or arabinose[9,10]
E. coli	Resistance to 5-methyltrytophan or streptomycin[11,12]
	Lack of galactose fermentation[13]
	Induction of prophage[14,15]
Fungi	
Streptomyces coelicolor	Resistance to streptomycin[16]
Saccharomyces cerevisiae	Adenine dependence[17–20]
Schysosaccharomyces pombe	Resistance to canavanine[21]
Neurospora crassa	Adenine biosynthesis[22–24]
Aspergillus nidulans	Resistance to 8-azaguanine[25]
Insects	
Drosophila melanogaster	Sex-linked recessive lethal[26–31]
	Autosomal recessive lethal[26–31]
Mammalian cells in culture	
Chinese hamster ovary[32] V79[33–41] Human fibroblast or[41–43] lymphoblast[44]	azaguanine, thioguanine resistance (hypoxanthine guanine phosphoribosyl transferase (HGPRT) locus)
L5178Y cells[45–47] P388F cells[48–50]	5,iodo-2-deoxyuridine, 5,bromo-2-deoxyuridine resistance (thymidine kinase (TK) locus)
L5178Y cells or P388F cells	Excess thymidine,[48–51] ouabain,[51] thioguanine[52] or alanine[53]

either high doses of applied chemical or high numbers of organisms are required. The great advantage of genetic marker organisms grown in culture is that extremely large numbers of cells may be used, facilitating the detection of low frequency events.

In vitro systems for detecting point mutations are of two types: forward or reverse mutation systems. Mutational events are detected in forward mutation systems when there is a loss of an enzyme function required for the biosynthesis of a critical chemical nutrient needed for the growth of the organism. This results in a new requirement for a growth factor to be included in the culture medium. Alternatively these events are detected as the acquisition of resistance to various toxic chemicals, substrate analogues or environmental agents such as heat. Examples of the organisms and the detection systems are given in Table 5.2. Systems which detect reverse mutations generally require the use of a mutant cell which contains a specific point mutation conferring a specific nutritional requirement on the organism. The point mutation in the mutant cells is generally characterised as either a

Table 5.3: Reverse Mutation Assays for Detecting Point Mutations

Organism	Marker
Bacteria	
S. typhimurium	Histidine independence[8,54–60]
E. coli	Arginine independence[61]
	Trytophan independence[62]
Yeast	
A. nidulans	Methionine independence[25]

base pair substitution or a frame shift mutation. The mutational event is detected as a re-establishment of normal function with loss of the particular nutritional requirement. This occurs as a consequence of a new substitution or deletion near the point mutation which characterised the mutant cell. Examples of the nutritional requirements and organisms which are commonly used in reverse mutation assays are given in Table 5.3. All the available reverse mutation assays depend on micro-organisms, but *in vitro* forward mutational assays are available for micro-organisms and mammalian cells in culture.

Chromosomal mutations cannot be detected in prokaryotic organisms because the DNA of prokaryotes is in a single circular strand within the cell and the cells multiply without undergoing meiosis or mitosis. Eukaryotic organisms have a nucleus in which the DNA is organised in characteristic sets of chromosomes, and are suitable for detecting chromosomal mutations. The most commonly used cells are mammalian cells, such as Chinese hamster ovary cells, or peripheral lymphocytes. Peripheral lymphocytes are stimulated to divide in culture with the use of a mitogen such as phytohaemagglutinin, and arrested in c-metaphase by the use of colchicine.[63] The metaphase figures can then be examined microscopically for the identification of changes in number or structure of the chromosomes. Although yeasts are eukaryotes with 16 chromosomes, the chromosomes do not organise themselves in a way which is suitable for microscopic examination. Hence, assays for chromosomal damage using yeasts are only capable of detecting certain mitotic events such as mitotic non-disjunction using classical genetic methods.[64,65]

In Vivo Mutagenicity Tests

The great advantage of *in vivo* mutagenicity assay systems is that they have an intact metabolic system which to a greater or lesser extent mimics the situation in man. They also have the advantage that either whole food, or components of food, or extracts of food, or pure chemicals can be tested in them. Their major disadvantage lies in the fact that low frequency events require enormous numbers of animals for their detection.

The Chemical. In order to enhance the probability of detecting low frequency events, the chemical, or extract, or food component is administered to the *in vivo* system at a maximum tolerated dose. In the case of pure chemicals or extracts, this can be determined with reasonable accuracy. If whole foods or components of food are to be administered to the animals there is a limitation of the maximum quantity that can be administered because of the need to maintain nutritional balance in the diet. Concerns expressed about the effects of mutagenic contaminants on the interpretation of *in vitro* tests apply equally to *in vivo* tests. Small quantities of potent mutagenic impurities can give positive results when the chemical or food component itself is not inherently mutagenic.

The Metabolic System. There are a variety of insects and mammals which are used in *in vivo* mutagenicity assays. Mammalian systems have the obvious advantage of the full metabolic competence of a mammalian organism. There are, however, many examples of differences in the response of different species to the same chemical and this complicates the problem of extrapolation of data derived from these systems to the human situation. These *in vivo* mammalian systems have the added advantage that observations in some cases are made directly on germ cell mutations. This allows the special metabolic features of the gonads to be taken into account during the extrapolation. The most commonly used insect systems are based on *Drosophila melanogaster*. This fly appears to have a fairly extensive range of enzyme systems, including mixed function oxidases, which are capable of metabolising xenobiotics in a way which is remarkably similar to that in mammals.[66]

The Genetic Marker. There are a variety of *in vivo* point mutation assays available, three of which are more commonly used.

(a) The *Drosophila* sex-linked recessive lethal assay. In this assay males of a specially selected strain of *Drosophila* are treated with the chemical concerned. Mutation results in the expression of lethality in the male offspring. This is detected by the absence of a particular phenotype (red-eyed; wild-type male) in the F_2 offspring with a consequent change in the sex ratio. The *Drosophila* sex-linked recessive lethal assay is considered to be a sensitive assay for point mutations in an *in vivo* system. It has the additional advantage of being relatively quick and cheap to carry out in comparison with the other *in vivo* point mutation assays.[26–31,66]

(b) The spot test.[67] This test detects somatic mutations. Mice which are heterozygous for various coat colour genes are treated *in utero* by administration to the mother. If a mutation occurs in a single cell during the development of the coat, a mosaic patch is observed in the mouse when it is fully grown. Thus, the female is treated during pregnancy and the offspring grown until the coat colour has fully developed. Each mouse is then examined for spots which are attributed to the clone arising from a single mutant cell. This assay system has the advantage of being one of the few assays for gene mutations in mammals. Its main disadvantages are that the somatic mutations for coat colour may occur as a consequence not only of gene mutations but also chromosomal mutations. These coloured spots can be very difficult to distinguish from the natural pigmentation zones of the coat.

(c) The specific locus test.[68,69] This test in mice is at present the only one available which can detect heritable gene mutations in mammalian germ cells. The test uses mouse strains, heterozygous for certain dominant, wild-type alleles and strains which are homozygous for recessive alleles at these loci. The dominant, wild-type strains are then treated and mated with the recessive strains. Changes induced at any of the loci cause visible changes in the offspring, particularly eye colour, ear size and coat colour. Although it is recognised that these mutational events are gene mutations, the exact nature of the event at the gene level cannot be identified. The main disadvantage of the specific locus test is that the spontaneous rate for these loci is low. Hence for a negative result to be recorded large numbers of offspring (up to 30,000) need to be scored, making the test extremely costly and laborious.

There are a number of biochemically based tests for detecting gene mutations in somatic and germ cells in mammals which are not yet extensively used. They depend on variations in proteins caused by gene mutations controlling the production of these proteins.[70–72] A variety of new approaches are also available in mice such as detection of changes in mandible shape,[73] skeletal mutations[74,75] histocompatibility loci[76] and recessive lethals measured by the inversion technique[77,78] and pigment loci.[79]

Genetic damage at the chromosomal level can be observed in *in vivo* systems, either by direct observation of chromosomes at metaphase or by indirect observations in various other systems.

(a) *In vivo* chromosome metaphase analysis.[80–85] This technique analyses the chromosomes in somatic cells in metaphase using similar techniques to those described for the *in vitro* system. The chemical or foodstuff is administered to the animal for a period of time and a few hours before death colchicine is administered to arrest cells in metaphase. Bone marrow is then recovered from the treated animals and chromosome aberrations observed by microscopy. It is obvious that this assay system does not detect chromosome mutations in germ cells and cannot be considered to provide evidence of heritability. It does have the advantage of direct observation on rapidly dividing cells in the mammalian system. Similar observations may be made on peripheral lymphocytes stimulated to divide by exposure to a mitogen or on spermatogonial cells.[99] In the latter cells the presence of translocations provide evidence of possible heritable changes which may be observed in mice using the heritable translocation test as partial sterility in the F_1 generation.

(b) Micronucleus assay.[86–89] During the maturation of the mammalian erythrocyte, the nucleus is extruded, producing the mature erythrocyte without a nucleus. In the event that chromosomal aberrations occur within that nucleus, a component of the chromosome can be left behind within the erythrocyte, creating the presence of a micronucleus. The observation of an increased number of micronuclei in maturing polychromatic erythrocytes from bone marrow smears is thus an indication of chromosome damage. In a typical assay, maximum tolerated doses of a chemical or components of the food are administered to the animal over a four- or five-day period. The animals are then killed and smears of the bone marrow made and stained. A number of micronuclei expressed as a proportion of maturing erythrocytes is determined by microscopy. Significant changes from the control are considered as positive results. The micronucleus assay has the great advantage of simplicity and speed of execution. However, the events that are observed occur in somatic cells and do not provide evidence of an heritable effect.

(c) The dominant lethal assay. This assay can be carried out in rodents[90–95] or in *Drosophila*.[66] The assay is based on the fact that chromosomal abnormalities can lead to abnormalities of the chromosome in the offspring which are incompatible with survival, resulting in death of the fetus. In a typical assay males are treated with the chemical or food component over a period which will cover the whole of spermatogenic maturation. In the case of mice this is for eight weeks. The males are then mated with virgin females and the number of fetal deaths counted by direct observation of the uterus after killing the animals, or by examination of the offspring. This assay has the advantage that it is an *in vivo* germ cell mutation assay but because the mutational event observed results in lethality, it cannot be considered as heritable. Evidence from the mouse dominant lethal assay suggests that it is insensitive in comparison to other assay systems for the detection of mutations, but it may reflect the true *in vivo* situation.

(d) Heritable translocation test.[96–101] This test has been carried out on mice and is basically an extension of the dominant lethal test. The principle of the test is that translocations occurring between adjacent chromosomes can result in partial sterility in the offspring. Thus, male mice are treated as in the case for a dominant lethal test and mated with virgin females. The F_1 progeny are allowed to grow to maturity and the males mated with a further set of virgin females to determine the incidence of sterility and partial sterility. Confirmation of partial sterility is required by remating schedules. Eventually confirmation of the translocation is obtained by examination of meiotic cells of the F_1 males which have been shown to be sterile or partially sterile. This test has the advantage that it is carried out on mammalian germ cells and demonstrates the heritable mutation. It is very time-consuming and expensive to carry out, and as a consequence relatively few chemicals have been tested in this system.

Tests for DNA Damage

There is a variety of tests which detect DNA damage. When a substance reacts with DNA and induces damage repair normally follows.[102,103] Tests which detect DNA damage are in general non-specific in terms of the genetic end-point which they measure and generally use indirect methods of observation. Two commonly used tests are as follows.

Differential Killing of Repair-deficient Strains.[104–108] A variety of mutant strains of micro-organisms have been developed which have specific deficiencies in DNA repair enzymes. The consequence is that when DNA damage is produced in these strains the organism will die because it cannot repair the damage. The assay system thus depends on differential killing between two strains of the bacteria, one of which is repair-proficient, and one of which is repair-deficient. Evidence of damage to DNA is produced when the repair-deficient strains are shown to be more susceptible to the toxic effects of the chemical than the repair-proficient strains. As in the case of other microbial assay systems, mammalian metabolic enzymes are absent and are added to the assay system in the form of an S9-mix. It has the advantage of being an extremely quick assay to carry out.

Unscheduled DNA Synthesis.[109–113] DNA damage can be detected by determining the incorporation of thymidine into DNA as a consequence of excision repair during the phase of the cell cycle when DNA synthesis does not normally occur. The incorporation of tritiated thymidine can be measured either by autoradiography or by scintillation counting techniques using mammalian cells in culture.

Tests for mitotic recombination or gene conversion in yeast[17–20] and for sister chromatid exchanges in mammalian cells[114–116] are also considered as tests for indirectly detecting DNA repair.

Many of the test systems discussed here have recently been reviewed in some detail by the Environmental Protection Agency Gene-Tox programmes.[253–274] Other drafts from the same source are awaiting publication.

Tests Which Can be Applied to Man

There are two basic types of observation which have been made in man exposed directly to chemicals and these could in principle be applied to populations eating certain foods. The first is the detection of active mutagenic chemicals or metabolites in human blood or urine by using microbial assays to detect the mutagenic activity.[117] The second involves the observations of non-heritable mutations in human tissues; for example detection of alkylation of body macromolecules,[117] electrophoretic variants of enzyme markers[71] and variants of haemoglobin.[118,119] Observation of sperm abnormalities[120–122] and the presence of YY bodies[123] gives indirect evidence of genetic events in germ cells, but cannot be considered to give unequivocal evidence of heritable effects.

Testing Strategy

There is obviously a wide variety of test systems measuring a variety of mutational events available to the laboratory wishing to test food or food components for mutagenic activity. Consideration of the following features may help guide the laboratory in selecting the tests most suitable for its task:

(a) It is generally believed that *in vivo* mammalian systems give information which is of more value in terms of assessment of risk to man than the systems dependent on lower organisms.

(b) There are two distinct genetic end-points in the various assays available (point mutation and chromosomal mutations). The selection of the assay systems must ensure that both genetic end-points are assayed.

(c) The assay system must be capable of having the compound or food component administered to it. *In vitro* systems are not wholly satisfactory for the inclusion of food components. Some of the *in vitro* systems however, under special circumstances (e.g. using extracts) and the *in vitro* systems can, be used for the administration of foods or food components.

(d) The total costs. Many of the assays, particularly the *in vivo* assays are extremely expensive and resource-consuming.

As a consequence of these considerations, any testing of foods or food components for mutation would require different assay systems. In general, when screening chemicals, at least a point-mutational system, usually a bacterial system with and without metabolic activation, and a chromosomal mutational system are used. Additional information can be obtained by using point-mutational assays in *Drosophila* and chromosomal aberration assays in *in vivo* systems. These latter two tests may be the only one which can be applied satisfactorily for whole foods or food components. If a positive result is obtained in any of these screening tests, further assays using the more extensive mammalian assay systems may be required before an accurate quantitation of the effect can be made.

Mutagens in Food

There are no reports of which we are aware that implicate food-borne

Table 5.4: Mutagenicity of Foods and Food Components

Nature and occurrence of mutagenic agent	Test systems in which mutagenicity was reported	Comments
Fungal metabolites		
Aflatoxin B_1 from *Aspergillus flavus*. Occurs in a variety of staples particular groundnuts, cottonseed and maize.[124,125]	Point mutation in *S. typhimurium*[127], *S. cerevisiae*[128] *Neurospora*,[129,130] *Drosophila*,[131] Chromosome aberrations in mammalian cells in culture[132] and in mice *in vivo*.[133]	The most potent mutagen[126,127] Also a carcinogen[134–138]
Aflatoxin B_2, G_1 and G_2,[134] metabolites of *A. flavus*	Similar organisms[136–137]	Less mutagenic and carcinogenic than Aflatoxin B_1[134,135]
Sterigmatocystin[137]	Similar organisms[138]	
Versicolorin A from *Aspergillus flavus* and *A. parasiticus*	*S. typhimurium* TA98[139]	
Dimethylsterigmatocystin, *S*-methylsterigmatocystin, Sterigmatin, Dimethylversicolorin A	*S. typhimurium*[138] TA98 and TA100	
PR toxin from *Pencillium roqueforti*	*S. typhimurium* TA100[138]	
Chaetoglobium from *Chaetomium globosium*	*S. typhimurium* TA100[138]	
Zearalenone from *Fusarium graminarium* found in mouldy grains	*B. subtilis* rec-assay[140]	
Luteoskyrin from *Penicillium islandicum*	*B. subtilis* rec-assay[140]	
Citrinin from *Pencillium citrinin*	*B. subtilis* rec-assay[140]	
Fungi		
Agaricus bisporous contains agaritine[141]	*S. typhimurium*[138]	Agaritine is broken down to 4-hydroxymethylhydrazine which causes lung tumours in mice[142]
Gyromitra esculenta contains acetaldehyde formylmethylhydrazone[143]	*E. coli* WP_2[144] *S. typhimurium* his G46 and TA92[138]	Also carcinogenic in mice and hamsters[145–146]
Plants		
Pteriduim aquilinum (bracken fern extract)	*S. typhimurium*[138,147]	The exact cause for the carcinogenicity and mutagenicity of bracken is not clearly established[138,148–152]

Table 5.4: Mutagenicity of Foods and Food Components *(contd)*

Mutagenicity possibly due to shikimic acid or	A dominant lethal mutagen[148]	
Kaempferol, a flavonoid	*S. typhimurium*[153]	
Cycad nuts containing cycasin	*S. typhimurium*[154]	Cycasin is metabolised to methyl-azoxy-methanol by gut bacteria[154] and is a carcinogen[155]
Pyrolidizine alkaloids from plant families[156] Leguminosae, Boraginaceae and Compositae	*Drosophilia,*[131,157] *Aspergillus,*[158] but not *Salmonella*[138]	Known as hepatocarcinogens[159,160]
Coltsfoot	Host-mediated[138] *S. typhimurium*	
Flavonoids found in roots and fruits particularly querticin and kaempferol.[153] Querticin found in onion skin[164] and the spice sumac[165]	*S. typhimurium*[138,161–163]	
Safrole	*S. typhimurium*[60,166,246]	Weakly carcinogenic[167]
Eugenol	*B. subtilis*[246]	
Cinnamon	*B. subtilis*[247]	
Food additives		
Nitrofuran derivatives AF_2 used for preserving fish and meat cakes	*S. typhimurium*[168,172] *E. coli*[168–170] Mammalian cells in culture[171]	Used as a food additive in Japan from 1965 until mutagenicity results led to retesting and a carcinogenic effect[173–178]
Saccharin	Non-mutagenic but some impurities may be mutagenic[6,7,179,180,251]	
Cyclamate	Chromosomal aberrations in mammalian cells[181–184] and rats[185] but not *Drosophila*[186]	
Amaranth	*Salmonella*[187] Host-mediated[188] for urine,[189] but no dominant lethal effects[190]	
Nitrite	*Salmonella*[127] Cultured mouse cells,[191] chromosomes,[192] mammals[193]	Reacts with amines and amides in the stomach[221]
Ascorbic acid	*Salmonella,*[194] mammalian sister chromatid exchange[195,196]	

Table 5.4: Mutagenicity of Foods and Food Components *(contd)*

Synthetic chemicals Pesticides	Bacterial mutation assays[197,198] *Drosophila*[199] Human lymphocytes[252]	Can be considered as indirect food additives. Some are positive in mutagenicity tests e.g. captan[200]
Food packaging materials,[201] Residual vinyl chloride was found in alcoholic beverage in PVC bottles	Microbial assays[202–205] including *S. typhimurium*[204] *E. coli*[205] Exposed workers[206]	Residual levels are now reduced so that migration into food is not measureable[201]
Food colours Azo, triphenylmethane and xanthene dyes	Various assay systems[235]	Most colours showing mutagenic activity are no longer permitted
Natural colours e.g. chlorophyll, carotene – curcumin, riboflavin	Anti-mutagenic[233,235] mutagenic in one or two tests, negative in others[235]	
Air and Water Pollution Contaminants of air and water. (Polycyclic aromatic hydrocarbon nitrosocompounds)[208,209]	Bacterial mutation tests[210]	Mutagens are found in water, shellfish and whisky[210]
Chlorination produces organochloride compounds[211]	*S. typhimurium*[212]	
Food processing Pyrolysis products Charred or broiled meat and fish	*S. typhimurium*[213,214,243–245]	Polycyclic aromatic hydrocarbons are not the only mutagens
Pyrolised protein giving products of tryptophan, serine, glutamic acid, ornithine and lysine	*S. typhimurium*[215]	
Pyrolised sugars (glucose, fructose, arabinose) and caramel	*S. typhimurium*[216]	
Pyrolised garlic, onion and weeds	*S. typhimurium*[138]	
Sodium nitrite and secondary amines producing nitrosamines in the stomach or in the food[217–222]	Bacterial mutation tests[219–220] (see also references 191–193)	
Irradiation of wheat	Some reports of polyploidy in children, monkeys and rats but others resport negative results[223–227]	Effect diminishes on storage
Irradiation of sucrose	*S. typhimurium*[228]	
Irradiation of whole diet	Dominant lethal effects have been seen but not in extensive multi-generation studies[229]	

Table 5.4: Mutagenicity of Foods and Food Components *(contd)*

Alcoholic drinks and beverages		
Wine, whisky and brandy citrus juices, milk, coffee, tea	*S. typhimurium*[234,236–242]	Flavonoids and their glycosides are often thought to be responsible but are poorly absorbed in mammals[250]

mutagens in the induction of mutagenic events in man (with the possible exception of the reports of polyploidy in children eating irradiated wheat (reference 223). Indeed, techniques for studying heritable effects in man require further development before definitive results could be expected. Information on the presence of mutagens in food is limited to experimental results using laboratory models.

There is a large body of data on the mutagenicity testing of food or chemicals in food. An outline of this range of chemicals, foods and food components which have been shown to be mutagenic in mutagenicity assays is given in Table 5.4.

There is good evidence that food from a variety of sources contains mutagenic agents. Some of these are natural components of food, such as the cycasin present in cycad nuts which are consumed in Guam.[154] Some species of mushrooms also contain mutagens.[141–146]

Interest in mycotoxins, initially as agents responsible for acute disease in poultry, but subsequently as the cause of a variety of chronic diseases, including cancer, has resulted in the isolation of a variety of mycotoxins. Recently many of these have been tested for mutagenic activity in laboratory tests. Notable among them are the aflatoxins because of their ubiquitous distribution in foodstuffs (such as nuts, grains and even milk) and because aflatoxin B_1 appears to be one of the most potent mutagens discovered. There are some mutagenic mycotoxins which are known to contaminate human food, and these are reported in Table 5.4.[124–140]

Man further contributes to the mutagenic components in his food by his attempts to improve various characteristics of the food. Improvements in the yield of crops are achieved by the use of a variety of chemicals designed to control insect, bacterial and fungal pests. Inevitably traces of some of these chemicals will persist into the food which is eaten and this is the subject of control by regulatory bodies. As mutagenicity testing is a fairly recent development, inevitably some of the chemicals used prior to their implementation have been found to be mutagenic (Table 5.4 gives the references to studies of a varicty of pesticides for mutagenicity).[197–200]

Once crops are harvested and foods prepared, additives are frequently used to improve texture, flavour, colour or storage. Once again, these chemicals were in use prior to mutagenicity testing becoming common, and several have been found to be mutagenic. One of the most interesting examples is AF_2 which was introduced as a food preservative in 1965 in Japan. Subsequently it was found to be a point mutagen in *E. coli* and *S. typhimurium*. This resulted in retesting for carcinogenicity, because of the link between mutagenicity and carcinogenicity. AF_2 was removed from the market in 1974 when it was found to be carcinogenic.[168–176, 236]

A large number of chemicals are or have been used as food dyes. Members of the three major classes (azo, tryphenylamethane and xanthene dyes) are known to be mutagenic.[235] With the exception of curcumin and riboflavin, all natural colours used for foods are non-genotoxic.[235]

The concern about the carcinogenic hazards of vinyl chloride as a consequence of reports of its carcinogenicity in rats and then in man, led to consideration of the problem posed by residual unreacted vinyl chloride in PVC used for food packaging.[201–206, 248] Manufacturing methods were altered and residual levels reduced to less than 1 p.p.m. Migration into foods is now considered insignificant ($< 1 \times 10^{-9}$). In this example and that of AF_2, regulatory action was based on knowledge of carcinogenicity, although both chemicals were also mutagenic.

The nitrosation of secondary amines as a consequence of their reaction with nitrites results in the formation of nitrosamines.[217–222] Many nitrosamines are known to be mutagenic and carcinogenic. Although nitrite occurs in foods, particularly some vegetables, it may also be produced by the action of bacteria in the saliva and in the gut on nitrates which also occur in foods. Sodium nitrite is also used as a food preservative in cooked meats. The nitrosation reaction is favoured by low pH and thus may occur in the stomach as well are more slowly in food. The result is that nitrosamines may be produced by nitrosation in the stomach or, as has been found, may occur at low concentrations in foods such as sausages.[243]

Man also contributes to the mutagenicity of food by cooking at high temperatures. Charring or broiling of meats, proteins and sugars results in the production of mutagenic components. These are limited to the outer charred portion of, for example, hamburgers.[230, 236] Beef[244] and boiled pork extracts[245] also contain mutagens. The chemical entities responsible for the mutagenic activity include polycyclics, such as 3,4-benzo(a)pyrene and the pyrolysis products of amino acids. In addition pyrolysis products, such as harman and nor-harman,[231] amplify the mutagenic activity of chemicals, probably by interfering with their metabolism.

Alterations in mutagenic activity may also result in a reduction in mutagenicity for example extracts of turnip and cabbage decreased the mutagenicity of the pyrolysis products of tryptophan.[232] It has been suggested that the anti-mutagenic properties of some common vegetables may be related to their chlorophyll content.[233] Vitamins A, C and E can inactivate or inhibit various mutagenic processes.[236]

The use of irradiation to preserve foods from microbial contamination by sterilisation has been studied extensively. Reports cover a wide range of studies of gene mutations, chromosomal abnormalities and dominant lethal effects in a variety of species including man.[223–229] Polyploidy has been observed in children, monkeys and rats fed irradiated wheat and hamsters fed irradiated diet.[223–227] Other studies of feeding irradiated wheat or foods irradiated for human consumption to rats have failed to demonstrate polyploidy. The induction of polyploidy appears to decrease when the food is stored and this, together with the variety of components of food, may result in the differences observed. Although dominant lethal effects have been observed in mice fed irradiated laboratory diets in one laboratory,[229] dominant lethal effects were not observed in several other studies in mice and rats. The absence of an effect was supported from multi-generation studies in both rats and mice. Thus although there are some reports implicating irradiated diet at mutagenic, there is a large body of evidence which shows that irradiation does not result in mutagenic food.[229]

Mutagens have been found in alcoholic drinks and various beverages. They have been detected in alcoholic spirits[236,237] including wine, whisky and brandy. Flavonoids and their glycosides account for some of the mutagenicity of red wine.[236] Flavonols such as kaempferol and quercitin have been detected in the extracts of the hydrolysates of citrus fruit juices.[238] Mutagenic activity has been detected in heat sterilised milk; the chemical specification of the mutagenic activity has not been defined, but was present in a chloroform extract of milk.[239] Equal mutagenic potential has been shown in brewed and instant coffee whether regular or decaffeinated.[236,240] Caffeine also has been shown to produce alterations in mutagenic activity for it is known to prevent the operation of the cell's self-repair mechanism.[234]

The mutagenic activity and black and green tea has also been partially accounted for by flavonoids and their glycosides.[236,240,241] Stoltz *et al.*[242] however, screened 28 widely consumed beverages including milk and tea and confirmed evidence of mutagenicity for only red wine, grape juice and instant coffee. Aeschbacher reports that flavonoids are poorly absorbed which might explain why they are not mutagenic/ carcinogenic in mammals.[250]

Extrapolation of Results from Laboratory Assays to Man

The first stage in any attempt to extrapolate results from mutagenicity tests to man must be the interpretation of results within the individual test system. Each assay has its own idiosyncrasies which make interpretation unique to that particular system. However, there are some characteristics of test systems which lead to the conclusion that some general principles can be applied to the interpretation of all mutagenicity assays. Positive results are on the whole more easy to interpret than are negative results. Generally a positive interpretation of the results of a mutagenicity assay requires the presence of an increase in the observed parameter over the control value (either on the basis of an arbitrary cut-off point or on the basis of a statistical analysis), the presence of a dose-response relationship and evidence that the observation is reproducible. One of the drawbacks to the *Salmonella* assay is that survival is not taken into account. Sometimes, for example with a toxic weak mutagen, experiments where survival is accounted for and mutation is expressed in terms of the surviving fraction may be required. Negative results from any assay system are never absolute and can only be described within the framework of the experimental system. In general terms it is more certain to describe the results as indicating no mutational effect at a given dose, rather than as negative results. There are a large number of substances with extremely weak activity in mutational assays, providing difficult decisions as to whether they are positive or negative. Once again, reproducibility is a key factor in the interpretation of negative results because there is a tendency to place more reliance on a single positive result than on a number of reproducible negative results.

Once a clear decision has been made on the interpretation of the results in the assay system, extrapolation to man can take a qualitative or a quantitative form. As one might expect, qualitative extrapolation tends to be easier, although even qualitative extrapolation is severely hampered by the fact that there are no good examples of human mutagens (except perhaps smoking). Extrapolation from the test systems is therefore theoretical in nature. In general terms, there are such large differences between the submammalian tests and the human situation, that these test systems can only be used for screening. Once an effect is observed in submammalian systems further experimentation in mammalian cells and *in vivo* systems would be required for confirmation that a mutagenic effect is observed. Even then, many assay systems are based on mutations in somatic cells and the extrapolation from somatic cell mutation to germ cell mutation is uncertain.

Quantitative extrapolation from assay systems to man is even less secure than is qualitative extrapolation. There are a number of reasons why this is so. Amongst them are the reasons for the difficulty in qualitative extrapolation, namely that a large number of assay systems are based on somatic cell mutations; there are no good models of human mutagens and there are large differences between the organisation of DNA in many of the organisms used in screening tests and in man. Apart from these variables it is well known that metabolic systems differ markedly between *in vitro* and various *in vivo* assay systems. The consequence is that quantitative results from assays of potential mutagens cannot be reliably extrapolated into other organisms with different metabolic systems. This is well illustrated by the large degree of variability between differing assay systems. Also, although the DNA is similar between the organisms used in assay systems and in man, the arrangement of the DNA into chromatin and chromosomes and the presence of repair systems may be vastly different between these assay systems, again rendering quantitative extrapolation unreliable.

Significance of Food-borne Mutagens to Man

There are insufficient data on which to form a clear view of the significance of the various mutagenic components of food to man. In general terms, evidence of mutagenicity of a food in a variety of prokaryotic and eukaryotic organisms would suggest that the food in question poses a mutagenic hazard to man. (In this context hazard, the potential for producing a mutagenic effect, is distinct from risk, or the probability that a mutagenic effect will be observed after consumption of a mutagenic food). Under these circumstances, some feel that it would be prudent to reduce the mutagenic risk by attempting to reduce consumption. When consumption is part of social behaviour (such as eating cycad flour and broiling hamburgers) such reduction could only be feasible by education. In other cases, such as AF_2, regulatory action may be feasible.

Assistance to regulatory decision making may be given by an assessment of the potency of the mutagenic hazard followed by an estimate of the risk. Such an analysis has been carried out by Bridges[200] for the pesticide captan. The residue limits for captan on vegetables and fruit recommended by FAO range from 5 to 40 p.p.m. with a mean of about 20 p.p.m. If the average person consumes 0.5 kg of fruit and vegetables containing 20 p.p.m. captan, the total daily consumption would be 10 mg or 0.125 mg/kg for an 80 kg person. It is reasonable to assume a 50 per cent reduction by cooking giving a daily intake of 0.0625 mg/kg. On the basis of observations of the dominant lethal effects of captan to mice, and using a log-linear extrapolation, 0.0625 mg/kg/day represents 0.066 rad-equivalents per day or 2.2 rad-equivalents per annum. If it is assumed that there is a threshold in the dominant lethal data this figure falls considerably to 0.058 rad-equivalents per annum. These figures were then compared with IRPC recommendation for acceptable annual exposure to radiation of 0.17 rad.

Calculations of this type made many assumptions (such as log-linear extrapolation, equivalent quantitative effects in mice and man, and equivalent effects from radiation and chemicals) and should be used with some caution.

In principle concern about the mutagenicity of food should start with the greatest mutagenic risk. There is insufficient information about the mutagenicity of foods, particularly in terms of quantitation and observed effects in mammalian germ cells, for a clear idea of the likely risks associated with mutagenic components to be formed. The priority in the immediate future should be for a more systematic study of this topic, particularly in mammalian test systems and in human populations. At this particular stage in the development of the science of mutagenesis, quantitative extrapolation for chemical and food substances is a difficult task. This should be contrasted with the situation from radiation and induced mutation where more experimentation has been carried out and extrapolation is better understood. Hence the best that can be achieved at this stage is to use qualitative information from sub-mammalian systems and quantitative information from mammalian systems to provide an indication of the likely effects of the substance. On the basis of this prioritisation can precede attempts to control mutagenic agents.

References

1. T. Sugimura, 'Let's be Scientific About the Problems of Mutagens in Cooked Food', *Mutat. Res., 55* (1979), p. 149.

2. FAO/WHO. Expert Committee on Food Additives: Evaluation of the Carcinogenic Hazards of Food Additives. WHO Tech. Rep., Ser. No. 220 (1961).

3. D. Anderson, 'Sub-mammalian Tests Other than the Ames Test for Mutagenesis' in G.E. Paget (ed.), *Topics in Toxicology, Mutagenesis in Sub-mammalian Systems. Status and Significance* (MTP Press, Lancaster, UK, 1979), pp. 95–134.

4. D. Anderson, 'Short-term Tests for Detecting Chemical Mutagens and Recommendations for their Use in Monitoring Exposed Populations' in K. Berg (ed.), *Genetic Damage in Man Caused by Environmental Agents* (Academic Press Inc., 1979), pp. 383–404.

5. Scientific Committee, Food Safety Council. Genetic Toxicology, *Fd Cosmet. Toxicol., 16* (1978), p. 35.

6. P.G.N. Kramers, 'The Mutagenicity of Saccharin', *Mutat. Res., 32* (1975), p. 81.

7. J. Ashby, J.A. Styles and D. Anderson, 'Saccharin: an Epigenetic Carcinogen/Mutagen?', *Fd Cosmet. Toxicol., 16* (1978), p. 95.

8. B.N. Ames, J. McCann and E. Yamasaki, 'Methods for Detecting Carcinogens and Mutagens with the Salmonella/Mammalian Microsome Mutagenicity Test', *Mutat. Res., 31* (1975), p. 347.

9. T.R. Skopek, J.L. Liber, J.J. Krowleski and W.G. Thilly, 'Quantitative Forward Mutation Assay in *Salmonella typhimurium* Using 8-Azaguanine Resistance as a Genetic Marker', *Proc. Natl Acad. Sci. USA, 75* (1978), p. 410.

10. C. Pueyo, 'A Forward Mutation Assay Using a 1-Arabinose Sensitive Strain', *Am. Envir. Mutagen Soc., Abstracts* (Ab-2), (1978).

11. G. Mohn, '5-Methyltryptophan Resistance Mutations in *Escherichia coli* K-12. Mutagenic Activity of Monofunctional Alkylating Agents Including Organophosphorus Insecticides', *Mutat. Res., 20* (1973), p. 7.

12. D. Wild, 'Chemical Induction of Streptomycin-resistant Mutations in *Escherichia coli*. Dose and Mutagenic Effects of Dichlorvos and Methyl-methanesulphonate', *Mutat. Res., 19* (1973), p. 33.

13. H. Saedler, A. Gullon, L. Fiethen and P. Starlinger, 'Negative Conuol of the Galactose Operon in *Escherichia coli*', *Mol. Gen. Genet., 102* (1968), p. 79.

14. P. Moreau, A. Bailone, R. Devoret, 'Prophage Induction in *Escherichia coli* K-12, enVA uvrB: a Highly Sensitive Test for Potential Carcinogens', *Proc. Natl Acad. Sci. USA, 73* (1976), p. 3700.

15. W.T. Speck, R.M. Santella and H.S. Rosenkranz, 'An Evaluation of the Prophage Induction (Inductest) for the Detection of Potential Carcinogens', *Mutat. Res., 54* (1978), p. 101.

16. A. Carere, G. Morpurgo, G. Cardamone, M. Biganami, F. Aulicino, G. Di Guisepie and C. Conti, 'Point Mutations Induced by Pharmaceutical Drugs', *Mutat. Res., 29* (1975), p. 235.

17. R.K. Mortimer and T.R. Manney, 'Mutation Induction in Yeast' in A. Hollaender (ed.), *Chemical Mutagens, Principles and Methods for their Detection* (Plenum Press, New York, London, 1971), vol. 1, pp. 209–237.

18. F.K. Zimmerman, 'Detection of Genetically Active Chemicals Using Various Yeast Systems' in A. Hollaender (ed.), *Chemical Mutagens, Principles and Methods for their Detection* (Plenum Press, New York, London, 1973), vol. 3, pp. 209–237.

19. F.K. Zimmerman, 'Procedures Used in the Induction of Mitotic Recombination and Mutation in the Yeast *Saccharomyces cerevisiae*', *Mutat. Res., 31* (1975), p. 71.

20. N. Loprieno, 'Use of Yeast as an Assay System for Industrial Mutagens' in A. Hollaender and F.J. de Serres (eds.), *Chemical Mutagens, Principles and Methods for their Detection* (Plenum Press, New York, London, 1978), vol. 5, pp. 25–54.

21. D.J. Brusick, 'Induction of Cyclohexamide Resistant Mutants in *Saccharomyces cerevisiae* with *N*-methyl-*N*-nitro-*N*-nitrosoguanidine and ICR-170', *J. Bacteriol., 109* (1972), p. 1134.

22. T.M. Ong and F.J. de Serres, 'Mutagenicity of Chemical Carcinogens in *Neurospora crassa*', *Cancer Res., 32* (1972), p. 1890.

23. F.J. de Serres and H.V. Malling, 'Measurement of Recessive Lethal Damage Over the Entire Genome and at Two Specific Loci in the Ad-3 Region of a Two Component Heterokaryon of *Neurospora crassa*', in A. Hollaender (ed.), *Chemical Mutagens, Principles and Methods for their Detection* (Plenum Press, New York, London, 1971), vol. 2, pp. 311–341.

24. T.M. Ong, 'Use of the Spot, Plate and Suspension Test Systems for the Detection of the Mutagenicity of Environmental Agents and Chemical Carcinogens in *Neurospora crassa*. (1983).

25. J.A. Roper, 'Aspergillis', in A. Hollaender (ed.), *Chemical Mutagens, Principles and Methods for their Detection* (Plenum Press, New York, London, 1971), vol. 2, pp. 343–363.

26. W.R. Lee, 'Chemical Mutagenesis' in M. Ashburner and E. Nivotski (eds.), *Genetics and Biology of Drosophila* (Academic Press, 1976), vol. IC, pp. 1299–1341.

27. S. Abrahamson and E.B. Lewis, 'The Detection of Mutations in *Drosophila*' in A. Hollaender (ed.), *Chemical Mutagens, Principles and Methods for their Detection* (Plenum Press, New York, London, 1971), vol. 2, pp. 461–489.

28. P. Mollet and E.E. Würgler, 'Detection of Somatic Recombination and Mutation in *Drosophila.* A Method for Testing Genetic Activity of Chemical Compounds', *Mutat. Res., 25* (1974), p. 131.

29. P. Mollet and W. Weilenmann, 'Characteristics of a New Mutagenicity Test: Induction of Somatic Recombination and Mutation in *Drosophila* by Different Chemicals', *Mutat. Res., 38* (1976), p. 131.

30. E. Vogel, 'Identification of Carcinogens by Mutagen Testing in *Drosophila.* The Relative Reliability for the Kinds of Genetic Damage Measured', in H.H. Hiatt, D. Watson and J.A. Winsten (eds.), *Origins of Human Cancer* (Cold Spring Harbour Laboratory, New York, 1977), vol. C, pp. 1483–1497.

31. B.S. Baker, J.B. Body, A.T.C. Carpenter, M.M. Green, T.D. Nguyen, P. Tipoll and P.D. Smith, 'Genetic Controls of Meiotic Recombination and Somatic DNA Metabolism in *Drosophila melanogaster', Proc. Natl Acad. Sci. USA, 73* (1976), p. 4140.

32. J.P. Neill, P.A. Brimer, R. Machanoff, G.P. Hirsch and A.W. Hsie, 'A Quantitative Assay of Mutation Induction at the HGPRT Locus in Chinese Hamster Ovary Cells: Development and Definition of the System', *Mutat. Res., 45* (1977), p. 91.

33. E. Huberman and L. Sachs, 'Cell-mediated Mutagenesis of Mammalian Cells with Chemical Carcinogens', *Int. J. Cancer, 13* (1974), p. 26.

34. D.F. Krahn and C. Heidelberger, 'Liver Homogenate-mediated Mutagenesis in Chinese Hamster V79 Cells in Polycyclic 'Aromatic' Hydrocarbons and Aflatoxins', *Mutat. Res., 46* (1977), p. 27.

35. E.H.Y. Chu, P. Brimer, K.B. Jacobson and E.V. Merriam, 'Mammalian Cell Genetics I. Selection and Characterization of Mutations Auxotrophic for 1-glutamine or Resistance to 8-azaguanine in Chinese Hamster Cells *in vitro*', *Genetics, 62* (1976), p. 359.

36. M. Fox, J.M. Boyle and B.W. Fox, 'Biological and Biochemical

Characteristics of Purine Analogue Resistant Clones of V79 Chinese Hamster Cells', *Mutat. Res., 35* (1976), p. 219.

37. O. Nikaido and M. Fox, 'The Relative Effectiveness of 6-thioguanine and 8-azaguanine in Selecting Resistant Mutants from Two V79 Chinese Hamster Clones *in vitro*', *Mutat Res., 35* (1976), p. 279.

38. C.F. Arlett, 'Mutagenicity Testing with V79 Chinese Hamster Cells' in B.J. Kilbey, M. Legator, W. Nichols and C. Ramel (eds.), *Handbook of Mutagenicity Test Procedures* (Elsevier Biomedical Press, North Holland, Amsterdam, 1977), pp. 175–191.

39. R.E. Duncan and P. Brooks, 'Enzyme Activities in Extracts of 8-azaguanine Resistant Mutants of Cultured Chinese Hamster Cells Induced by the Carcinogen 7-bromomethyl benz(a)anthracene', *Mutat. Res., 26* (1974), p. 37.

40. F.D. Gillin, D.J. Roufa, A.L. Beaudet and T. Caskey, '8-azaguanine Resistance in Mammalian Cells. 1-hypoxanthine Guanine Phosphoribosyl Transferase', *Genetics, 72* (1972), p. 239.

41. B.J. Dean and K.R. Senner, 'Detection of Chemically Induced Mutation in Chinese Hamsters', *Mutat. Res., 46* (1977), p. 403.

42. L. Jacobs and R. de Mars, 'Chemical Mutagenesis with Diploid Human Fibroblasts' in B.J. Kilbey, M. Legator, W. Nichols and C. Ramel (eds.), *Handbook of Mutagenicity Test Procedures* (Elsevier Biomedical Press, North Holland, Amsterdam, 1977), pp. 193–220.

43. R. de Mars, 'Resistance of Cultured Human Fibroblasts and Other Cells to Purine and Pyrimidine Analogues in Relation to Mutagenesis Detection', *Mutat. Res., 24* (1974), p. 335.

44. W.G. Thilly, J.G. de Luca, I.V.H. Hoppe and B.W. Penmann, 'Mutation of Human Lymphoblasts by Methyl Nitrosourea', *Chem. Biol. Interact., 15* (1976), p. 33.

45. D.W. Clive and J.F.S. Spector, 'Laboratory Procedure for Assessing Specific Locus, Mutations at the TK Locus in Cultured L5178Y Mouse Lymphoma Cells', *Mutat. Res., 31* (1975), p. 17.

46. D.W. Clive, 'Recent Developments with the L5178Y TK Heterozygote Mutagen Assay System', *Environ. Hlth Perspect., 6* (1973), p. 119.

47. D.W. Clive, W.G. Flamm, M.R. Machesko and N.H. Bernheim, 'Mutational Assay System Using the Thymidine Kinase Locus in Mouse Lymphoma Cells', *Mutat. Res., 16* (1972), p. 77.

48. D. Anderson and M. Fox, 'The Induction of Thymidine and 1UdR Resistant Variants in P388 Mouse Lymphoma Cells by X-rays, UV and Mono- and Bifunctional Alkylating Agents', *Mutat. Res., 25* (1974), p. 107.

49. D. Anderson, 'The Selection and Induction of 5-iodo-2-deoxyuridine and Thymidine Variants of P388 Mouse Lymphoma Cells with Agents which are Used for Selection', *Mutat. Res., 33* (1975), p. 399.

50. M. Fox and D. Anderson, 'Induced Thymidine and 5-iodo-2-deoxyuridine Resistant Clones of Mouse Lymphoma Cells', *Mutat. Res., 25* (1974), p. 89.

51. J. Cole and C.F. Arlett, 'Ethyl Methanesulphonate Mutagenesis with L5178Y Mouse Lymphoma Cells. A Comparison of Ouabain, Thioguanine and Excess Thymidine Resistance', *Mutat. Res., 34* (1976), p. 507.

52. A.G.A.C. Knaap and J.W.I.M. Simons, 'A Mutational Assay System for L5178Y Mouse Lymphoma Cells Using Hypoxanthine Guanine Phosphoribosyl Transferase Deficiency as a Marker. The Occurrence of a Long Expression Time for Mutations Induced by X-rays and EMS', *Mutat. Res., 30* (1975), p. 97.

53. J. Nakamura, N. Susuki and S. Okada, 'Mutagenicity of Furylfuramide, a Food Preservative Tested by Using Alanine – requiring Mouse L5178Y Cells *in vitro* and *in vivo*', *Mutat. Res., 46* (1977), p. 355.

54. B.N. Ames, F.D. Lee and W.E. Durston, 'An Improved Bacterial Test

System for the Detection and Classification of Mutagens and Carcinogens', *Proc. Natl Acad. Sci. USA, 70* (1973), p. 782.

55. B.N. Ames, W.E. Durston, E. Yamasaki and F.D. Lee, 'Carcinogens are Mutagens: a Simple Test System Combing Liver Homogenates for Activation and Bacteria for Detection', *Proc. Natl Acad. Sci. USA, 70* (1973), p. 2281.

56. J. McCann, N.E. Spingarn, J. Kobori and B.N. Ames, 'The Detection of Carcinogens as Mutagens: Bacterial Tester Strains with R Factor Plasmids', *Proc. Natl Acad. Sci. USA, 72* (1975), p. 979.

57. D. Maron, J. Katzenellenbogen and B.N. Ames, 'Compatibility of Organic Solvents with the *Salmonella*/microsome test', *Mutat. Res., 88* (1981), pp. 343–50.

58. J. McCann and B.N. Ames, 'Detection of Carcinogens as Mutagens in the *Salmonella*/microsome Tests. Assay of 300 Chemicals. Part II', *Proc. Natl Acad. Sci. USA, 73* (1976), p. 950.

59. M.M. Coombs, C. Dixon and A.M. Kissonerghis, 'Evaluation of the Mutagenicity of Compounds of Known Carcinogenicity Belonging to the Benz(a)-anthracene, Chrysene and Cyclopenta(a)phenanthrene Series, Using Ames's Tests', *Cancer Res., 36* (1976), p. 4525.

60. I.F.H. Purchase, E. Longstaff, J. Ashby, J.A. Styles, D. Anderson, P.A. Lefevre and F.R. Westwood, 'Evaluation of Six Short-term Tests for Detecting Organic Chemical Carcinogens and Recommendations for their Use', *Nature, 264* (1976), p. 624.

61. E. Mohn, J. Ellenberger and D.B. McGregor, 'Development of Mutagenicity Tests Using *Escherichia coli* K-12 as an Indicator Organism', *Mutat. Res., 23* (1974), p. 187.

62. M.H.L. Green and W.J. Muriel, 'Mutagen Testing Using Tryp + Reversion in *Escherichia coli*', *Mutat. Res., 38* (1976), p. 3.

63. J.H. Evans and M.L. O'Riordan, 'Human Peripheral Blood Lymphocytes for the Analysis of Chromosome Aberrations in Mutagen Tests', *Mutat. Res., 31* (1975), p. 135.

64. M. Bignani, G. Morpurgo, R. Pagliani, A. Carere, G. Conte and G. Di Guiseppe, 'Non-disjunction and Crossing-over Induced by Pharmaceutical Drugs in *Aspergillus nidulans*', *Mutat. Res., 26* (1974), p. 159.

65. J.M. Parry, 'The Detection of Chromosome Non-disjunction in the Yeast *Saccharomyces cerevisiae*' in D. Scott, B.A. Bridges and F.H. Sobels (eds.), *Progress in Genetic Toxicology* (Elsevier Biomedical Press, North Holland, Amsterdam, 1977), pp. 223–229.

66. F.E. Würgler, F.H. Sobels and E. Vogel, '*Drosophila* as an Assay System for Detecting Genetic Changes' in B.J. Kilbey, M. Legator, W. Nichols and C. Ramel (eds.), *Handbook of Mutagenicity Test Procedures* (Elsevier Biomedical Press, North Holland, Amsterdam, 1977), pp. 335–375.

67. R. Fahrig, 'The Mammalian Spot-test. A Sensitive *In vivo* Method for the Detection of Genetic Alterations in Somatic Cells of Mice' in A. Hollaender and F.J. de Serres (eds.), *Chemical Mutagens, Principles and Methods for their Detection* (Plenum Press, New York, London, 1978), vol. 5,pp. 151–176.

68. W.L. Russel, 'Specific Locus Mutations in Mice', *Cold Spring Harbour Quant. Biol., 16* (1951), p. 327.

69. A.G. Searle, 'The Specific Locus Test in the Mouse', *Mutat. Res., 31* (1975), p. 277.

70. L.R. Valcovic and H.V. Malling, 'An Approach to Measuring Germinal Mutations in the Mouse', *Environ. Hlth Perspec., 6* (1973), p. 201.

71. J.V. Neel, 'Some Trends in the Study of Spontaneous and Induced Mutation in Man', *Genetics, 92*, No. 1, Part 1 (1977), p. S25.

72. J. Mays, J. McAninch, R'J. Feurs, J. Burkhart, H. Mohrenweiser and

D.A. Casciano, 'Enzyme Activity as a Genetic Marker to Detect Microlesions in Mammals', *Mutat. Res., 53* (1978), No. 1 Abstract 98.

73. M.F.W. Festing and G.L. Wolff, 'Quantitative Characters of Potential Value in Studying Mutagenesis', *Genetics, 92* No. 1 Part 1 (1977), p. S173.

74. U.H. Ehling, 'Dominant Mutations Affecting the Skeleton in Offspring of X-irradiated Male Mice', *Genetics, 54* (1966), p. 1381.

75. P.B. Selby, 'Induced Skeletal Mutations', *Genetics, 92* No. 1 (1972), p. 127.

76. H.I. Kohn, 'H Gene Mutations Induced by TEM in the Mouse', *Mutat. Res., 20* (1973), p. 235.

77. J.H. Roderick, 'Chromosomal Inversion in Studies of Mammalian Mutagenesis', *Genetics, 92* No. 1 Part 1 (1977), p. S121.

78. E.P. Evans and R.J.S. Philips, 'Inversion Hetrozygosity and the Origin of XO Daughters of Bpa/+ Female Mice', *Nature, 256* (1975), p. 40.

79. A.G. Searle, 'The Use of Pigment Loci for Detecting Reverse Mutations in Sometic Cells in Mice', *Arch. Toxicol., 38* (1977), p. 105.

80. W.W. Nichols, P. Moorhead and G. Brewen, 'Chromosome Methodologies in Mutation Testing', *Toxicol. Appl. Pharmacol., 22* (1972), p. 269.

81. M.S. Legator, K.A. Palmer and I.D. Adler, 'Collaborative Study *in vivo* Cytogenetic Analysis', *Toxicol. Appl. Pharmacol., 24* (1973), p. 337.

82. W. Schmid, D.T. Arakaki, N.A. Breslau and J.C. Culberston, 'Chemical Mutagenesis. The Chinese Hamster Bone Marrow as an *in vivo* Test System', *Humangenetik, 11* (1971), p. 103.

83. W. Schmid, 'Chemical Mutagen Testing on *in vivo* Somatic Mammalian Cells', *Agents and Actions, 3* (1973), p. 77.

84. J.H. Tjio and J. Whang, 'Chromosome Preparations of Bone Marrow Cells Without Prior *in vitro* Culture of *in vivo* Colchicine Administrations', *Stain Technol., 37* (1962), p. 17.

85. B.J. Dean, 'Chemical Induced Chromosome Damage', *Lab. Anim., 3* (1969), pp. 159–174.

86. J.A. Heddle, 'A Rapid *in vivo* Test for Chromosomal Damage', *Mutat. Res., 18* (1973), p. 187.

87. W. Schmid, 'The Micronucleus Test for Cytogenetic Analysis' in A. Hollaender (ed.), *Chemical Mutagens* (Plenum Press, New York, London, 1976), vol. 4, pp. 31–53.

88. P. Maier and W. Schmid, 'Ten Model Mutagens Evaluated by the Micronucleus Test', *Mutat. Res., 40* (1976), p. 325.

89. W. Schmid, 'The Micronucleus Test', *Mutat. Res., 31* (1975), p. 9.

90. A.J. Bateman, 'The Dominant Lethal Assay in the Mouse' in B.J. Kilbey, M. Legator, W. Nichols and C. Ramel (eds.), *Handbook of Mutagenicity Test Procedures* (Elsevier Biomedical Press, North Holland, 1977), pp. 325–335.

91. A.J. Bateman and S.S. Epstein, 'Dominant Lethal Mutations in Mammals' in A. Hollaender (ed.), *Chemical Mutagens: Principles and Methods for their Detection* (Plenum Press, New York, London, 1971), vol. 2, p. 541.

92. S.S. Epstein, 'The Use of the Dominant Lethal Test to Detect Genetic Activity of Environmental Chemicals', *Environ. Hlth Perspect., 6* (1973), p. 23.

93. S.S. Epstein and G. Röhrborn, 'Recommended Procedures for Testing Genetic Hazards from Chemicals Based on the Induction of Dominant Lethal Mutations in Mammals', *Nature, 230* (1971), p. 459.

94. S. Green, F.M. Moreland and W.E. Flamm, 'A New Approach to Dominant Lethal Testing', *Toxicol. Appl. Pharmacol., 39* (1977), p. 549.

95. S. Green and J.A. Springer, 'The Dominant Lethal Test. Potential Limitations and Statistical Considerations for Safety Evaluation', *Environ. Hlth Perspect., 6* (1973), p. 73.

96. N.L.A. Cachiero, L.B. Russell and M.S. Swartout, 'Translocation, the Predominant Cause of Total Sterility in Sons of Mice Treated with Mutagens', *Genetics, 76* (1974), p. 73.

97. W.M. Generoso, K.T. Cain and S.W. Huff, 'Inducibility by Chemical Mutagens of Heritable Translocations in Male and Female Germ Cells in Mice', in W.G. Flamm and M.A. Mehlman (eds.), *Advances in Modern Toxicology* (Halsted Press, 1978), vol. 5, pp. 109–129.

98. W.M. Generoso, K.T. Cain, S.W. Huff and D.G. Gosslee, 'Heritable Translocation Test in Mice' in A. Hollaender (ed.), *Chemical Mutagens: Principles and Methods of their Detection* (Plenum Press, New York, 1977), vol. 5, pp. 55–78.

99. A. Léonard, 'Observations on Meiotic Chromosomes of the Male Mouse as a Test of the Potential Mutagenicity of Chemicals' in A. Hollaender (ed.), *Chemical Mutagens: Principles and Methods for their Detection* (Plenum Press, New York, 1973), vol. 3, p. 21.

100. A. Léonard, 'Tests for Heritable Translocations in Male Mammals', *Mutat. Res., 31* (1973), p. 291.

101. R. Lang and I.D. Adler, 'Heritable Translocation Test and Dominant Lethal Assay in Male Mice with Methyl Methanesulphonate', *Mutat. Res., 48* (1977), p. 75.

102. J.E. Cleaver, 'Methods for Studying Excision Repair of DNA Damaged by Physical and Chemical Mutagens' in B.J. Kilbey, M. Legator, W. Nichols and C. Ramel (eds.), *Handbook of Mutagenicity Test Procedures* (Elsevier Biomedical Press, North Holland, Amsterdam, 1977), pp. 19–47.

103. J.E. Cleaver, 'Methods for Studying Repair of DNA Damaged by Physical and Chemical Carcinogens' in H. Busch (ed.), *Methods in Cancer Research* (Academic Press, New York, 1975), vol. 9, pp. 123–165.

104. E. Slater, M.D. Anderson and H.S. Rosenkranz, 'Rapid Detection of Mutagens and Carcinogens', *Cancer Res., 31* (1971), p. 970.

105. T. Kada, M. Morija and Y. Shirasu, 'Screening of Pesticides for DNA Interactions by REC-assay and Mutagenic Testing and Frameshift Mutagens Detected', *Mutat. Res., 26* (1974), p. 243.

106. D. Inchinotsubo, H.F. Mower, J. Setliff and M. Mandel, 'The Use of REC-Bacteria for Testing Carcinogenic Substances', *Mutat. Res., 46* (1977), p. 53.

107. J. Tanooka, 'Development and Applications of *Bacillus subtilis* Test System for Mutagens Involving DNA Repair Deficiency and Suppressible Auxotrophic Mutations', *Mutat. Res., 48* (1977), p. 367.

108. R.B. Kelly, M.R. Atkinson, J.A. Huberman and A. Kornberg, 'Excision of Thymine Dimers and Other Mismatched Sequences by DNA Polymerases of *E. coli*', *Nature, 224* (1969), p. 495.

109. H.F. Stitch, R.H.C. San and Y. Kawazoe, 'DNA Repair Synthesis in Mammalian Cells Exposed to a Series of oncogenic and non-oncogenic Derivatives of 4-nitroquinoline-1-oxide', *Nature, 229* (1971), p. 416.

110. H.F. Stich, R.H.C. San, P. Lam, J. Koropatnick and L. Lo, 'Unscheduled DNA Synthesis of Human Cells as a Short-term Assay for Chemical Carcinogens', in H.H. Hiatt, J.D. Watson and J.A. Winsten (eds.), *Origins of Human Cancer* (Cold Spring Harbor Laboratory, New York, 1977), vol. C, pp. 1499–1512.

111. D. Stoltz, L.A. Poirier, C.C. Irving, H.F. Stitch, J.H. Weisburger and H.C. Grice, 'Evaluation of Short-term Tests for Carcinogenicity', *Toxicol. Appl. Pharmacol., 29* (1974), p. 157.

112. G.A. Sega, 'Unscheduled DNA Synthesis in the Germ Cells of Male Mice Exposed *in vivo* to EMS', *Proc. Natl Acad. Sci. USA, 71* (1974), p. 4955.

113. J.A. Swenberg, G.L. Petzold and P.R. Harbach, '*In vitro* DNA Damage/Alkaline Elution Assay for Predicting Carcinogenic Potential', *Biochem. Biophys.*

Res. Commun., 72 (1976), p. 732.

114. P. Perry and H.J. Evans, 'Cytological Detection of Mutagen – Carcinogen Exposure by Sister Chromatid Exchange', *Nature, 258* (1975), p. 121.

115. D.G. Stetka and S. Wolff, 'Sister Chromatid Exchange as an Assay for Genetic Damage Induced by Mutagens-carcinogens. I. *In vivo* Test for Compounds Requiring Metabolic Activation', *Mutat. Res., 41* (1976a), p. 333.

116. D.G. Stetka and S. Wolff, 'Sister Chromatid Exchange as an Assay for Genetic Damage Induced by Mutagens-carcinogens. II. *In vitro* Test for Compounds Requiring Metabolic Activation', *Mutat. Res., 41* (1976b), p. 343.

117. M.S. Legator, S. Zimmering and T.H. Connor, 'The Use of Indirect Indicator Systems to Detect Mutagenic Activity in Human Subjects and Experimental Animals' in A. Hollaender (ed.), *Chemical Mutagens* (Plenum Press, New York, London, 1976), vol. 4, pp. 171–191.

118. P.E. Nute, W.G. Wood, G. Stammatoyannopoulos, C. Olivery and P.J. Failkow, 'The Kenya Form of Hereditary Persistence of Foetal Haemoglobin: Structural Studies and Evidence for Homogeneous Distribution of Haemoglobin F Using Fluorescent Anti-haemoglobin F Antibodies', *Br. J. Haematol., 32* (1976), p. 55.

119. R.A. Popp, G.P. Hirsch and B.S. Bradshaw, 'Amino Acid Substitution: Its Use in Detection and Analysis of Genetic Variants', *Genetics, 92*, Part 1, (1979), p. S39.

120. A.J. Wyrobek and W.R. Bruce, 'Chemical Induction of Sperm Abnormalities in Mice', *Proc. Natl Acad. Sci. USA, 72* (1975), p. 4425.

121. J. Topham, 'The Detection of Carcinogen – Induced Sperm Head Abnormalities in Mice', *Mutat. Res., 69* (1980), p. 149.

122. J. Topham, 'Chemically Induced Transmissable Abnormalities in Sperm Head Shape', *Mutat. Res., 70* (1980), p. 109.

123. The Working Group of the Sub-Committee on Environmental Mutagenesis. Approaches to Determining the Mutagenic Properties of Chemicals: Risk to Future Generations. L.R. Valcŏvic DHEW-NIH-NIEHS. Prepared for the Department of Health, Education and Welfare Committee to Co-ordinate Toxicology and Related Programmes (1976).

124. K. Sargernt, A. Sheridan, J. O'Kelly and R.P.A. Carnagha, 'Toxicity Associated with Certain Samples of Grounduts', *Nature, 192* (1961), p. 1095.

125. C.W. Hesseltine, 'Natural Occurrence of Mycotoxins in Cereals', *Mycopathol. Mycol. Appl., 53* (1974), p. 141.

126. J.J. Wong and D.P.H. Hsieh, 'Mutagenicity of Aflatoxins Related to their Metabolism and Carcinogenic Potential', *Proc. Natl Acad. Sci. USA, 73* (1976), p. 2241.

127. J. McCann, E. Choi, E. Yamasaki and B.N. Ames, 'Detection of Carcinogens as Mutagens in the *Salmonella*/Microsome Test: Assay of 300 Chemicals', *Proc. Natl Acad. Sci. USA, 72* (1975), p. 5135.

128. M.H. Kuczak, P.M. Benson, H. Heath and A.W. Hayes, 'Evaluation of the Mutagenic Potential of Mycotoxins Using *Salmonella typhimurium* and *Saccharomyces cerevisiae*', *Mutat. Res., 53* (1978), p. 11.

129. T. Ong, 'Mutagenicity of Aflatoxins in *Neurospora crassa*', *Mutat. Res., 9* (1970), p. 615.

130. T.M. Ong and F.J. de Serres, 'Mutagenicity of Chemical Carcinogens in *Neurspora crassa*', *Cancer Res., 32* (1972), p. 1890.

131. E. Vogel and F.H. Sobels, 'The Function of *Drosophila* in Genetic Toxicology Testing' in A. Hollaender (ed.), *Chemical Mutagens, Principles and Methods for their Detection* (Plenum Press, New York, London, 1976), vol. 4, pp. 93–142.

132. H. Umeda, T. Tsutsui and M. Saito, 'Mutagenicity and Inducibility of

DNA Single Strand Breaks and Chromosome Aberrations by Various Mycotoxins', *Gann, 68* (1972), p. 619.

133. S.S. Epstein and H. Schafner, 'Chemical Mutagens in the Human Environment', *Nature, 219* (1968), p. 385.

134. W.H. Butler, M. Greenblatt and W. Lijinsky, 'Carcinogenesis by Aflatoxin B_1 G_1 and B_2', *Cancer Res., 29* (1969), p. 2206.

135. G.N. Wogan, G. Edwards and P.M. Newkerne, 'Structure Activity Relationship in Toxicity and Carcinogenicity of Aflatoxins and Analogs', *Cancer Res., 31* (1971), p. 1936.

136. J.M. Essigmann, R.G. Croy, A.M. Nadzan, W.F. Busky, V.N. Beinhold, G. Buchi and G.N. Wogan, 'Structural Identification of the Major DNA Adduct Formed by Aflatoxin B_1 *in vitro*', *Proc. Natl Acad. Sci. USA, 74* (1977), p. 1870.

137. M. Manabe and L. Tsuruta, 'Mycological Damage of Domestic Brown Rice During Storage in Warehouse Under Natural Conditions', *Trans. Mycol. Soc. Jpn, 16* (1975), p. 399.

138. M. Nagao and T. Sugimura, 'Environmental Mutagens and Carcinogens', *Ann. Rev. Genet., 12* (1978), p. 117.

139. J.J. Wong, R. Singh and D.P.H. Hsieh, 'Mutagenicity of Fungal Metabolites Related to Aflatoxin Biosynthesis', *Mutat. Res., 44* (1972), p. 447.

140. Y. Ueno and K. Kubota, 'DNA Attaching Ability of Carcinogenic Mycotoxins in Recombination – Deficient Mutant Cells of *Bacillus subtilis*', *Cancer Res., 36* (1976), p. 445.

141. E.G. Daniels, R.B. Kelly and J.W. Hinmann, 'Agaritine: An Improved Isolation Procedure and Confirmation of Structure by Synthesis', *J. Am. Chem. Soc., 83* (1961), p. 3333.

142. B. Toth, D. Nagel, K. Patil, J. Erickson and K. Antonson, 'Tumour Induction with the *N'*-acetyl Derivatives of 4-hydroxymethylphenylhydrazine, a Metabolite of Agaritine of *Agaricus bisporus*', *Cancer Res., 38* (1978), p. 177.

143. B. Toth, 'Synthetic and Naturally Occurring Hydrazines as Possible Cancer Causative Agents', *Cancer Res. 35* (1975), p. 3693.

144. A.V. Wright, A. Niskanen and H. Pyysalo, 'The Toxicities and Mutagenic Properties of Ethylidene Gyromitrin and *N'*-methylhydrazine with *Escherichia coli* as Test Organisms', *Mutat. Res., 56* (1977), p. 105.

145. B. Toth, 'Hydrazine, Methylhydrazine and Methylhydrazine Sulphate Carcinogenesis in Swiss Mice: Failure of Ammonium Hydroxide to Interfere in the Development of Tumours', *Int. J. Cancer, 9* (1972), p. 109.

146. B. Toth and H. Shimizu, 'Methyl Hydrazine Tumorigenesis in Syrian Golden Hamster and the Morphology of Malignant Histiocytomas', *Cancer Res., 33* (1973), p. 2744.

147. A.M. Pamukcu, E. Etrück, S. Yalciner, U. Milli and G.T. Bryan, 'Carcinogenic and Mutagenic Activities of Milk from Cows Fed Bracken Fern (*Pteridium aquilinum*)', *Cancer Res., 38* (1978), p. 1556.

148. I.A. Evans and M.A. Osman, 'Carcinogenicity of Bracken and Shikimic Acid', *Nature, 250* (1974), p. 343.

149. I.A. Evans and J. Mason, 'Carcinogenic Activity of Bracken', *Nature, 208* (1965), p. 913.

150. I.A. Evans, R.S. Jones and R. Mainwaring-Burton, 'Passage of Bracken Fern Toxicity into Milk', *Nature, 237* (1972), p. 107.

151. I.A. Evans, 'The Bracken Carcinogen in Chemical Carcinogens', *American Society Monograph, 173* (1976), p. 690.

152. I. Hirono, K. Fushimi and N. Matsubara, 'Carcinogenicity Test of Skikimic Acid in Rats', *Toxicol. Lett., 1* (1977), p. 9.

153. J.P. Brown, 'A Review of the Genetic Effects of Naturally Occurring

Flavonoids, Anthraquinones and Related Compounds', *Mutat. Res., 75* (1980), p. 243.

154. H. Matsumoto, A. Shirai, M. Sawamura, T. Matshushima and T. Sugimura, 'Mutagenicity of Cycasin and its Derivatives', *Proc. Jpn Cancer Assoc.*, 35th Ann. Meet. (1976), p. 33.

155. G.L. Laqeuer, 'Carcinogenic Effects of Cycad Meal and Cycasin, Methyl-Azoxymethanol Glycoside in Rats and Effects of Cycasin in Germgree Rats', *Fed. Proc., 23* (1964), p. 1386.

156. L.B. Bull, C.C.J. Culvenor and A.T. Dick, *The Pyrrolizidine Alkaloid* (Elsevier, North Holland, Amsterdam, 1968).

157. A.M. Clark, 'The Mutagenic Activity of Some Pyrrolizidine Alkaloids in *Drosophila*', *Z. Vererbungls, 91* (1960), p. 74.

158. T. Alderson and A.M. Clark, 'Interlocus Specificity Chemical Mutagens in *Aspergillus nidulans*', *Nature, 210* (1966), p. 593.

159. IARC. Monographs on the *Evaluation of Carcinogenic Risk of Chemicals to Man* (Int. Agency Res. Cancer, Lyon, 1976), vol. 10, pp. 333–43.

160. R. Schoental, 'Carcinogens in Plants and Micro-organisms', in C. Searle (ed.), *Chemical Carcinogens* Am. Chem. Soc. (Washington DC, 1976), pp. 629–89.

161. A.A. Haridgree and J.L. Epler, 'Mutagenicity of Plant Flavonols in Microbial Systems', *Mutat. Res., 53* (1978), p. 89.

162. L.F. Bjeldanes and G.W. Chang, 'Mutagenic Activity of Quercetin and Related Compounds', *Science, 197* (1977), p. 577.

163. J.P. Brown, P.S. Dietrich and R.T. Brown, 'Frameshift Mutagenicity of Certain Naturally Occuring Phenolic Compounds in the *Salmonella*/microsome Test. Activation of Anthraquinone and Flavonol Glycoxides by Gut Bacterial Enzymes', *Biochem. Soc. Trans., 5* (1977), p. 1489.

164. K. Hermann, 'Flavonols and Flavons in Pod Plants. A Review', *J. Food Technol., 11* (1976), p. 433.

165. Y. Seino, M. Nagao, T. Yahagi, T. Sugimura, T. Yasuda and S. Nishimira, 'Identification of a Mutagenic Substance in a Spice Sumac, as Quercetin', *Mutat. Res., 58* (1978), p. 225.

166. J.L. Dorange, P. Janiaud, M. Delarfirge, P. Levi and P. Padieu, 'Comparative Study of Microsomal Activation Systems for Mutagenic Studies of Safrole', *Abstr. 2nd Int. Conf. Environ. Mutagens, 128* (1977).

167. IARC Monographs on the *Evaluation of Carcinogenic Risk of Chemicals to Man* (Int. Agency Res. Cancer, Lyon, 1962), vol. 1, p. 169.

168. T. Yahagi, M. Nagao, K. Hara, T. Matsushima, T.Sugimura and G.T. Bryan, 'Relationships Between the Carcinogenic and Mutagenic or DNA Modifying Effects of Nitrofuran Derivatives, Including 2-(2-furyl)-3-(5-nitro-2-furyl) acrylamide, a Food Additive', *Cancer Res., 34* (1974), p. 2266.

169. S. Kondo and H. Ichikawa-Ryo, 'Testing and Classification of Mutagenicity of Furylfuramide in *Escherichia coli*', *Jpn J. Genet., 48* (1973), p. 295.

170. T. Kada, '*Escherichia coli* Mutagenicity of Furyl Furamide', *Jpn J. Genet., 48* (1973), p. 301.

171. A. Tonomura and M.S. Sasaki, 'Chromosome Aberration and DNA Repair Synthesis in Cultured Human Cells Exposed to Nitrofurans', *Jpn J. Genet., 48* (1973), p. 291.

172. T. Yahagi, T. Matsushima, M. Nagao, Y. Seino, T. Sugimura and G.T. Bryan, 'Mutagenicity of Nitrofuran Derivatives on a Bacterial Tests Strain with an R Factor Plasmid', *Mutat. Res., 40* (1976), p. 9.

173. Y. Ideda, S. Horiuchi, T. Furuija, K. Suzuki and J. Azagami, 'Interim Report. Induction of Gastric Tumours in Mice by Feeding of Furyl Furamide. Food Sanitation Study', Council Ministry of Health and Welfare, Japan (1974).

174. T. Nomura, 'Carcinogenicity of the Food Additive Furyl Furamide in Foetal and Young Mice', *Nature, 258* (1975), p. 610.

175. K. Yokoro, H. Kajihara, Y. Kodama, K. Nagao, K. Hamada and A. Kinomura, 'Chronic Toxicity of 2-(2-furyl)-3-(5 nitro-2-furyl) acrylamide (AF-2) in Mice with Special Reference to Carcinogenicity in the Forestomach', *Gann, 68* (1977), p. 825.

176. T. Sano, T. Kawachi, N. Matsukura, K. Sasajima and T. Sugimura, 'Carcinogenicity of a Food Additive, AF-2 in Hamster and Mice', *Z. Krebsforsch, 89* (1977), p. 61.

177. S. Tackayama and N. Kuwabara, 'The Production of Skeletal Muscle Atrophy and Mammary Tumours in Rats by Feeding 2-(-2-furyl)-3-(5-nitro-2-furyl) acrylamide', *Toxicol. Lett., 1* (1977), p. 11.

178. Y. Taxima, T. Kada and A. Murakami, 'Mutagenicity of Nitrofuran Derivatives Including Furylfuramide, a Food Preservative', *Mutat. Res., 32* (1975), p. 55.

179. R.P. Batzinger, S.Y.L. Ou and E. Bueding, 'Saccharin and Other Sweeteners. Mutagenic Properties', *Science, 198* (1977), p. 944.

180. D.R. Stoltz, B. Stavric, R. Klassen, R.D. Bendall and J. Craig, 'The Mutagenicity of Saccharin Impurities. 1 Detection of Mutagenic Activity', *J. Environ. Pathol. Toxicol., 1* (1977), p. 139.

181. D.E. Stone, E. Lamson, Y.S. Chang and K.W. Pickering, 'Cytogenetic Effects of Cyclamates on Human Cells *in vitro*', *Science, 164* (1969), p. 568.

182. T. Tokumitsu, 'Some Aspects of Cytogenetic Effects of Sodium Cyclamate on Human Leucocytes *in vitro*', *Proc. Japan. Acad., 47* (1971), p. 635.

183. S. Green, K.A. Palmer and M.S. Legator, '*In vitro* Cytogenetic Investigation of Calcium Cyclamate, Cyclohexylamine and Triflupromazine', *Fd Cosmet. Toxicol., 8* (1970), p. 617.

184. D.R. Stoltz, K.S. Khera, R. Bendall and S.W. Gunner, 'Cytogenetic Studies with Cyclamate and Related Compounds', *Science, 167* (1970), p. 1501.

185. M.S. Legator, K.A. Palmer, S. Green and K.W. Petersen, 'Cytogenetic Studies in Rats of Cyclohexylamine a Metabolite of Cyclamate', *Science, 165* (1969), p. 1139.

186. A.G.A.C. Knapp, G. Pieter, N. Kramers and F.H. Sobels, 'Lack of Mutagenicity of the Cyclamate Metabolites in *Drosophila*', *Mutat. Res., 21* (1973), p. 341.

187. R.C. Garner and C.A. Nutman, 'Testing of Some Azo Dyes and Their Reaction Products for Mutagenicity Using *Salmonella typhimurium* TA1538', *Mutat. Res., 44* (1977), p. 9.

188. T. Kada, K. Tutkawa and Y. Sadaie, '*In vitro* and Host-mediated "Rec-assay" Procedures for Screening Chemical Mutagens: and Phloxine, a Mutagenic Red Dye, Detect', *Mutat. Res., 16* (1972), p. 165.

189. R. Münzner, 'Mutagenicity Testing of the Urine of Rats Treated with Amaranth', *Fd Cosmet. Toxicol., 17* (1979), p. 563.

190. K.A. Palmer, C.W. Sheu and S. Green, 'Mutagenicity Studies of R-amino Salt, a Metabolite of Amaranth (FD & C Red No. 2) in Mouse Lymphoma Cells Heterozygous at the Thymidine Kinase Locus and in the Rat Dominant Lethal Test', *Fd Cosmet. Toxicol., 17* (1979), p. 1.

191. F. Kodama, M. Umeda and T. Tsutsui, 'Mutagenic Effect of Sodium Nitrite on Cultured Mouse Cells', *Mutat. Res., 40* (1976), p. 119.

192. H. Tsuda and K. Kato, 'High Rate of Endoreplications and Chromosomal Aberrations in Hamster Cells Treated with Sodium Nitrite *in vitro*', *Mutat. Res., 56* (1977), p. 69.

193. L. Knudsen and O.A. Meyer, 'Mutagenicity Studies in Rats and Mice

Given Canned, Heated, Nitrite-treated Pork', *Mutat. Res., 56* (1977), p. 177.

194. H.F. Stich, J. Karim, J. Koropatnick and L. Lo, 'Mutagenic Action of Ascorbic Acid', *Nature (Lond.), 260* (1976), p. 722.

195. S.M. Galloway and R. Painter, 'Vitamin C is Positive in the DNA Synthesis Inhibition and Sister-chromatid Exchange Tests', *Mutat. Res., 60* (1979), p. 321.

196. W.D. MacRae and H.F. Stich, 'Induction of Sister Chromatid Exchanges in Chinese Hamster Cells by the Reducing Agents Bisulfite and Ascorbic Acid', *Toxicology, 13* (1979), p. 167.

197. Y. Shirasu, M. Moriya, K. Kato, F. Lienard, H. Tezuka, S. Teramoto and T. Kada, 'Mutagenicity Screening of Pesticides in the Microbial System', *Mutat. Res., 40* (1977), p. 19.

198. D. Wild, 'Mutagenicity Studies of Organophosphorus Insecticides', *Mutat. Res., 32* (1975), p. 133.

199. E. Vogel, 'Identification of Carcinogens by Mutagen Testing in *Drosophila*. The Relative Reliability for the Kinds of Genetic Damage Measured' in H.H. Hiatt, J.D. Watson and J.A. Winsten (eds.), *Origins of Human Cancer* (Cold Spring Habour NY, Cold Spring Harbour Lab., 1977), pp. 1483–97.

200. B.A. Bridges, 'The Mutagenicity of Captan and Related Fungicides', *Mutat. Res., 32* (1975), p. 3.

201. Food Surveillance Paper No. 2. Survey of Vinyl Chloride Content of Polyvinyl Chloride for Food Contact and of Foods (Ministry of Agriculture, Fisheries and Food, 1978).

202. H. Bartsch and R. Montesano, 'Mutagenic and Carcinogenic Effects of Vinyl Chloride', *Mutat. Res., 32* (1975), p. 93.

203. L. Fishbein, 'Industrial Mutagens and Potential Mutagens I. Halogenated Aliphatic Derivatives', *Mutat. Res., 32* (1976), p. 267.

204. Y. Rannug, A. Johansson, C. Ramel and C.A. Washmeister, 'The Mutagenicity of Vinyl Chloride After Metabolic Activation', *Ambio, 3* (1974), p. 194.

205. H. Greim, E. Bonse, Z. Radwan, D. Reichert and D. Henschler, 'Mutagenicity *in vitro* and Potential Carcinogenicity of Chlorinated Ethylenes as a Function of Oxirane Formation', *Biochem. Pharmacol., 24* (1975), p. 2013.

206. I.F.H. Purchase, C.R. Richardson and D. Anderson, 'Chromosomal and Dominant Lethal Effects of Vinyl Chloride', *Lancet, ii* (1975), p. 410.

207. J.P. Brown, G.W. Roehm and R.J. Brown, 'Mutagenicity Testing of Certified Food Colours and Related Azo, Xanthene and Triphenylmethane Dyes with the *Salmonella*/microsome System', *Mutat. Res., 5* (1978), p. 249.

208. D.H. Fine, D.P. Rounbehler, T. Fan and R. Ross 'Human Exposure to *N*-nitroso Compounds in the Environment' in H.H. Hiatt, J.D. Watson and J.A. Winsten (eds.), *Origins of Human Cancer* (Cold Spring Harbour NY, Cold Spring Harbour Lab., 1977), p. 309.

209. T. Page, R.H. Harris and S.S. Epstein, 'Drinking Water and Cancer Mortality', *Louisiana Science, 193* (1976), p. 55.

210. M. Nagao and T. Sugimura, 'Mutagenesis: Microbial System' in P.O.P. ts'O, and H.V. Gelboin (eds.), *Polycyclic Hydrocarbons and Cancer* (Academic Press, New York, 1978), vol. 2, pp. 19–121.

211. R.H. Harris, T. Page and N.A. Reiches, 'Carcinogenic Hazards of Organic Chemicals in Drinking Water' in H.H. Hiatt, J.D. Watson and J.A. Winsten (eds.), *Origins of Human Cancer* (Cold Spring Harbour NY, Cold Spring Harbour Lab., 1977), p. 309.

212. T. Wlodrawski and H.S. Rosenkranz, 'Mutagenicity of Sodium Hypochlorite for *Salmonella typhimurium*', *Mutat. Res., 31* (1975), p. 39.

213. M. Nagao, M. Honda, Y. Seino, T. Yahagi and T. Sugimura,

'Mutagenicites of Smoke Condensates and the Charred Surface of Fish and Meat', *Cancer Lett., 2* (1977), p. 221.

214. W. Lijinsky and P. Shubik, 'Benzo(a)pyrene and Other Polynuclear Hydrocarbons in Charcoal-Broiled Meat', *Science, 145* (1964), p. 53.

215. T. Matsumoto, D. Yoshida, S. Mizusaki and H. Okamoto, 'Mutagenic Activity of Amino Acid Pyrolysates in *Salmonella typhimurium* TA98', *Mutat. Res., 48* (1977), p. 279.

216. M. Nagao, T. Yahagi, T. Kawachi, Y. Seino, M. Hondo, N. Matsukura, T. Sugimura, K. Wakabayashi, K. Tsuji and T. Kosuge, 'Mutagens in Foods and Especially Pyrolysis Products of Protein' in D. Scott, B.A. Bridges and F.H. Sobels (eds.), *Progress in Genetic Toxicology* (Elsevier, North-Holland, Amsterdam, 1977), pp. 259–265.

217. P.N. Magee and J.M. Barnes, 'Carcinogenic Nitroso Compounds', *Adv. Cancer Res., 10* (1967), p. 163.

218. T. Yahagi, M. Nagao, Y. Seino, T. Matsushima, T. Sugimura and M. Okada, 'Mutagenicities of *N*-nitrosamines on *Salmonella*', *Mutat. Res., 48* (1977), p. 121.

219. K. Lee, B. Gold and S.S. Mirvish, 'Mutagenicity of 22 *N*-nitrosamides and Related Compounds for *Salmonella typhimurium* TA1535', *Mutat. Res., 48* (1977), p. 131.

220. T.K. Rao, A.A. Hardigree, J.A. Young, W. Lijinsky and J.L. Epler, 'Mutagenicity of *N*-nitrosopiperidines with *Salmonella typhimurium*/microsomal Activation System', *Mutat. Res., 56* (1977), p. 131–45.

221. J. Sander and G. Burkle, 'Induction of Malignant Tumours in Rats by Simultaneous Feeding of Nitrite Secondary Amines', *Z. Krebsforsch., 73* (1969), p. 54.

222. W. Lijinsky and S.S. Epstein, 'Nitrosamines as Environmental Carcinogens', *Nature, 225* (1970), p. 21.

223. C. Bhashkaram and G. Sadasivan, 'Effects of Feeding Irradiated Wheat to Malnourished Children', *Am. J. Clin. Nutr., 28* (1975), p. 130.

224. Vijayalaxmi, 'Cytogenetic Studies in Monkeys Fed Irradiated Wheat', *Toxicology, 9* (1978), p. 181.

225. Vijayalaxmi, 'Cytogenetic Studies in Rats Fed Irradiated Wheat', *Int. J. Radiat. Biol., 27*, No. 3 (1975), p. 283.

226. Vijayalaxmi and G. Sadasivan, 'Chromosomal Aberrations in Rats Fed Irradiated Wheat', *Int. J. Radiat. Biol., 27*, No. 2, (1975), p. 135.

227. Vijayalaxmi and K. Visweswara Rao, 'Dominant Lethal Mutations in Rats Fed on Irradiated Wheat', *Int. J. Radiat. Biol. 29*, No. 1 (1979), p. 93.

228. A.S. Aiyar and V. Subba Rao, 'Studies on Mutagenicity of Irradiated Sugar Solutions in *Salmonella typhimurium*', *Mutat. Res., 48* (1977), p. 17.

229. D. Anderson, M.J.L. Clapp, M.C.E. Hodge and T.M. Weight, 'Irradiated Laboratory Animal Diets: Dominant Lethal Studies in the Mouse', *Mutat. Res., 80* (1981), pp. 333–345.

230. B. Commoner, A.J. Vithayathil and Dolera Piero, 'Mutagenic Analysis as a Means of Detecting Carcinogens in Foods', *J. Fd Protect., 41*, No. 21, (1978), p. 996.

231. K. Umezawa, A. Shirai, T. Matsushima and T. Sugimura, 'Co-Mutagenic Effect of Norharman and Harman with 2-acetyl aminofluorene Derivatives', *Proc. Natl Acad. Sci. USA, 75* (1978), p. 928.

232. T. Kada, K. Morita and T. Inoue, 'Antimutagenic Action of Vegetable Factor(s) on the Mutagenic Principle of Trypotophan Pyrolysate', *Mutat. Res., 53* (1978), p. 351.

233. L. Chui-Nan, M.A. Butler and T.S. Matney, 'Antimutagenic Activities of Common Vegetables and Their Chlorophyll Content', *Mutat. Res., 77* (1980), p. 245.

234. J. Timson, 'How Harmful is Your Daily Caffeine?', *New Scientist, 78* (1978), p. 1107.

235. R.D. Coombes and R.B. Haveland-Smith, 'A review of the genotoxicity of food, drug and cosmetic colours and other azo, triphenylmethane and xanthene dyes', *Mutat. Res., 98* (1982), 101–248.

236. T. Sugimura, 'Mutagens, Carcinogenesis and Tumour promotors in our daily food', *Cancer, 49* (1982), 1970–1984.

237. M. Nagao, Y. Takahashi, K. Wakabayashi, T. Sugimura, 'Mutagenicity of alcoholic beverages', *Mutat. Res., 88* (1981) 147–54.

238. Mazaki Masanori, Ishii Takao and Uyeta Masahike (Kochi), 'Mutagenicity of hydrolysates of citrus fruit juices', *Mutat. Res., 101* (1982), 283–91.

239. M. Green, E. Ben Hur, E. Riklis, S. Gordin and I. Rosenthal, 'Application of Mutagenicity Test for Milk', *J. Dairy Sci., 63* (1980), 358–61.

240. M. Nagao, Y. Takahashi, H. Yamanaka, and T. Sugimura, 'Mutagens in Coffee and Tea', *Mutat. Res., 68* (1979), 101–6.

241. M. Uyeta, S. Taue and M. Mazaki, 'Mutagenicity of hydrolysates of tea infusions', *Mutat. Res., 88* (1981), 233–40.

242. D. Stoltz, B. Stavric, D. Krewski, R. Klassen, R. Bendall and B. Junkins, 'Mutagenicity Screening of Foods. 1. Results with Beverages', *Environmental Mutagenesis, 4* (1982), 477–92.

243. E. Crocke, M. Eckhardt, T. King and D. Wild, 'Mutagenicity study of fried sausages in Salmonella, Drosophila and Mammalian cells *in vitro* and *in vivo*', *Mutat. Res., 101* (1982), 293–304.

244. P. Dolaro, 'Beef extract mutagens and rat liver S9, increase mutation frequency in the Ames Test', *Mutat. Res., 104* (1982), 345–6.

245. J.Y. Lin, H. Lee and H.I. Huang, 'Formation of mutagens in boiled pork extract', *Fd. Chem. Toxic., 20* (1982), 531–3.

246. J. Sekizawa and T. Shibamoto, 'Genotoxicity of Safrole – related chemicals in microbial test systems', *Mutat. Res., 101*, 127–40.

247. M. Ungsurungsie, O. Suthienkul and C. Paovalo, 'Mutagenicity screening of popular Thai spices', *Fd. Chem. Toxic., 20* (1982), 527–30.

248. L. Rossi and J.B.H. von Lierop, 'Repeatability and reproducibility of determinants of vinly chloride in foodstuffs', *Fd. Chem. Toxic., 20* (1982), 603–10.

250. H.V. Aesbacher, 'The significance of mutagens in food', *Abstracts 12th European Environmental Mutagen Society*, (1982), p. 73.

251. R. Fahrig, 'Effects of food additives in the Mammalian Spot test', *Abstracts 12th European Environmental Mutagen Society*, (1982), p. 74.

252. B. Debuyst and Van Larabeke, 'Induction of sister chromatid exchanges in human lymphocytes by Aldicarb, thiofanox and methomyl', *Abstracts 12th European Environmental Mutagen Society*, (1982), p. 75.

253. D.J. Brusick, V.F. Simmon, H.S. Rosenkranz, V.A. Ray and R.S. Stafford, 'An Evaluation of the *Escherichia coli* WP2 and WP2 uvrA Reverse Mutation Assay', *Mutat. Res., 76* (1980) 169–90.

254. W.M. Generoso, J.B. Bishop, D.G. Gosslee, G.W. Newell, Ching-Ju Sheu and E.S. von Halle, 'Heritable Translocation Test in Mice', *Mutat. Res., 76* (1980), 191–215.

255. A.W. Hsie, D.A. Casciano, D.B. Couch, D.F. Krahn, J.P. O'Neill and B.L. Whitfield, 'The use of Chinese Hamster ovary cells to quantify specific locus mutation and to determine mutagenicity of chemicals: a report of the Gene-Tox program', *Mutat. Res., 86* (1981), 193–214.

256. M.O. Bradley, B. Bhuyan, M.C. Francis, R. Langenbach, A. Peterson and E. Huberman, 'Mutagenesis by chemical agents in V-79 Chinese Hamster cells: a review and analysis of the literature, a report of the Gene-Tox program', *Mutat.*

Res., *87* (1981) 81–142.

257. R.J. Preston, W. Au, M.A. Bender, J.G. Brewen, A.V. Carrano, J.A. Heddle, A.F. McFee, S. Wolff and J.S. Wassom, 'Mammalian *in vivo* and *in vitro* cytogenetic assays: a report of the Gene-Tox program', *Mutat. Res.*, *87* (1981), 143–88.

258. Z. Leifer, T. Kada, M. Mandel, E. Zeiger, R. Stafford and H.S. Rosenkranz, 'An evaluation of tests using DNA repair-deficient bacteria for predicting genotoxicity and carcinogenicity: a report of the US EPA's Gene-Tox program', *Mutat. Res.*, *87* (1981), 211–97.

259. L.B. Russell, P.B. Selby, E. Von Hallé, W. Sheridan and L. Valcovic, 'The mouse specific-locus test with agents other than radiations: interpretation of data and recommendations for future work', *Mutat. Res.*, *86* (1981) 329–54.

260. L.B. Russell, P.B. Selby, E. Von Hallé, W. Sheridan and L. Valvolic, 'Use of the mouse spot test in chemical mutagenesis: Interpretation of past data and recommendations for future work', *Mutat. Res.*, (1981), 355–79.

261. S.A. Latt, J. Allen, S.E. Bloom, A. Carrano, E. Falke, D. Kram, E. Schneider, R. Schreck, R. Tice, B. Whitfield and S. Wolff, 'Sister-chromatid exchanges: a report of the Gene-Tox program', *Mutat. Res.*, *87* (1981), 17–62.

262. E. Kafer, B.R. Scott, G.L. Dorn and R. Stafford, 'Aspergillus nidulans: systems and results of tests for chemical induction of mitotic segregation and mutation. I. Diploid and duplication assay systems: a report of the US EPA Gene-Tox program', *Mutat. Res.*, *98* (1982), 1–148.

263. B.R. Scott, G.L. Dorn, E. Kafer and R. Stafford, 'Aspergillus nidulans: systems and results of tests for induction of mitotic segregation and mutation. II. Haploid assay systems and overall response of all systems: a report of the US EPA Gene-Tox program', *Mutat. Res.*, *98* (1982), 49–94.

264. M.S. Legator, E. Bueding, R. Batzinger, T.H. Connor, E. Eisenstadt, M.G. Farrow, G. Ficsor, A. Hsie, J. Seed and R.S. Stafford, 'An evaluation of the host-mediated assay and body fluid analysis: a report of the US Environmental Protection Agency Gene-Tox program', *Mutat. Res.*, *98* (1982), 319–74.

265. K.H. Larsen, D. Brash, J.E. Cleaver, R.W. Hart, V.M. Maher, R.B. Painter and G.A.Sega, 'DNA repair assays as tests for environmental mutagens: a report of the US EPA Gene-Tox program', *Mutat. Res.*, *98* (1982), 287–318.

266. M.J. Constantin and R.A. Nilan, 'The chlorophyll-deficient mutatant assay in barley (Hordeum vulgare): a report of the US environmental protection agency Gene-Tox program', *Mutat. Res.*, *99* (1982), 37–49.

267. M.J. Constantin and R.A. Nilan, 'Chromosome aberration assays in barley (Hordeum vulgare): a report of the US Environmental Protection Agency Gene-Tox program', *Mutat. Res.*, *99* (1982), 13–36.

268. M.J. Constantin and E.T. Owens, 'Introduction and perspectives of plant genetic and cytogenetic assays: a report of the US environmental protection agency Gene-Tox program', *Mutat. Res.*, *99* (1982), 1–12.

269. Te-Hsiu Ma (Macomb), 'Vicia cytogenetic tests for environmental mutagens. A report of the US Environmental Protection Agency Gene-Tox program', *Mutat. Res.*, *99* (1982), 257–71.

270. W.F. Grant (Ste. Anne de Bellevue), 'Chromosome aberration assays in Allium. A report of the US Environmental Protection Agency Gene-Tox program', *Mutat. Res.*, *99* (1982), 273–91.

271. Te-Hsiu Ma (Macomb), 'Tradescentia cytogenetic tests (root-tip mitosis, pollen mitosis, pollen mother-cell meiosis). A report of the US Environmental Protection Agency Gene-Tox program', *Mutat. Res.*, *99* (1982), 293–302.

272. J. Van't Hof and L.A. Schairer (Upton), 'Tradescantia assay system for gaseous mutagens. A report of the US Environmental Protection Agency

Gene-Tox program', *Mutat.Res., 99* (1982), pp. 303–15.

273. M.J. Plewa (Urbana), 'Specific-locus mutation assays in Sea mays. A report of the US Environmental Protection Agency Gene-Tox program', *Mutat. Res., 99* (1982), 317–37.

274. B.K. Vig (Reno), 'Soybean (Glycine max [L.] merrill) as a short-term assay for study of environmental mutagens. A report of the US Environmental Protection Agency Gene-Tox program', *Mutat. Res., 99* (1982), 339–47.

275. H.F. Stich, M.P. Rosin, C.H. Wu and W.D. Powrie, 'Mutagenicity testing to valuate food' in J.A. Heddle (ed.), *Mutagenicity – New Horizons in Genetic Toxicology* (Academic Press, New York), pp. 117–42.

276. F.J. de Serres and Ashby, 'Evaluation of short-term tests for carcinogens', *Progress in Mutation Res., Vol. 1*, Report of the International Collaborative programme (Elsevier, North Holland), 1981.

6 THE GASTROINTESTINAL TRACT IN FOOD TOXICOLOGY

I.R. Rowland and R. Walker

Introduction

In the context of food toxicology the gastrointestinal tract is of major importance in that it is the primary site of exposure to potential toxicants and the concentrations involved are likely to be higher than those in other tissues. This can lead to localised lesions due to irritation or to more specific toxic effects. Furthermore, dietary components may undergo metabolism in the gut by endogenous mammalian enzymes or by the gut microflora and this may significantly influence the toxicological sequelae. Pre-existing enteropathies or enzyme deficiencies can alter the course of both these events.

Role of Gut Microflora in Intestinal Metabolism

The Gastrointestinal Flora

The characterisation of the gastrointestinal floras of animals and man is a very complex and difficult task, and presents the investigator with several problems. Firstly sampling the gut contents, especially the human gut, can be difficult. Secondly, the predominant types of bacteria in the mammalian gut are oxygen sensitive and require relatively sophisticated techniques for their isolation and subcultures.[1,2] Thirdly, enormous numbers of species of bacteria are found in the gut (over 400)[3] many of which are difficult and tedious to identify.

Major Types of Bacteria in the Mammalian Gut. The bacteria of the mammalian gut can be divided into transient and indigenous types although often it can be difficult to distinguish the two. The subject has been extensively discussed by Savage,[4] who has described several criteria that should be fulfilled before an organism isolated from the gut can be considered indigenous to that area of the gut. It must be noted that an organism indigenous to one area of the alimentary tract may be a transient component of another region, e.g. bacteria indigenous to the mouth can be washed down into the stomach where they may

Table 6.1: Major Groups of Micro-organism Found in Human Intestine (Healthy Adults)

Family and genera	Common species	Description
Bacteridaceae		
Bacteroides	*B. fragilis* ss *vulgatis* ss *thetaiotaomicron* ss *distasonis* *B. melaninogenicus*	Gram negative rods strictly anaerobic, non-spore forming.
Fusobacterium	*F. prausnitzii* *F. fusiforme*	
Propionobacteriaceae		
Propionobacterium	*P. acnes*	Gram positive rods, strictly anaerobic, non-sporing.
Eubacterium	*E. aerofaciens* *E. rectale*	
Lactobacillaceae		
Bifidobacterium	*B. adolescentis* *B. infantis*	Gram positive branched rods, strictly anaerobic non-sporing.
Peptostreptococcus	*P. productus*	Gram positive cocci strictly anaerobic.
Lactobacillus	*L. leishmanii* *L. acidophilus* *L. catenaforme* *L. salivarius*	Gram positive non-sporing rods strictly anaerobic or microaerophilic
Streptococcus	*S. faecalis* *S. faecium*	Gram positive cocci facultatively anaerobic.
Enterobacteriaceae		
Escherichia *Klebsiella* *Proteus*	*E. coli*	Gram negative rods facultatively anaerobic.
Neisseriaceae		
Veillonella	*V. parvula*	Gram negative cocci strictly anaerobic.
Micrococcaceae		
Peptococcus *Sarcina*		Gram positive anaerobic cocci.
Staphylococcus		Gram positive facultatively anaerobic cocci.
Bacillaceae		
Clostridium	*C. leptum*	Gram positive rods, spore forming, strictly anaerobic.
Yeast		
Candida	*C. albicans*	

survive for short periods but are unable to become established in that region.

Table 6.1. lists the major groups of bacteria indigenous to the normal human gut with examples of the most prominent genera and species within the groups. This subject has been reviewed in detail by Brown.[5] Somewhat paradoxically, the most extensively studied of the intestinal organisms (because of their clinical importance and their ease of culture), namely the enterobacteria, streptococci and staphylococci, are numerically the least important members of the flora.

The anaerobes have, until recently, been seriously neglected partly because of the difficulty in isolation and culture and partly because of the enormous diversity of types and lack of simple methods of distinguishing these types. Even today the taxonomy of certain groups especially the Gram positive non-sporing rods is the subject of considerable controversy. Holdeman and Moore[1] have published a comprehensive scheme for the identification of anaerobes.

Distribution of Bacteria Along the Gastrointestinal Tract: Interspecies Differences.

General. There are large interspecies differences in the numbers and types of bacteria in the gut and also in the distribution of organisms along the alimentary canal, reflecting morphological and physiological differences in the various regions of the gut in different animal species.

Mouth and Saliva. The mouth harbours a complex flora probably due to the variety of ecological niches present and saliva contains facultative and strict anaerobes including streptococci, *Neisseria, Veillonella*, bacteroides, lactobacilli and yeasts.[6]

Stomach. Although considerable numbers and types of microorganism pass into the stomach in the saliva and food, in man the low pH (1-2) of the gastric juice quickly kills most of the bacteria. Immediately after a meal up to 10^5 organisms/g are found, but the count declines concomitantly with the decrease in pH and after about 1 h, as the pH falls to 2-3, the numbers of bacteria are only $10\text{–}10^3$/g.[6] Thus it is clear that the human stomach has a transient flora but it should be noted that even minor changes in gastrointestinal physiology e.g. slight rise in pH of gastric juice can result in increased numbers of bacteria (see section on disorders of the gastrointestinal tract below).

In contrast to the human and the rabbit, which also secretes gastric juices of very low pH, the stomach of the mouse and rat harbours a complex flora dominated by lactobacilli, bacteroides and bifidobacteria.[6,8] The most likely explanation of this colonisation of the rodent stomach is the higher pH of the gastric contents (pH 4–5 in the rat, Rowland, unpublished observation).

Figure 6.1: Distribution of Bacteria in Gastrointestinal Tracts of Various Laboratory Animals and Man. ●, man; ○, rat; △, mouse; ▲, rabbit; □, guinea pig.

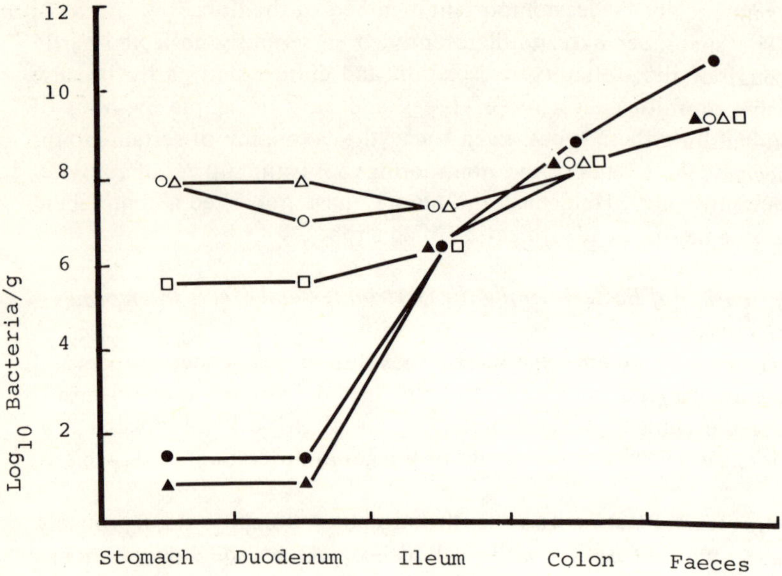

Duodenum and Jejunum, and Ileum. In the human, the extent of colonisation of the small intestine increases with the distance from the stomach rising from 10^2–10^5 organisms/g in the jejunum[5,6,9] to 10^6–10^8/g in the distal ileum (Figure 6.1).

Various factors may be involved in limiting the bacterial populations of the proximal ileum, such as gastric acid, limited availability of nutrients and perhaps most importantly, peristaltic movement. It is of note that in people inhabiting tropical areas of the world such as India and South America, the duodenum and ileum harbour a much larger and more diverse flora.[10,11] However, it is not known whether this reflects physiological differences between tropical residents and those in temperate regions or nutritional status and health of the individual.[12–14]

With the exception of the rabbit most laboratory animals harbour an abundant flora in the proximal ileum although usually in slightly lower numbers than in the stomach.[6,8] (Figure 6.1.)

Caecum, Large Intestine and Faeces. In virtually all laboratory animals and man the large intestine and faeces contain the highest numbers of bacteria (Figure 6.1; Table 6.2).

The predominant components of the faecal floras of various laboratory animals and man are in most cases non-sporing anaerobic genera such as *Bacteroides, Eubacterium, Bifidobacterium, Fusobacterium, Peptostreptococcus* and *Propionobacterium* which constitute over 90 per cent of the total faecal flora. In recent studies *Bacteroides fragilis* has been found to be the major species comprising 22–52 per cent of the total viable count in man.[15–17] Thus although over 400 species of bacteria have been identified from the human gut, the majority of the Gram negative organisms can be assigned to one species and with the Gram positive non-sporing anaerobes, the majority belong to five types *Bifidobacterium adolescentis, Eubacterium aerofaciens, E. rectale, Peptostreptococcus productus* and *Ruminococcus bromii*.[18]

There is considerable variation both quantitative and qualitative[15] in faecal flora between individuals and a certain amount of daily variation although the latter is much less than the former. Such variations have also been seen in the rat gut flora.[19]

There are, however, marked quantitative and qualitative differences between the gut floras of some animals and man. For example, the large intestine of the rabbit is devoid of lactobacilli and has very few enterobacteria by comparison to the human gut flora (Table 6.2). The baboon and dog differ from most other species of laboratory animal in having a gut flora which is not dominated by bacteroides, instead large numbers (about 10^{11}/g) of lactobacilli are found and, in the dog, facultative anaerobes such as *streptococci* and *staphylococci* form a significant proportion of the flora.[20,21] The rat gut flora is similar to that of the human in being dominated by *Bacteroides fragilis*, with the subspecies *vulgatus* and *thetaiotaomicron* the most common.[19]

In the mouse, however, although the strictly anaerobic bacteria are predominant, the species found are different from those in the human gut and, most notably, *Bacteroides fragilis* is not found.

In rats and mice, bacteria are found intimately attached to the intestinal mucosa virtually throughout the gastrointestinal tract. These organisms include various spiral-shaped bacteria,[22,23] fusiform organisms, and a species of yeast.[4,24] In the human, bacteria have been found associated with the colon and caecum walls only.[25]

In summary it is clear from Figure 6.1 that there are large quantitative differences in gut flora between man and several laboratory animals. These differences in bacterial flora between different regions

Table 6.2: Predominant Organisms in Large Intestine or Faeces of Various Animals and Man

	Log$_{10}$ no. of organisms/g intestinal contents or faeces							
	Man (faeces)[15–17]	Mouse[58–60]	Rat[19,49]	Dog[20]	Baboon[21]	Rabbit[60]	Guinea pig[60]	Pig[60]
Bacteroides	10.0–11.2	9–10	7–9	9–10	10.3	7–9	7–9	0–8
Eubacterium	10.3–10.8	9–11	ND	10	10.5	ND	ND	ND
Bifidobacterium	10.0–11.0	ND	10.1	10–11	NP	7–9	-9	ND
Fusobacterium	10.2–10.3	9–11	9.9	-8	9.9	ND	ND	ND
Peptostreptococcus	9.6–10.5	9	10.9	9–10	10.1	ND	ND	ND
Propionobacterium	8.6–10.1	9	ND	-7	NP	ND	ND	ND
Lactobacillus	7.0–9.7	8–10	8–5	10–11	11	3	ND	6–9
Streptococcus	4.0–9.2	6–8	9.6	9–10	10.4	2–4	3–4	6–8
Clostridium	5.0–10.0	7–8	7.0	10–11	NP	-3	-4	3–6
Enterobacteria	4.0–8.4	4–6	6.9	6	ND	-2	-3	5–8
Staphylococcus	-8	-2	-3	-9	ND	–	-4	ND
Veillonella	-6	-6	10.2	-8	ND	–	-4	ND
Total	10.0–11.5	9–11	11.0	11	11.3	9–10	9–10	8–9

ND = Not determined; NP = Not present (below detection limit).
Numbers shown are the range of means of the viable count.
– number indicates a range of values from below the limit of detection to the number.

of the gut should be considered in conjunction with the sites of absorption of nutrients and anutrients in the gut,[6] especially when studying the metabolism of a xenobiotic substance either *in vitro* or *in vivo* in a laboratory animal. The large numbers of bacteria in the rodent stomach and small intestine mean that a substance will have a greater chance of being modified by the gut flora before being absorbed than in the human, where a compound may well be absorbed in the duodenum and jejunum before it ever reaches the regions of the gut heavily populated by micro-organisms. It should be remembered however, that gastrointestinal disorders which result in bacterial colonisation of the stomach and upper small intestine in man are by no means uncommon, and may modify the metabolism of ingested compounds.

Factors Affecting Flora

Age. The mammalian gut is sterile at birth, but is rapidly colonised by bacteria from the mother's skin, vaginal and faecal microflora and from air and food.

In the human, the type of flora which becomes established depends on whether the baby is breast fed or bottle fed. In breast-fed infants, a fairly stable flora develops within the first week consisting predominantly of bifidobacteria at 10^8–10^9/g faeces (99 per cent of total viable count), with coliforms, enterococci and lactobacilli contributing the remaining 1 per cent of the flora.[26,27] Organisms such as bacteroides and clostridia are rarely found. In contrast, in bottle-fed babies, the faecal pH is higher than in breast-fed children and the faeces have a different microflora consisting of bacteroides, *Streptococcus faecium* and *E. coli* each at about 10^8 organisms/g.[27] The stools contain amines, ammonia and a wide spectrum of bile acids.[26] When an infant is changed from breast milk to mixed feeding the counts of bifidobacteria fall steadily, numbers of *E. coli* and *S. faecium* rise to 10^8/g as do clostridia and bacteroides[27] and a flora qualitatively similar to that of the adult becomes established at 4–12 months.[28] Over the period 13–48 months most of the faecal organisms reach numbers similar to those of the adult flora except for lactobacilli which may take up to eight years to reach adult levels.

Disorders of Gastrointestinal Tract. A wide variety of gastrointestinal disorders (Table 6.3) are accompanied by bacterial colonisation of the stomach and/or duodenum and proximal ileum – regions which in the healthy human are almost sterile.[29] The presence of bacteria in the upper intestinal tract can have profound metabolic, clinical and

Table 6.3: Some Disorders, Conditions and Diseases Which Influence Gastrointestinal Flora

Condition	Change in flora	References
Polya partial gastrectomy Blind loop syndrome in small intestine caused by strictures Acute intestinal obstruction Small intestinal diverticula	Change from virtually sterile conditions in stomach, jejunum and duodenum to a complex flora qualitatively similar to the faecal flora, i.e. strict anaerobes, lactobacilli, enterobacteria Numbers 10^4–10^7 organisms/ml	29
Gastric operations	Slight increase in gastric and jejunal viable count	29
Crohns disease	Colonisation of jejunum and duodenum (10^4–10^7 organisms/ml by *E. coli, P. acnes*, streptococci) *C. albicans*	61
Achlorhydria	*E. coli, S. faecalis* and throat commensals found in stomach and jejunum, sometimes faecal-type flora found in stomach	62 63
Tropical Sprue	Jejunal mucosa colonised by anaerobes, *Alcaligines, Enterobacter, Hafnia*. No anaerobes. *E. coli*. occasionally isolated	64
Protein-calorie malnutrition	Stomach and jejunum colonised	13
Chronic diarrhoea	Increase in facultative anaerobes in faeces. Numbers of strict anaerobes depressed	
Ulcerative colitis	Increase in streptococci and reduction in bifidobacteria	65
Acute diarrhoea	Enteric pathogen occurs in large numbers in faeces. Reduction in numbers of strict anaerobes	66 67

toxicological sequelae such as anaemia, vitamin deficiency, weight-loss, steatorrhoea, diarrhoea, methaemoglobinaemia caused by nitrate reduction, and nitrosamine formation (p. 239). Many of these clinical consequences are attributable to the interference by the bacteria in the absorption of nutrients and vitamins by the intestine.

Diet. Studies on the influence of diet on the gut flora in general fall into two categories: (a) comparisons of the faecal floras of people inhabiting different areas of the world; and (b) investigations of the effect on faecal flora of different diets in the same individuals or population groups. Many of these investigations have been conducted to determine factors involved in cancer of the large bowel and rectum (see page 210).

Several investigations of the faecal floras of populations in different regions of the world have revealed significant differences in the number and proportions of types present. Much of the work has concentrated on comparisons of the developed western nations (USA and Western Europe) and underdeveloped countries such as Uganda and India (Table 6.4).

Table 6.4: Faecal Flora of People Inhabiting Different Countries

	Log_{10} no. of organisms/g faeces Country			
Bacterial group	England	India	Uganda	Japan
Bacteroides	9.8	9.2	8.2	9.4
Gram+ve non-sporing anaerobes	9.9	9.6	9.3	9.7
Clostridia	5.7	5.7	5.1	5.6
Lactobacilli	6.5	7.6	7.2	7.4
Enterococci	5.8	7.3	7.0	8.1
Enterobacteria	7.9	9.9	8.0	9.4

Source: From Aries *et al.*[30] and Drasar.[31]

The predominant bacteria in all the populations studied were the non-sporing anaerobes, but the western populations had thirty times more bacteroides than the Ugandans, although the differences between the western and Indian or rural Japanese populations were not so marked[30,31] (Table 6.4). The facultative anaerobes and lactobacilli were in lower numbers in the faeces of the western population. There are obvious and large differences in diet between the groups studies by Aries *et al.*[30] and Drasar[31] notably the English diet (in common with that of most western countries) is rich in animal protein and fat whereas the other population groups had diets high in carbohydrate, and hence the gut flora differences may well be attributable to differences in diet. However there is clearly a variety of other environmental and cultural factors which could be involved as well as genetic differences between the groups. In an attempt to minimise the other environmental and genetic factors, Crowther *et al.*[32] studied the faecal floras of various economic groups in Hong Kong. Members of the highest income group had higher counts of bacteroides (although similar counts of facultative anaerobes) than the lower economic groups. The former group was said to have a more 'westernised' diet than the lower economic group.

In a recent study, Moore and Holdeman[33] investigated the faecal bacteria of Japanese–Hawaiians, North Americans, rural Japanese and Africans and were unable to confirm the results of Aries *et al.*[30] mainly due to the complexity of the gut flora and the large degree of individual variation. Moore and Holdeman[33] found that in the rural Japanese and Africans, the range of predominant species in each individual was small which they suggest may be due to a more uniform intestinal physiology, reflecting a more routine existence and less varied diet. Although by comparison a more heterogeneous faecal flora was found in the North American individuals, the average composition of the floras of the native populations was similar to that of the American group.

Attempts to alter the human faecal flora by dietary changes have generated conflicting data. For example, several studies have been performed using no-residue, completely absorbable diets with radically different results. Winitz *et al.*[34] using chemically defined diets found that the flora changed dramatically after only short periods on the diet. However, Attebury *et al.*[35] using an almost identical no-residue diet to that of Winitz *et al.*[34] found few changes in flora and the total anaerobic count remained constant at 10^{11} organisms/g after transfer of the subjects to the experimental diet despite an increase in transit time (1–2 days on control diet, 5–7 days on the defined diet) and a large reduction in bulk of the stools. Attebury *et al.*[35] also reported that patients on starvation diets for extreme obesity and who have only one bowel movement each month, possess a faecal flora similar to their pre-diet flora. Other workers[33,36,37] have also reported only minor changes in faecal flora of individuals consuming a largely absorbable diet.

As might be expected more moderate alterations to the diet have little effect on the faecal flora. For example the gut floras of strict vegetarians or of people fed a vegetarian diet experimentally, were very similar to those of people eating a mixed diet.[38]

An investigation of the influence of dietary fibre on the faecal flora revealed no differences in the numbers and types of bacteria in faeces of subjects on high and low fibre diets.[39] Furthermore, although the predominant species of anaerobe was different for each subject, its predominance was not affected by changing the fibre content of the diet.

Recently, interest in the influence of diet on gut flora has centred on the effect of diets high in animal protein and fat, which are thought to be related to the incidence of colon cancer.[40] Conflicting results have emerged when attempts have been made to change the flora by altering the protein and/or fat content of the diet. Reddy *et al.*[41] reported that the consumption of a diet rich in animal protein and fat – a 'mixed western diet', increased numbers of bacteroides, bifidobacteria, peptococci and lactobacilli by comparison with their numbers in the subjects consuming a non-meat diet. This increase was reflected in a rise in β-glucuronidase[42] and an increase in the faecal excretion of bacterially modified bile acids such as deoxycholic acid[41] (see p. 210).

In contrast, in a similar study in which subjects were given a high-beef diet, a non-meat diet, and a control mixed diet,[43] no significant differences were found in total counts of facultative or strict anaerobes on the different diets and only occasional differences in species of bacteria occurred. The discrepancies between the two studies may be related to differences in food consumed even though both diets were described as high-meat or high-beef. For example Reddy *et al.*[41] maintained the protein content of the diet constant (although changing its source from vegetable to animal) and altered the fat content, both in total amount and in extent of saturation. Whereas in the study by Hentges *et al.*[43] the fat content was kept fairly constant and the protein varied. Cummings *et al.*[44] also reported no significant changes in numbers of faecal bacteria when a low-animal-fat diet was replaced by one high in animal fats and cholesterol but of otherwise similar composition.

The effects of dietary alterations on the floras of laboratory animals has not been extensively studied although Chung *et al.*[45] showed that rats fed a meat diet had larger numbers of facultative anaerobes and clostridia in their faeces than rats given a standard laboratory diet.

In general, it would appear that even major changes in diet do not greatly influence the numbers and types of bacteria in human faeces. This suggests that the human faecal flora is an extremely stable ecosystem with a high degree of organisation in its bacterial composition. This is not unexpected since evolutionary processes would tend to eliminate from the gut ecosystem organisms which could not easily co-exist. The resistance of the gut flora to change may also be attributable to the constant environment in the lower regions of the gut. Virtually all dietary nutrients are absorbed in the upper region of the human gut, so it is likely that the bacteria in the colon obtain their nutrients from desquamated mucosal cells, intestinal mucin and other mammalian secretions and breakdown products. Unless a change in diet radically altered the intestinal physiology, few changes in faecal flora would be expected. Poorly absorbed anutrients which reach the lower gut may however cause changes in the microflora of that region, for example saccharin (see p. 236). In addition, the abundant stomach and duodenal floras in the mouse and rat may be highly sensitive to changes in diet.

Although the numbers of the major types of organisms in the gut do not alter greatly after dietary changes, there is evidence that metabolic changes do occur, e.g. in cyclamate metabolism (p. 237). Thus conventional methods of identifying bacteria may not be relevant to investigations of dietary changes, and enzymatic parameters should be studied. For example, Goldin and Gorbach[46] showed that the activity of β-glucuronidase, nitroreductase and azoreductase in faeces could be increased up to 2.5-fold by changing rats from a grain-based to a meat-based diet. In addition, Wise *et al.* (unpublished observation) demonstrated an increase in nitrate reductase activity in rats given a pectin-containing semi-synthetic diet by comparison with animals given a fibre-free diet.

Metabolism by the Gut Flora

Methodology

Comparison of Oral versus Parenteral Administration. Some clues to the extent of the involvement of intestinal enzymes in the metabolism of a compound are obtained by comparing the rates or products of metabolism when the compound is administered orally and by another route, intravenous or intraperitoneal. It should be noted however, that the above technique does not distinguish bacterial metabolism from metabolism by mammalian enzymes in the gut and that more importantly, giving a substance intraperitoneally or intravenously does not by-pass the gut entirely, since the substance can be excreted into the gut in the bile.

Use of Germ-free or Antibiotic-treated Animals. Animals lacking a gut flora can be obtained by antibiotic treatment or by germ-free isolation. The use of antibiotics is a very convenient means of obtaining a 'pseudo' germ-free animal and does not necessitate the use of sophisticated germ-free isolators. Various antibiotics can be used, for example a mixture of bacitracin, neomycin and streptomycin in the drinking water eliminates the gut flora of the rat within three or four days and can maintain the 'germ free' status for several weeks until the gut becomes contaminated with a yeast or antibiotic-resistant bacteria.[47]

Care must be taken to ensure that the compound studied docs not react or bind to the antibiotics and it must be remembered that the antibiotics themselves may interfere with the physiology and metabolic processes of the animal, either directly or via their effect on the gut flora.

Germ-free animals can be employed to investigate gut flora metabolism in a manner similar to antibiotic-treated animals, except of course, the former can be used for extended periods up to the lifetime of the animal which enable carcinogenicity studies to be performed e.g. with cycasin (see. p. 228). However, any differences in the metabolism or toxicity of a compound in conventional and germ-free animals must be interpreted against the background of the many physiological disparities between the two types of animal. The absence of microbial activity in the gut of germ-free animals renders the pH and especially E_h higher than in animals with an intact flora.[48,49] In germ-free rodents the caecum is grossly enlarged and there are morphological differences in the intestinal mucosa. Furthermore, the absence of bacteria results in a slower turnover of mucosal cells in the small intestine which could influence mucosal enzyme activity in the gut. In addition, the transport of several low-molecular-weight compounds such as sugars and vitamins is higher in germ-free animals[50] and there are significant changes in hepatic enzymes.[51]

In vitro Methods

Suspensions of gut contents. A specimen of gut contents or faeces, usually collected under an oxygen-free gas to preserve strict anaerobes, is suspended in a suitable medium and incubated with the test compound under oxygen-free CO_2 or N_2. Various modifications of this basic procedure can be used and several more sophisticated apparatuses are extant.[52,53]

Although *in vitro* methods of studying metabolism by the gut flora are extensively used and have yielded important information, there are limitations to this type of study. For example the transfer of bacteria from the gut ecosystem to an *in vitro* situation alters a multitude of biotic and abiotic conditions such as pH, E_h nutrient availability and the influence of the host, which could significantly affect the metabolic activity of the flora. Also the conditions in this type of culture are constantly changing leading to a succession of bacterial types and hence a continuous change in overall metabolic activity.

It must also be noted, that unless a suspension of gut contents is fractionated it will contain a great many other sources of enzymes such as intestinal mucosal cells and soluble enzymes of mammalian origin.

Continuous culture of gut micro-organisms. The indigenous microbial population in the intestinal tract is at a steady state, i.e. the numbers of bacteria remain more or less constant because their rate of multiplication balances their outflow from the gut. Thus the gut can be considered as a continuous culture system or chemostat. The rumen flora[54] and, more recently, the rat caecal flora[19] have been modelled using a chemostat and this may provide a more realistic means of investigating metabolism by the gut flora. Indeed in a preliminary study, a high degree of similarity was found in the enzymatic activity of bacteria in the gut and the simulated flora in a chemostat (Mallett and Rowland, unpublished results).

Pure cultures. A further step removed from the *in vivo* ecosystem in the study of metabolism is the use of pure cultures of strains of bacteria isolated from the gut, nevertheless such studies can be extremely useful in the investigation of specific reactions, for example the investigation of steroid dehydrogenation by clostridia from human faeces.[55]

Cell-free extracts. Methods have also been developed to prepare cell-free extracts of the intestinal flora which have been used to study the metabolism by the gut flora of azo-dyes to mutagenic products.[56,57]

A final factor to be considered when attempting to extrapolate information gained from *in vitro* experiments to the intact animal is the site of absorption of the substance being studied from the intestine and the relationship of this site to the regions of the gut where metabolism is likely to occur. Such considerations are especially important in man, who possesses very sparse populations of bacteria in the stomach and upper small intestine where absorption of many compounds occurs. It is only in the lower regions of the healthy human alimentary canal that sufficient bacteria are present to effect metabolism.

Metabolic Transformations Effected by the Gut Microflora

Introduction. The importance of the gut microflora in the metabolism of orally ingested compounds has been recognised by nutritionists for very many years but it is only in about the last 15 years that any systematic study of the metabolism of xenobiotics by intestinal micro-organisms has been undertaken. Largely due to the pioneering work by Scheline[68,69] and Williams,[70] it is now clear, however, that the metabolic potential of the microbial inhabitants of the mammalian gastrointestinal tract is comparable with that of the liver and considerably more diverse. Furthermore, metabolism by the gut flora is not only important for orally ingested compounds since compounds entering the body by other routes (or their metabolites) may enter the gut via the bile.

The many types of molecular transformations which are effected by the gut microflora have been reviewed comprehensively by Scheline[69] and, with some additions, these are summarised in Table 6.5.

Table 6.5: Metabolism by Intestinal Microflora

Hydrolyses	
1.	Glucuronides
2.	Other glycosides
3.	Amides
4.	Ethereal sulphates
5.	Sulphamates
6.	Organic nitrates
Reductions	
1.	–CH=CH– bonds
2.	Nitro-compounds
3.	Azo compounds
4.	*N*-oxides, *N*-hydroxy-compounds
5.	Carbonyl compounds
6.	Alcohols, phenols (dehydroxylations)
7.	Arsonic acids
8.	Inorganic nitrate (to nitrite)
9.	Reductive dehalogenation
Degradations	
1.	Decarboxylation
2.	*O*-dealkylation
3.	*N*-dealkylation
4.	Dealkylation of organometallic compounds
5.	Deamination
6.	Decomposition of nitrosamines
7.	Heterocyclic ring fission
Syntheses	
1.	Esterification
2.	Acetylation
3.	Formation of nitrosamines
4.	Alkylation of metals
5.	Hydration
Miscellaneous	
1.	Aromatisation
2.	Isomerisation
3.	Oxidation

Hydrolyses. Hydrolysis of glucuronides is one of the best known examples of intestinal microbial metabolism and one of the most significant. Many compounds are excreted in the bile conjugated with glucuronic acid and are thus subjected to the hydrolytic activity of the gut microflora. These include endogenous compounds e.g. thyroxine,[71] plant constituents, e.g. ferulic acid,[72] terpene alcohols,[73] food additives, e.g. *p*-hydroxybenzoic acid[74] and drugs, e.g. morphine,[75] diethylstilboestrol,[76] chloramphenicol,[77] salicylic acid.[74] Intestinal microorganisms have been shown to be involved in the hydrolysis of glucuronide conjugates of alcohols, phenols, carboxylic and *N*-hydroxy compounds.

Glucuronidase activity is found mainly in the heavily populated regions of the gut while activity is low in the stomach of most species except ruminants.[78] Comparative experiments in conventional and germ-free animals indicate that the glucuronidase activity of intestinal tissue is much less significant than the microbial activity *in vivo*.[79,80]

Glycosides other than glucuronides may also be hydrolysed in the gut and many plant glycosides which are ingested in the normal diet are hydrolysed in this way, sometimes, as in the case of cycasin or cyanogenic glycosides, with toxic consequences (see p. 228).

The ability of the gut microflora to hydrolyse amides has been shown to be significant both in the metabolism of *N*-acyl compounds, such as *N*-acetyl derivatives of sulphonamides,[81] and amide conjugates of xenobiotics with amino acids such as glycine, taurine and glutamic acid. Amide hydrolysis is also a feature of the metabolism by intestinal microflora of penicillins, methotrexate and chloramphenicol.

Comparative studies in normal, antiobiotic-treated and germ-free rats and mice have shown that the gut flora are involved in the hydrolysis of methotrexate to 4-amino-4-deoxy-N^{10}-methylpteroic acid,[82–84] while amide hydrolysis of chloramphenicol to 1-*p*-nitrophenyl-2-amino-1,3-propanediol is observed in the metabolism of this drug by rats *in vivo*.[77]

Hydrolysis of sulphate esters by intestinal microflora has been reported to occur to a small extent with sodium amylopectin sulphate,[85] cyclohexylphenyl-2-sulphate[86] and some phenyl sulphate derivatives[87] but Scheline[69] concludes that this reaction is of limited occurrence and importance.

The classic example of hydrolysis of sulphamates by gut flora is the conversion of the synthetic sweetener, cyclamate, into cyclohexylamine and, in view of the toxicological implications, this is discussed in more detail later (see p. 237).

Hydrolysis of organic nitrates in the gut has been demonstrated with the anti-anginal drug, pentaerythritol tetranitrate,[88] where pentaerythritol and its mono-, di- and tri-nitrates have been identified as hydrolysis products.

Intestinal hydrolysis of esters frequently is associated with the esterases of intestinal secretions or of the mucosa, but some examples are known which are dependent on microbial activity. These include the pyrollizidine alkaloid, lasiocarpine, which can be degraded by sheep rumen contents to yield 7-α-hydroxy-1-α-methyl-8-α-pyrollizidine,[89] and the anti-ulcer drug, carbenoxolone, which is hydrolysed extensively by rat gastrointestinal microflora but to a much lesser degree in man.[90,91]

Reductions. In view of the low redox potential in the lumen of the caecum and colon it is not surprising that reductions form a large class of metabolic reactions characteristic of the gut microflora and in this chapter it is only possible to mention a few typical examples.

Reduction of carbon–carbon double bonds has long been recognised as a characteristic reaction in the metabolism of dietary unsaturated fatty acids by ruminants and other species (see p. 211) but other unsaturated compounds may also be reduced both in ruminants and other species. Thus caffeic acid, cinnamic acid[72] and some unsaturated monoterpenes (Rahman and Walker, unpublished), all of which occur free or as derivatives in plant foods, are reduced by intestinal microflora in the rat. The metabolism of lasiocarpine by sheep rumen flora was also observed to involve a reductive step.

Reduction of aromatic nitro compounds has been extensively studied since this reaction was shown to be important in the metabolism and toxicity of chloramphenicol. The major proportion of the intestinal metabolites of this drug in rats consists of aryl amines formed by reduction of the nitro group.[77] Similarly, the gut microflora have been shown to effect reduction of nitrobenzene, chlorinated nitrobenzenes, isomeric nitrobenzoic acids and *m*-nitrophenol in rabbits, rats and dogs.[92,93]

The pesticide, parathion, is reduced to the corresponding amine by bovine rumen contents[94] and it was suggested that this might explain the relative lack of toxicity of parathion to cattle. This is somewhat unusual since the amines produced by reduction of nitro compounds are frequently more toxic than the parent compound.

Reductive fission of azo compounds has been studied in considerable detail, primarily in relation to their use as food colours, although early studies had shown the importance of azo-reduction in the metabolic activation of the sulphonamide drugs, prontosil rubrum and neoprontosil.[95,96] It is now recognised that, in many mammalian species, water-soluble sulphonated azo-dyes are reduced primarily in the gut by the intestinal microflora.[97,98] The toxicological consequences of these conclusions will be discussed in more detail later.

The reduction of *N*-oxides by gastrointestinal microflora has been demonstrated in a limited number of cases. *In vitro* studies have shown that 4-nitroquinoline-*N*-oxide is reduced to 4-aminoquinoline by some common intestinal micro-organisms,[99] which involved reduction both of the nitro group and of the *N*-oxide. Similarly, heliotrine-*N*-oxide is reduced to heliotrine by sheep rumen contents *in vitro.*[100] Evidence for reduction of *N*-oxides *in vivo* has been provided for man where (–)-nicotine-*N*-oxide is quantitatively excreted in urine after intravenous administration but is excreted as nicotine or cotinene after oral dosing.[101]

Reduction of *N*-hydroxy compounds has already been mentioned in that microbial reduction of nitro-compounds proceeds via the hydroxylamine intermediate which is further reduced to the corresponding amine. A similar reduction occurs in the intestinal metabolism of *N*-hydroxy-*N*-2-fluorenylacetamide.[80,102]

In several instances, reduction of carbonyl compounds has been found to be a consequence of intestinal microbial metabolism. Thus, a number of benzaldehyde derivatives were reduced to the corresponding benzyl alcohols by rat intestinal organisms[103] and the monoterpene, carvone, was reduced to carveol and its saturated derivatives by rat caecal microflora *in vitro* and in the animal *in vivo* (Walker, unpublished results).

Reductive dehydroxylation by the gut microflora has been observed in the metabolism of both phenols and alcohols. The metabolism of plant phenolics has been extensively studied (see review by Scheline).[69] Caffeic acid and a series of related catechols, including *L*-dopa were observed to undergo *para*-dehydroxylation to a significant extent in the gut of conventional animals but this was suppressed by antibiotics or in germ-free animals. Protocatechuic acid was only dehydroxylated to a small extent and catechol itself was not dehydroxylated to phenol. In some cases, *meta*-dehydroxylation was also observed as a minor pathway. Similar dehydroxylations have also been observed with trihydric phenols and phenolic acids, e.g. pyrogallol was dehydroxylated to resorcinol (but not to catechol) while 3,4,5-trihydroxyphenyl-acetic and -propionic acids were similarly dehydroxylated preferentially in the

4-position.[104–106] Dehydroxylation reactions are also frequently encountered in the metabolism of flavonoids and isoflavonoids. All these compounds cited contain two hydroxyl groups in *ortho*-positions but this is not obligatory for aromatic dehydroxylation to occur since both phloretic acid (*p*-hydroxyphenyl propionic acid) and tyrosine are dehydroxylated by rumen micro-organisms.[107,108] Similarly, heterocyclic hydroxyl compounds kynurenic acid, xanthurenic acid and the 8-methyl ether of xanthurenic acid are all dehydroxylated in the 4-position by the gut microflora.[109–112]

Probably the best-known example of dehydroxylation of alcohols by the gut microflora is the reduction of *p*-hydroxybenzyl alcohol to *p*-cresol[103] and a similar reduction occurs in the metabolism of vanillin and vanillyl alcohol.[113] Reduction of arsonic acids has been studied in relation to the intestinal metabolism of arsenicals used as growth promoters and in disease control in poultry, e.g. 4-nitrophenyl arsonic acid. In this case, both the corresponding arsenoxide and metabolites in which the nitro group was also reduced to the amine were detected.[114]

The reduction of nitrate to nitrite *in vivo* in mammals is due almost exclusively to the microflora which populate the alimentary tract. This reduction has important consequences in relation to the acute toxicity of ingested nitrite and the formation of carcinogenic nitrosamines *in vivo*, and is discussed in more detail in the section on Nitrate, Nitrite and Nitrosamines.

Both reductive dehalogenation (i.e. replacement of Cl by H) and dehydrochlorination have been observed in the intestinal metabolism of DDT. Thus, DDD has been shown to be a metabolite of DDT in rats[115] and several common representatives of the intestinal bacteria have been implicated.[116,117]

It will be noted that none of the reactions illustrated in the metabolism of DDT involve aromatic dechlorination but this has been observed in the metabolism of polychlorinated benzenes in the rabbit.[118,119]

Degradations. Following degradation a fragment of the original molecule may be utilised as a carbon or nitrogen source by the organisms involved.

These degradations include decarboxylation and a number of amino acids are converted into amines by the gut microflora *in vivo.*[120] Decarboxylation is involved in the conversion of tyrosine and *L*-dopa to simple phenols[121,122] and orally ingested *L*-dopa is largely decarboxylated prior to absorption in man.[123] Decarboxylation has also been shown to be an important step in the intestinal metabolism of phenolic carboxylic acids from plant sources which are excreted as simple phenols, e.g. resorcinol, catechol derived from plant tannins.[72,124] Decarboxylation appears to occur primarily with *p*-hydroxylated aromatic carboxylic acids and to a much lesser extent with *m*-hydroxylated acids.

O-demethylation is a common pathway in the degradation of methoxylated isoflavones and flavonoids, e.g. in the metabolism of hesperidin, 3,4-dihydroxy phenyl propionic acid and *m*-hydroxyphenylpropionic acid are produced from the B ring and a demethylation step is involved. The involvement of the gut microflora in this reaction has been demonstrated.[72,125] Many other methoxy compounds of this type are dealkylated but *O*-dealkylation usually does not appear to occur with higher ethers.

N-dealkylation reactions have been reported in studies on a number of drugs, e.g. imipramine which is converted into desmethyl imipramine by the gut microflora of rats and man.[126]

Dealkylation of organometallic compounds has been found to occur with diethyl tin, methyl mercury and organo lead compounds (see p. 241).

Deamination represents another class of degradations which are commonly effected by intestinal microflora. This reaction is encountered in the intestinal metabolism of dopa,[122] triiodo-L-thyronine[127] and the conversion of 5-fluorocytosine into 5-fluorouracil.[128]

Heterocyclic ring fission has been observed with both *N*- and *O*-heterocycles and frequently results in the formation of metabolites bearing little structural resemblance to the parent compound. The food colours indigo carmine and tartazine both undergo metabolism by the gut flora with ring fission occurring to a small extent. Thus indigo carmine is degraded to isatine-5-sulphonic acid and 5-sulphoanthranilic acid, steps which involve the cleavage of both N-C and C-C bonds. Tartrazine is primarily metabolised by reduction of the azo groups but some of the sulphanilic acid which is excreted is derived from the *p*-sulphophenylpyrazolone moiety by fission of the pyrazolone ring.[129]

Ring fission in *O*-heterocycles has been extensively studied in relation to the intestinal metabolism of coumarins,[130,131] flavanoids and isoflavanoids such as quercetin,[132] rutin, hesperitin and hesperidin (see review by Scheline).[69] In all cases, extensive fission of the heterocyclic ring occurs.

Syntheses. Synthetic reactions are much less commonly encountered than the other reactions mentioned in the metabolism of xenobiotics by intestinal microflora. Where syntheses do occur they are sometimes the converse of the hydrolytic reactions already mentioned. Thus, esterification has been demonstrated with some sterols.[133] *N*-acetylation has been observed with a number of amines including histamine,[134] sulphanilic acid and *p*-phenylenediamine sulphonic acid (metabolites of Acid Yellow)[135] and metanilic acid and 4-amino diphenylamine (metabolites of metanil yellow; Walker and Khanna, unpublished results).

Inorganic mercury may be methylated by some enteric bacteria[136] the fate of mercurials in the gut is discussed on p. 241.

Hydration is a relatively minor reaction which has been reported for some fatty acids, e.g. the conversion of oleic acid into hydroxystearic acid by intestinal organism *in vitro*[137,138] (see p. 211).

Nitrosamines may be formed spontaneously from nitrite and secondary or tertiary amines in the acid conditions of the stomach, but this reaction may be carried out at neutral pH by the gut microflora.[139,140] Many strains of bacteria isolated from the human gastrointestinal tract may effect this synthesis, including *Escherichia coli*, enterococci, clostridia, bacteroides and bifidobacteria and some of the organisms involved are also able to reduce nitrate to nitrite (see p. 242).

Miscellaneous. Among the miscellaneous group of reactions are included isomerisations, such as the conversion of linoleic acid into its cis-, trans-, cis-isomer by rumen microflora,[141] and oxidations which are rare. The degradation of indigo carmine referred to above involves an oxidation and a minor pathway in the metabolism of some substituted benzaldehydes involved oxidation to the corresponding benzoic acid, though the major pathway was by reduction to the benzyl alcohol.

Dehydrogenation represents a special case of oxidation and, in some cases, this can lead to aromatisation. Thus, quinic acid is extensively converted into benzoic acid by the gut microflora in many species.[142] Dehydrogenation of steroids by nuclear dehydrogenating clostridia has been reported to lead to aromatisation of both A and B rings in bile acids.[143]

Comparison of Intestinal and Hepatic Metabolism. From the foregoing it can be seen that the reactions performed by the gut microflora are mainly hydrolytic, reductive and degradative, whereas metabolism of xenobiotics in the liver usually involves oxidations and syntheses (conjugations). These are contrasted in Table 6.6. Clearly this gives rise to many possibilities of competition, sequential reactions and

Table 6.6: Comparison Between Mammalian and Microbial Metabolism

Mammalian	Microbial
Phase 1 – Oxidations	Reductions very common
Phase 2 – Syntheses	Hydrolyses
(a) Conjugation with glucuronic acid	(a) Hydrolysis of glucuronides
(b) Conjugation with sulphate	(b) Hydrolysis of sulphates
(c) Conjugation with amino acids	(c) Hydrolysis of amino acid conjugates
(d) Acetylation	(d) Hydrolysis of amides (*N*-acetyl) Hydrolysis of esters (*O*-acetyl)

Table 6.7: Consequences of Intestinal Metabolism

1. Lethal synthesis (including production of carcinogens).
2. Enterophatic circulation.
3. Species, strain and interindividual differences in metabolism of xenobiotics and toxicity.
4. Changes in pattern of metabolism of anutrients due to:
   (a) adaptation
   (b) dietary modification
   (c) drugs

enterohepatic circulation. Many of the references cited provide evidence of intestinal metabolites being further metabolised in the liver and other tissues after absorption. Conversely, there are many examples of hepatic metabolites being excreted in bile and then being further modified by the gut microflora – and this may result in parenterally administered compounds being metabolised to a significant extent in the gut.

Consequences of Metabolism by the Gut Microflora. There are several important consequences of metabolism by the gut microflora which are summarised in Table 6.7 and which are discussed in more detail in the following sections.

Lethal Synthesis. In the current context, lethal synthesis is taken to mean the formation of metabolites which are more toxic than the original compound. A few examples only will be cited to illustrate lethal synthesis by intestinal microflora but these typify a much wider range of possibilities.

Reduction of azo compounds by the gut microflora commonly results in the production of amines which are more toxic than the parent compound. This is the case with the food colours Brown FK, Red 2G, Red 10B, Orange G and Orange RN (see p. 232). Similarly, reduction of nitrate to nitrite leads to an increase in acute toxicity (methaemoglobinaemia) and an increased likelihood of nitrosamine formation (see p. 242).

Hydrolysis of cyanogenic glycosides markedly increases the acute toxicity while hydrolysis of the glycoside, cycasin, is an obligatory activation step leading to the formation of a carcinogen (see p. 228).

Many more examples of lethal synthesis in the gut are given in subsequent sections.

Enterohepatic circulation. It has been mentioned that, due to the different sequential reactions which may occur in the liver and gut, an enterohepatic circulation may be established. Many xenobiotics are conjugated with glucuronic acid, acetic acid, amino acids or sulphuric acid and excreted in the bile. These polar conjugates are generally poorly absorbed from the gut but are there hydrolysed by the micro-flora. The liberated, deconjugated parent compound may then be absorbed into the portal system and returned to the liver, thus establishing the enterohepatic cycle delaying excretion.[144]

Figure 6.2: Enterohepatic Circulation

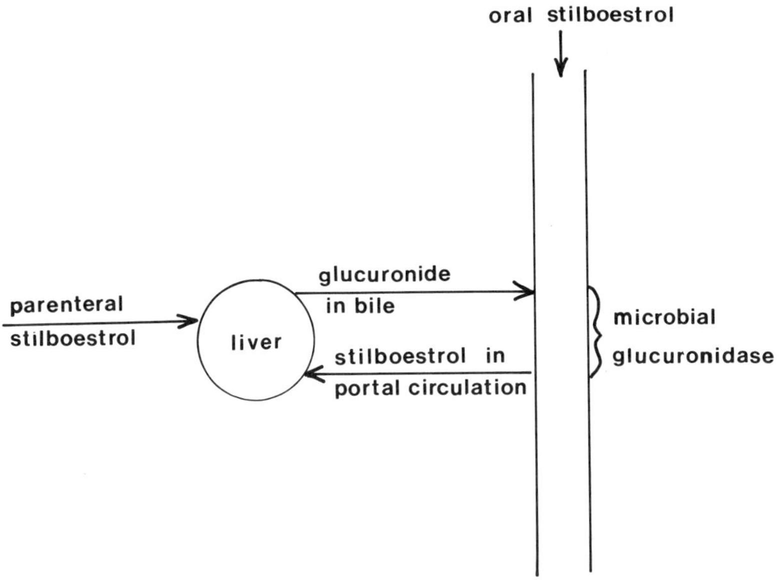

Consider the classic case of stilboestrol (Figure 6.2). Whether administered parenterally or orally, much of the dose ultimately reaches the liver and is excreted in bile as the glucuronide conjugate. It has been estimated that 70 per cent of the glucuronide is hydrolysed and reabsorbed on the first pass.[145] If this proportion is constant for each turn of the cycle, it will take 10 cycles to reduce the amount in circulation to 3 per cent of its original value. Clearly this delays excretion giving rise to prolonged exposure and protracted pharmacological (and possibly toxic) activity.

Enterohepatic circulation has been reported to be significant in the metabolism of antioxidants like BHT[146] and food flavours, including linalool[73] and is obviously a quite common phenomenon when conjugates are excreted in bile.

Species Differences. Inter-species differences in the metabolism and toxicity of compounds may arise as a result of quantitative and qualitative differences in the composition and distribution of the gastrointestinal microflora (p. 185) as well as from differences in gut structure and motility.

The most extensive study yet reported is that on the aromatisation of quinic acid by 22 species, including man.[142] Aromatisation varied from zero in dog, cat and mouse to 60 per cent in man and striking differences were observed between Old World and New World monkeys.

Differences such as these in the metabolism of drugs and food additives may be important in interpreting results of toxicological studies and extrapolating to man.

Strain and Interindividual Differences. These factors have not been extensively studied except in comparisons between metabolism in germ-free and conventional animals, where the differences may be profound. Cycasin is carcinogenic in conventional, but not germ-free rats[147] as a result of differences in the metabolism of this glycoside. However, this represents an extreme case and more comparative studies are needed where the differences in the gut microflora are more subtle. Strain and interindividual differences have also been observed in the intestinal conversion of cyclamate into cyclohexylamine in man and rat (see p. 237).

Changes in the Pattern of Intestinal Metabolism.

Adaptation. In chronic studies on the metabolism of cyclamate in rats, Renwick and Williams[148] found that prolonged exposure to dietary cyclamate resulted in the acquisition of the ability to convert the compound into cyclohexylamine. This adaptation to dietary cyclamate has also been observed in man and is reversible.

Adaptation may result from selection pressures resulting in qualitative differences in the gut microflora, or from induction of the enzymes involved. It seems likely that the metabolic differences observed in human subjects and referred to above, arose from differences in the qualitative composition of the gut microflora.

Regrettably, little further work has been reported on adaptation to dietary and other environmental factors although this may be important in chronic toxicity studies and in day-to-day exposure of man to food additives over prolonged periods.

Dietary modification. The extent to which diet influences intestinal metabolism of anutrients has received little attention even though this may help to explain interlaboratory differences in metabolic studies. It is known that there are dramatic changes in gut flora on weaning, these being particularly striking in the case of ruminants, and and there are also enormous differences in the gut population between herbivores and carnivores (see p. 188).

It has been hypothesised[149] that dietary differences, particularly in fat content, are important in their effects on the ability of the gut microflora to dehydrogenate bile acids and hence in the aetiology of colonic cancer (see p. 213). It has also been observed that dietary riboflavin moderates the carcinogenicity of butter yellow, possibly by stimulating the intestinal azo-reductase. Possibilities such as these are clearly worthy of much more attention than has been the case hitherto.

Drugs. Drugs may modify their own metabolism or that of other compounds due to effects on the gut microflora (see also p. 194). In a number of cases suppression of the gut microflora with antibiotics modified the toxicity of orally administered compounds and this is a factor to be borne in mind both in experimental work and in relation to human exposure to foreign compounds in the course of medication.

Conclusion

It has been shown that the metabolism of drugs, food additives and other dietary anutrients by the gut microflora is important, whether these compounds enter the gut directly after oral ingestion or via the bile. The factors which might influence the course and extent of

Table 6.8: Factors Reflecting Intestinal Metabolism of Xenobiotics

A.	Nature of xenobiotics
	1. Polarity
	2. Formulation and dose
	3. Structure and stability
B.	Nature of exposure
	1. Time and route of administration
	2. Chronic dosing and adaptation
C.	Nature of animal
	1. Species differences in structure of the G.I. tract
	2. Gut motility
	3. Pathological conditions (enteropathies, diarrhoea)
D.	Nature and distribution of gut flora, influenced by
	1. Age
	2. Diet
	3. Exposure to infection ('germ-free', SPF, etc.)
	4. Drugs
	5. Coprophagy

intestinal metabolism are summarised in Table 6.8. Some of these factors are speculative because of insufficient experimental data but others have been more clearly demonstrated.

In general it can be concluded that intestinal metabolism is particularly important for polar compounds which are poorly absorbed from the gut and for the many compounds which are excreted, free or conjugated, in the bile.

Role of Mammalian Enzymes in Intestinal Metabolism

Table 6.9 summarises the major reactions occurring in the gut, a subject reviewed by Hartiala.[150] Many of the enzymes found in the liver which detoxify drugs and other foreign compounds are present, albeit in lower activities, in the small intestine, including those dependent on cytochrome P450.[151,152] Furthermore several of the enzymes are induced by pretreatment of the animal with the classical hepatic enzyme inducers, phenobarbital and 3-methylcholanthrene.[153]

Mammalian Enzymes and Deficiencies

Lipid Metabolism: Lipase Deficiency. In addition to congenital lipase deficiency a variety of clinical conditions are associated with a decreased synthesis or output of pancreatic lipase, including pancreatitis, cystic fibrosis, Schwachman syndrome, pancreatic resection and vagotomy.[165]

Table 6.9: Reactions Catalysed by Mammalian Enzymes in the Gut

1. Hydrolytic enzymes involved in metabolism and absorption of food components	(a) Amylase (b) Disaccharidases: lactase trehalase isomaltase sucrase heat-resistant maltases I and II β-galactosidase (c) enterokinase (d) pancreatic proteases: trypsin chymotrypsin elastase carboxypeptidase (e) pancreatic lipases
2. Hydrolysis	(a) Esterases– specific and non-specific[154,155] (b) Deacetylase[150] (c) β-glucuronidase[156] (d) Sulphatases[150]
3. Oxidation	(a) Alcohol dehydrogenase[157] (b) Cytochrome P450/P488 – dependent mixed function oxidases e.g. benzpyrene hydroxylases[158] (c) Amine oxidases[154] (d) Sulphur oxidation desulphuration[159]
4. Reduction	(a) Dehydroxylation of flavanoids (b) Reduction of *N*-hydroxyacetylamino-fluorene[160]
5. Conjugation	(a) Steroid reduction[150] Glucuronide formation: UDP-glucuronyl transferase UDP-glucuronyl dehydrogenase[150,161] (b) Acetylation[150] (c) Sulphate conjugation of phenols and steroids[162] (d) Gluthathione-*S*-transferase[163]
6. Miscellaneous	(a) Deamination (b) Dealkylation[164]

Under these conditions fat digestion is impaired resulting in steatorrhoea characterised by increased amounts of faecal triglycerides.

Steatorrhoea and apparent lipase deficiency also occur when the secreted lipase is inactivated in the gut by an abnormally low jejunal or duodenal pH in diseases such as gastroma and cystic fibrosis.[166]

Protein Metabolism. Disorders of intestinal protein metabolism can occur as a result of deficiencies in endogenous enzymes or the presence of dietary protease inhibitors (see p. 230).

Enterokinase Deficiency. The pancreatic proteases are produced and secreted as proenzymes (trypsinogen, chymotrypsinogen, proelastase, procarboxypeptidase) which are converted into their active forms in the duodenum. Enterokinase, a glycoprotein, catalyses the first stage in the activation sequence converting trypsinogen into trypsin by the removal of specific peptide fragments, trypsin in turn activates the other three proenzymes. Thus enterokinase plays a key role in protein digestion and its deficiency results in several clinical symptoms including diarrhoea, hypoproteinaemia, oedema, anaemia, failure to thrive and vomiting.[167] The disease, which would appear to be congenital[168] and of greatest clinical importance in childhood, responds well to oral administration of pancreatic extracts.

Carbohydrate Metabolism. The disaccharidases located in the brush border membranes of the intestinal epithelium convert the poorly absorbed disaccharides such as sucrose, lactose and maltose into rapidly absorbed monomers, glucose, galactose and fructose, and deficiencies in one or other of these disaccharidases leads to malabsorption of the corresponding disaccharide.

Disaccharidase Deficiency. The most well-known clinical manifestation of an enzyme deficiency is lactose intolerance. Hypolactasia is characterised by a very low jejunal lactase concentration (although other disaccharidases are normal) which is insufficient to hydrolyse a physiological lactose load. The symptoms after ingestion of lactose usually include diarrhoea and abdominal pain. the former thought to be due to osmotic effects caused by fermentation of the unabsorbed lactose by colonic bacteria.[169]

Lactose intolerances appear to be due to at least three different mechanisms. The first of these, congenital lactase deficiency, is very rare and gives pronounced clinical symptoms during infancy. The disease is inherited recessively and is apparently due to a mutation in the structural gene for lactase, leading to an inactive enzyme.[170]

The second class of lactase deficiency occurs in the majority of adult populations of the world but is not usually seen in infants of these populations.[169] It has been found that in children, lactase activity in the jejunum reaches a peak at parturition, and then in most populations declines to about 5–10 per cent of the level in newborn infants over the first few years of life when the child is no longer completely dependent on breast feeding.[171] This decline occurs whether or not lactose continues to be consumed.[172] In adults therefore, hypolactasia is the normal state for the majority of the world's population and it is thought that the high lactose concentrations which persist in adults in Europe, North America and in North Eastern Africa are the result of a mutation in a regulatory gene for lactase, which has survived due to natural selection in populations where consumption of dairy products has been high for centuries.[169]

The third class of disaccharidase deficiency occurs in patients with diseases or conditions which cause damage to the small intestinal mucosa such as Crohn's disease, ileostomy,[173] bacterial overgrowth in the duodenum, jejunum or small intestine.[174,175] Marked depression of disaccharidase activities (lactase and sucrase) is also seen in alcoholics.[176] Since the symptoms of lactase deficiency can usually be reversed in adults by using lactose-free diets, nutritional deficiencies associated with lactose intolerance are only likely to be of significance in very young children, especially those in populations such as in Thailand, Peru, Nigeria and Uganda in whom lactase activity falls rapidly within the first few months of life.[169,177]

Malabsorption of other sugars due to deficiencies in specific disaccharidases, such as sucrase and trehalase, is rare.

Glucose-galactose Malabsorption. In patients with glucose-galactose malabsorption, which is a rare autosomal recessive congenital disorder resulting in severe diarrhoea in infancy, the brush border disaccharidase activity is normal. It has been suggested that such patients possess fewer carrier sites for monosaccharide translocation in the intestine.[178] The clinical symptoms can be alleviated by substituting the usual dietary carbohydrate by fructose which is absorbed normally.

Toxicological Aspects of Intestinal Metabolism

Natural Dietary Constituents

Lipids.

Unsaturated Fatty Acids. Dietary fatty acids are more unsaturated than the fatty acids in faeces and this difference has been ascribed to metabolism during passage through the gastrointestinal tract.[179] The most probable agents of fatty acid reduction are members of the gut flora since *in vitro* studies using micro-organisms from human intestine and the rumen have demonstrated the conversion of the triene acid, linolenic acid, into the diene, linoleic acid[180] and the further reduction of the latter to oleic acid.[181,182] Thomas[138] showed that several intestinal bacteria can convert oleic acid into hydroxystearic acid and Watson[183] demonstrated the intestinal reduction of ricinoleic acid to hydroxystearic acid. Patients with steatorrhoea have 10-hydroxystearic acid as the major hydroxyacid in the faeces, which presumably is derived from unabsorbed dietary oleic acid.[184] It has been suggested that the hydroxystearic acid may contribute to steatorrhoea, since many purgative oils such as castor oil contain long-chain hydroxy acids.[184]

Lipids and Large Bowel Cancer.

Bile acids. In addition to the metabolism of unabsorbed fatty acids there is a more indirect interaction of the gut flora with dietary fat, although potentially of greater significance to man than the former, namely the ability of the gut flora to metabolise bile salts and bile acids. In humans the faecal excretion of acid and neutral steroids appears to be related to the intake of fat. The relationship of faecal steroid excretion to the fat content of the diet has been incorporated into a theory of the aetiology of cancer of the large bowel, a disease which, from epidemiological studies (discussed below) is known to involve one or more components of man's diet, although the precise component has yet to be determined. The fat theory of colon cancer as proposed by Hill and colleagues in Britain and Wynder and colleagues in the USA postulates the following:

1. The incidence of colon cancer and fat content of the diet are highly correlated.
2. The amount of fat in the diet governs the concentration of steroids in the faeces.
3. The diet influences the numbers and types of bacteria in the gut and also the physiological conditions in the gut.
4. Gut bacteria can produce carcinogenic or co-carcinogenic compounds from biliary steroids.

1. Demographic studies of a large number of populations throughout the world have revealed that the incidence of colon cancer correlates highly (correlation coefficient about 0.85) with the dietary intake of fat and protein,[40] in particular with 'combined fat', that is the fat associated with meat. No correlation between fibre or vegetable consumption and colon cancer was found.[40]

Studies of populations which have migrated from areas of low colon cancer incidence to areas of high incidence and in doing so have acquired a high colon cancer incidence, have reinforced the view that diet is the most important factor in colon cancer and that dietary fat in particular is implicated.[185–187]

2. Hill,[188] showed that a change in diet from a normal, high-fat diet to one low in fat led to a fall in faecal bile acid concentration to 25–30 per cent of the control level within a few days. The concentration of neutral steroids (derived from cholesterol) fell more slowly to approximately 40 per cent of the control value.

These results have been in general confirmed by other studies including that of Crowther *et al.*[189] and Cummings *et al.*[44] It has also been found that the type of dietary fat is important in determining faecal bile acid excretion, which is increased by a high intake of polyunsaturated fat[190] but unaffected by the level of cholesterol in the diet.[191,192] Studies of the faecal constituents of patients with bowel cancer have revealed a high concentration of faecal bile acids in 82 per cent of the cancer patients compared with only 17 per cent of the controls.[193]

3. As has already been discussed above major differences in the composition of the gut flora have been found in people living in different areas of the world, especially when western populations (high colon cancer incidence) are compared with populations of underdeveloped countries.[30,31] However, not only are the dietary differences between these populations extremely diverse in terms of the type and quantity of the various food components, but there are numerous cultural differences and disease processes which make it difficult to form any conclusions as to the influences of diet on the gut flora. Attempts to change the gut flora by altering the diet have also given rise to ambiguous results (see p. 192).

However, it is quite feasible that changes in the diet can lead to alterations in the metabolic activity of the flora. For example Reddy *et al.*[42] demonstrated an increase in β-glucuronidase activity in faeces of people fed diets high in animal fats and meat, and Goldin and Gorbach[194] found higher levels of nitroreductase, azoreductase and β-glucuronidase in faeces of rats whose diet was changed from one based on grain to one consisting predominantly of beef. It has also been shown that not only is total faecal steroid concentration higher in populations experiencing a high incidence of colon cancer, but also the extent of metabolism (by the gut flora) of those steroids is increased.[32,37,195,196] Similar changes in faecal steroid concentration and steroid degradation can be induced in man using diets high and low in dietary fat or meat.[41,44,189]

It should be noted, however, that in a recent comparison of two Scandinavian populations with high and low incidences of colon cancer, no differences were found in faecal steroid excretion or in extent of degradation of neutral steroids.[197]

4. The postulated involvement of acid and neutral steroids in large bowel cancer has received support from studies which show that carcinogenic and co-carcinogenic actvity is associated with some of these steroids, e.g. lithocholic acid and deoxycholic acid.[37,198,199] Short-term tests utilising micro-organisms and mammalian cells in culture have been used to study the potential mutagenicity and carcinogenicity of bile acids and their metabolites. In the *Salmonella*/mammalian microsome mutagenicity test none of the 30 bile acids tested was mutagenic, but lithocholic acid and 3-keto-5-β-cholanic acid increased the mutagenicity of 2-aminoanthracene.[200] Furthermore, lithocholic acid induced low numbers of transformed hamster embryo cells *in vitro* which is indicative of carcinogenic activity.[201]

Bile acid metabolism by intestinal bacteria has been studied in detail and the reactions can be divided into two groups, the first in which the polar substituents on the steroid nucleus are metabolised and the second involving the nucleus itself.

Reactions involving substituent groups. Modifications to the substituents are common, for example all bile acids in faeces are deconjugated indicating high activity of the hydrolase enzyme. Hydroxycholanoyl dehydrogenases can act at the 7α, 3α and 12α positions and are found in that order of abundance, i.e. the 7α-hydroxycholanoyl dehydrogenase is produced by the most bacterial types in the human faeces.[202] The oxidation reactions however are not of great significance *in vivo* in the gut since the enzymes are reversible and the reverse reaction is favoured under the pH and E_h conditions of the gut. Hydroxycholanoyl dehydroxylases, especially the 7α-dehydroxylase appear to be widely distributed in the human gut flora and greater than 80 per cent of the faecal bile acids are known to be dehydroxylated, for example deoxycholic acid and lithocholic acids are the dehydroxylated products of cholic and chenodeoxycholic acids respectively.[203] The 7α-dehydroxylase is an inducible enzyme having a pH optimum of 7–8 and for its production requires strict anaerobiosis.[202]

Nuclear dehydrogenation reactions. In a series of papers, Hill and his co-workers have demonstrated several dehydrogenations of the steroid nucleus by organisms isolated from the gut.[143,202,203] These reactions are oxidative and require an oxidising agent, such as menadione which was shown to be present in human faeces.[204] However the significance of these reactions *in vivo* is questionable since Lombardi *et al.*[205] have shown that the *in vitro* metabolites of steroids are dependent on the concentration of faeces used, at high concentrations reductive reactions predominate, while at low faecal concentrations oxidative reactions were also found, but only in the presence of electron acceptors.

Hill has provided some evidence that the incidence of colon cancer is related to the faecal deoxycholate concentration, the number of organisms possessing 7α-dehydroxylase and the number of nuclear-dehydrogenating clostridia per gram of faeces.[206] He has suggested that the production of unsaturated steroid derivatives by the nuclear-dehydrogenating clostridia may be involved in the aetiology of colon cancer and in high risk populations one might expect to have a combination of high faecal bile acid concentration and the presence of large numbers of clostridia capable of nuclear dehydrogenation.[206]

There is some supportive evidence for the implication of clostridia in colorectal cancer notably that a high proportion (82 per cent) of patients with the cancer carried nuclear-dehydrogenating clostridia in their faeces compared with only 40 per cent of control patients.[193] However, other studies have failed to confirm these results[207,208] and two recent epidemiological studies, one in Scandinavia[197] and the other in Hong Kong,[32] have revealed no significant differences in numbers of nuclear-dehydrogenating clostridia in high-risk and low-risk colon cancer populations.

Cholesterol. In human faeces, unchanged cholesterol accounts for less than 30 per cent of the total neutral steroids, the remainder being mainly the reduction product, coprostanol (5β-cholestan-3β-ol), with coprostanone providing a further 5–10 per cent.[37] Studies with germ-free rats have demonstrated that the intestinal flora is entirely responsible for this reduction. Even though coprostanol is less readily absorbed from the gut than cholesterol, it is unlikely that the gut flora metabolism of cholesterol affects faecal excretion of neutral steroids since the metabolism occurs in non-absorptive regions of the gut.[37]

The relevance of this microbial transformation of cholesterol to the aetiology of colorectal cancer has not been fully elucidated. Studies of faecal neutral steroid concentrations in populations experiencing widely different incidences of colon cancer have revealed a correlation between the two parameters.[30] Similarly Reddy and Wynder[195] showed that the total neutral steroid excretion in Americans eating a conventional western diet was much higher than in American Seventh Day Adventists and Japanese (who have lower colon cancer incidence) and furthermore the faecal cholesterol was more extensively metabolised by the Americans on the western diet. In a study of the effects of low-risk and high-risk diets for colon cancer on faecal parameters in man, Reddy *et al.*[41] demonstrated a reduction in excretion of microbially modified neutral steroids when the subject transferred from a high-risk to the low-risk diet. However, in a recent epidemiological study no differences in faecal concentration, total excretion and microbial degradation of neutral steroids were found between two populations experiencing different colon cancer incidences.[197]

Patients with colon cancer or with other colon diseases known to predispose them to cancer of the colon such as familial polyposis, adenomatous polyps and ulcerative colitis often have higher faecal concentrations of cholesterol and its metabolites than control patients,[209–211] although other studies have shown that the colon cancer patients and patients with polyposis coli have a reduced ability to degrade cholesterol bacterially.[212–214] Furthermore, although most healthy North Americans could convert cholesterol into coprostanol, approximately 20 per cent were unable to do so[215] and low-risk populations of rural Africans have been found to be capable of bacterial cholesterol metabolism.[216] Thus the metabolism of cholesterol by the gut flora may be related to colon cancer risk, but the relationship may be highly complex and the conflicting nature of much of the evidence is presumably due to large variations in diet and gut flora between individuals and population groups (see p. 190).

Cruse *et al.*[217] have suggested that cholesterol promotes a carcinogenic process which has been previously initiated by, for example, other dietary carcinogens or by genetic factors. Hill *et al.*[218] have proposed a hypothesis to explain the aetiology of the adenoma-carcinoma sequence of large bowel cancer, in which a tumour-promoting agent is implicated. Hill *et al.*[218] on the basis of their extensive studies suggest that the identity of the tumour promoter is a bile acid metabolite, whereas Cruse *et al.*[217] postulate cholesterol as an alternative. The latter's theory, that cholesterol is a co-carcinogen, is supported by their studies of the induction of colon carcinoma by dimethylhydrazine in rats fed various diets.[219] Carcinogen-treated animals eating a cholesterol-free, chemically defined diet survived longer and took longer to develop colon tumours than similarly treated rats fed a standard formula diet. Furthermore, addition of cholesterol to the chemically defined diet reduced both survival and time to tumour presentation in dimethylhydrazine-treated rats and also increased the incidence of metastases, (which were normally uncommon in this animal model). In a recent study utilising an *in vitro* cell transformation test,[201] cholesterol α-epoxide (a peroxidation metabolite) but not cholesterol itself, was found to possess transforming activity indicating its carcinogenic potential.

Proteins

Amino Acid Metabolism. Amino acids are initially metabolised by the gut flora by two major classes of reaction, deamination and debarboxylation. These two processes can lead to highly reactive and toxic products such as amines, ammonia and, in the case of amino acids such as tyrosine, phenols. Although in the healthy individual these substances are usually detoxified by mammalian enzymes, especially those in the liver, in certain disease states such as hepatic cirrhosis these detoxification mechanisms can break down. Amino acid decarboxylases are widely distributed amongst members of the gut flora. The products of decarboxylation of several amino acids are pharmacologically active monoamines and diamines[220,221] such as histamine from histidine, tryptamine from trytophan and tyramine from tyrosine. The monoamines are in general pressor substances which in very small quantities can cause vasoconstriction and an increase in blood pressure and have in certain cases been implicated in migraine attacks.[222] Decarboxylation of lysine and ornithine results in, respectively, the diamines, cadaverine and putrescine which have weak pressor activity. The diamine histamine however is a powerful vasodilator and depressor and is associated with the severe symptoms of nausea, vomiting and headache of scombroid poisoning.[223] There is good evidence from experimental animals that amino acid decarboxylation occurs *in vivo* and that such reactions are responsible for the major proportion of amines and diamines in the urine.[37]

It has also been found that histamine excretion is increased after consumption of meat[224] presumably due to decarboxylation of the histidine after proteolysis of the meat.

The potential toxicity of these pharmacologically active amines is ameliorated in the healthy individual by deamination by monoamine oxidase and diamine oxidase in mammalian tissues. Administration of monoamine oxidase inhibitors increases both the urinary concentration of amines and the range of amines found in the urine in humans.[120] In patients with hepatic cirrhosis, monoamine oxidase may have reduced activity or may be prevented from interacting with its substrates, resulting in increased blood levels of the toxic amines which may contribute to the 'hepatic coma' induced in these patients after a high protein meal.[220] The ability to deaminate amino acids and other amines is possessed by many enterobacteria and clostridia – the only groups of gut bacteria which have been extensively investigated for this reaction. There is evidence however that deamination also occurs in bacteroides, bifidobacteria, and staphylococci. Drasar and Hill[37] have reviewed the various pathways involved in the release of ammonia from amino acids, the major pathway being reductive deamination resulting in a saturated fatty acid and ammonia.

In the case of some amino acids, notably tyrosine and tryptophan, deamination, decarboxylation and other reactions lead to the production of potentially toxic and carcinogenic products.

Tyrosine Metabolism. The gut bacteria can metabolise tyrosine derived from dietary protein to several phenolic compounds including phenol, *p*-cresol and 4-ethylphenol.[225] There is evidence that phenols are co-carcinogens and that they possess convulsant activity[226,227] and although they are normally conjugated in the liver prior to excretion, under conditions of hepatic dysfunction (e.g. cirrhosis) blood levels of toxic phenols may be elevated.[37,228]

Metabolism of Tryptophan. Tryptophan can be metabolised by a complex series of reactions, some bacterial others mammalian, and of these reactions, one of the most studied is the production of indole, the 'tryptophanase' reaction, the enzyme being distributed widely in the gut flora.[229] Tryptophan has been implicated in the aetiology of bladder cancer by Dunning *et al.*[230] who showed that rats fed 2-acetylaminofluorene (2-AAF) together with tryptophan or indole had a high incidence of bladder tumours which were not produced by 2-AAF alone.

It was subsequently found that 2-AAF stimulates tryptophan metabolism[231] and that several tryptophan metabolites of both bacterial and mammalian origin, including kynurenine, 3-hydroxyanthranilic acid and xanthurenic acid, induce tumours on implantation in the bladder.[232] Furthermore, indole given orally or subcutaneously induced lymphoreticular tumours and leukaemia, although no bladder tumours were reported.[233,234]

An earlier report that bladder cancer patients have an increased rate of urinary excretion of tryptophan metabolites[235] was subsequently shown to be due to the presence of the tumour since the excretion of several metabolites, including 3-hydroxyanthranilic acid, kynurenine and xanthurenic acid, was reduced in patients treated for bladder cancer.[236]

Two tryptophan metabolites, 3-hydroxykynurenine and 3-hydroxyanthranilic acid were mutagenic in mammalian cells in culture[37] although they were negative in the *Salmonella* mutagenicity test of Ames.[237]

Attempts have been made to implicate tryptophan and its metabolites in the aetiology of large bowel cancer. Epidemiological studies have shown that colon cancer is closely correlated with the consumption of diets high in animal protein and fat.[40,238] Animals fed an all-meat diet have significantly higher concentrations of tryptophan in their faeces than animals on mixed diets[45] and furthermore the activity of tryptophanase was higher in the faeces of meat-fed animals. However, there is as yet no evidence that indole is a carcinogen or co-carcinogen using a short-term predictive test.[237] In addition, in a recent epidemiological study of colon cancer in two Scandinavian populations, faecal tryptophan concentrations were higher in the population with a low incidence of bowel cancer even though the high incidence group had a 50 per cent higher intake of meat.[197] Further evidence that meat is not involved in colon cancer comes from a recent retrospective study of the diets of colon cancer patients compared with control groups[239] which found no increase in risk related to consumption of meat but a decrease in risk associated with a higher intake of vegetables.

Protein Intolerance. A variety of dietary proteins may elicit an idiosyncratic response in sensitive individuals. This is exemplified by the best characterised of these intolerances, namely gluten intolerance or coeliac disease. Coeliac disease is an intestinal malabsorption syndrome in which most nutrients are poorly absorbed including fats, sugars, amino acids, minerals and vitamins.[240] Diarrhoea, bulky stools, weight loss and abdominal cramps are the commonly encountered symptoms.[241] Histologically, villous atrophy and changes in the structure of epithelial cells of the jejunum are seen in the coeliac patient.[242,243] Wheat flour and more specifically the protein fraction (or gluten) of the flour was shown to be implicated in the development of the disease[244] and coeliac patients appear to possess a specific intolerance for protein-bound glutamine and are unable to metabolise certain peptides.[243] The toxic fraction of the gluten is thought to comprise peptides rich in glutamine and proline.[242]

The gastric and pancreatic enzyme activities appear to be normal in coeliac patients[245] and there do not seem to be any deficiencies in mucosal peptidases.[246] However, Cornell and Rolles[247] have demonstrated that small intestinal mucosa from coeliac patients was less active in digesting a fraction of wheat gliadin than mucosa from healthy individuals. It seems probable that such a mucosal deficiency together with an abnormal immunological response to the toxic fraction of gluten may be responsible for gluten toxicity.[241,247]

Carbohydrates

Starches. It is commonly asserted that large amounts of starch in the diet can be digested and well utilised by most species of laboratory and domestic animal, and by man. Indeed, semi-synthetic diets used in some toxicological studies are commonly based on starch as the major, or even sole, carbohydrate source. However, it has long been known that raw potato starch is less digestible than maize or wheat starches[248] and Beazell *et al.*[249] concluded that raw potato starch is almost completely resistant to hydrolysis by gastrointestinal analyses in humans. Jelinek *et al.*[250] reported that raw potato starch was poorly utilised in the rat and caused enlargement of the caecum but gelatinisation of the starch, or damage to the granule by ball-milling abolished or greatly reduced this effect. Subsequent studies[251] confirmed these observations and demonstrated that raw potato starch could cause such gross caecal enlargement in rats, that respiratory function was impaired and death ensued in some cases (Figure 6.3). The caecal enlargement does not appear to be unique to the rat and Whittemore *et al.*[252] recorded that pigs fed raw potato have much enlarged caeca containing particles of undigested potato.

Comparative studies with raw starches from a variety of plant sources[253,254] have shown that there is a correlation between resistance of a starch to hydrolysis by pancreatic amylase and caecal size in rats fed the starch; raw cassava or sorghum starches were less readily hydrolysed than maize starch and caused an increase in caecal size, though less dramatic than raw potato starch. Ketiku and Oyenuga[255] have reported that the starch in yams is poorly utilised and in this respect resembles potato. A postulated mechanism by which starches may influence caecal size suggests that when the starch is not hydrolysed and absorbed from the small intestine, it may be partially broken down by the caecal microflora to lower molecular weight, osmotically

Figure 6.3: Caecal Enlargement in Rats Caused by Raw Potato Starch. Comparison between rats receiving a diet containing 71% maize starch (A4) and 71% raw potato starch (B4). Note the secondary effects of compression of the diaphragm and distension of the rib cage leading to respiratory distress.

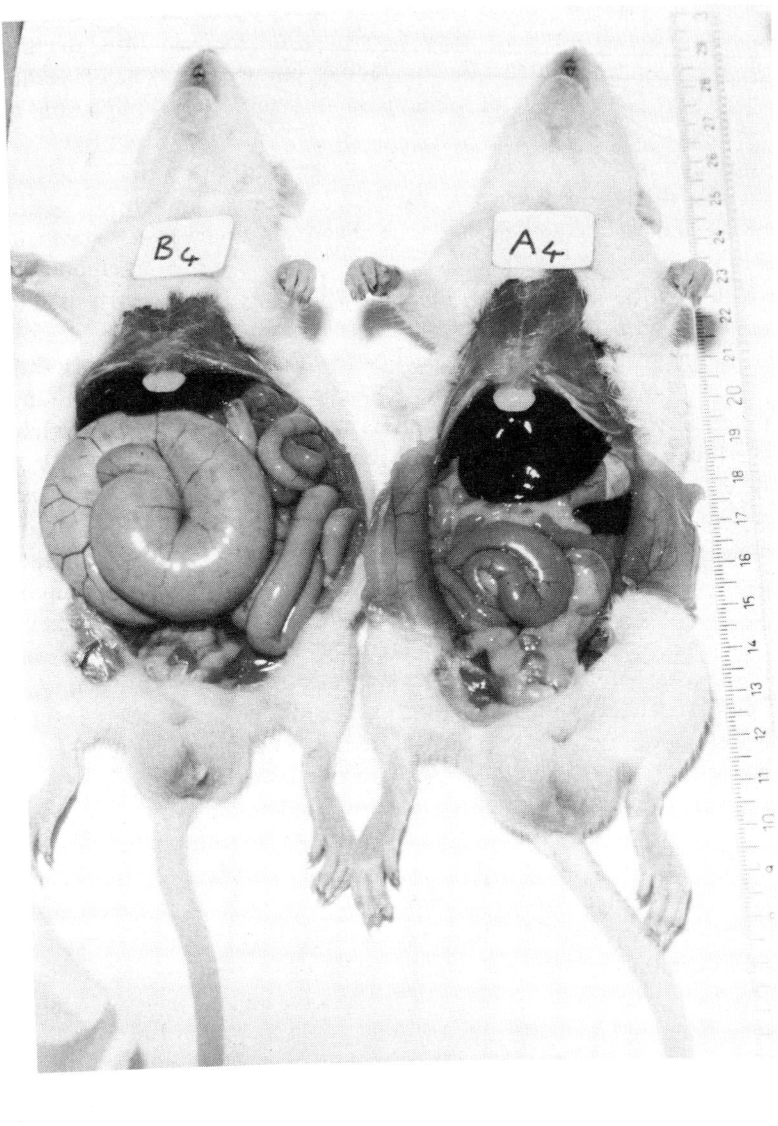

active oligosaccharides which results in water retention in caecal contents. A number of other poorly absorbed, osmotically active compounds have also been observed to cause caecal enlargement (see p. 246).

In cases where the degree of caecal enlargement is small, there was only a slight increase in tissue mass and the tissues appeared histologically normal. In contrast, the gross caecal enlargement caused by raw potato starch was accompanied by a marked increase in tissue mass with some oedema in the mucosa and sub-mucosa[256] (Figure 6.4): evidence for some degree of hyperplasia and cell hypertrophy was also presented.

In addition to the changes in caecal size raw potato starch caused both quantitative and qualitative changes in the caecal microflora, with increased counts dominated by lactic anaerobes, and a lowering of the pH of caecal contents.

Although the effects of raw potato starch were most dramatic on the caecum, they were not confined to this organ. Increased occurred in the weight of all sections of the gastrointestinal tract except the stomach, and duodenal contents had depressed amylase activity.[253] Associated with these changes, there is increased water retention in the faeces, increased faecal mass (on a dry weight basis) and impaired food efficiency and nitrogen utilisation.[253,257]

Oligosaccharides. Oligosaccharides which are poorly hydrolysed and absorbed from the upper gastrointestinal tract, either because of resistance to endogenous hydrolytic enzymes or as a result of enzyme deficiency, can present both toxicological and nutritional problems. The oligosaccharides reaching the lower gut may induce a diarrhoea due to osmotic effects and, in extreme cases, this may be associated with malabsorption syndromes, electrolyte depletion and dehydration. Additionally, unabsorbed oligosaccharides may undergo fermentation by the microflora leading to gas production and flatulence.

The disaccharide, lactose, exemplifies the gastrointestinal effects resulting from endogenous enzyme deficiency and lactose intolerance has been discussed in detail on p. 210. When lactose is included in the diet of normal rats it causes caecal enlargement[258] again probably due to osmotic effects, and similar results have been obtained when higher oligosaccharides contained in glucose syrup fractions were administered in the diet.[259,260]

Consumption of some plant foods, particularly legumes, frequently results in flatulence. Extensive studies on beans showed that

Figure 6.4 (a): Sections of Caecum from Rats Fed Raw Maize Starch

(a)

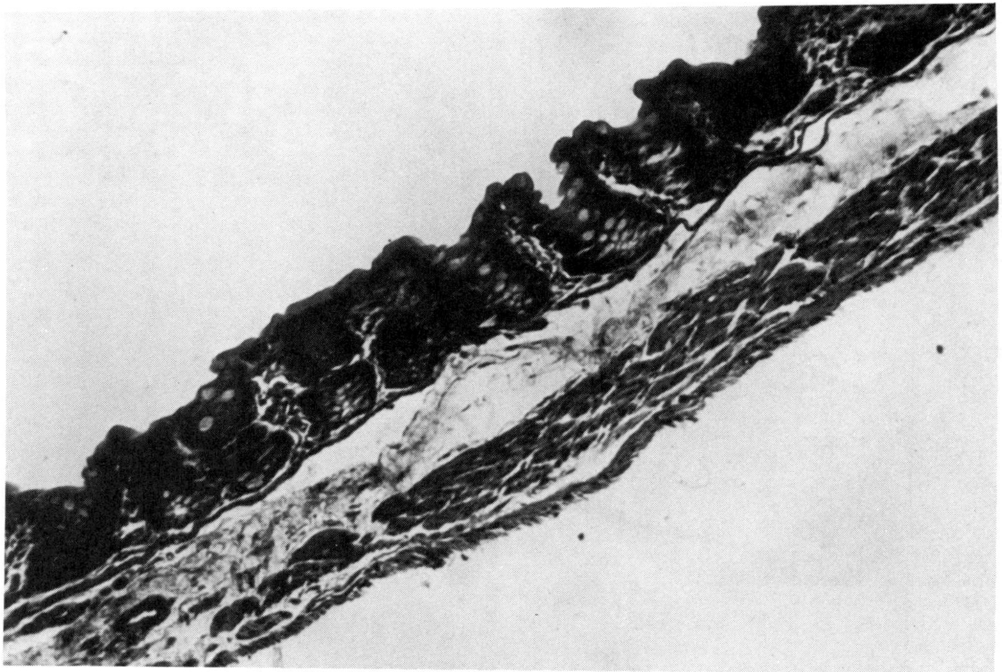

Figure 6.4 (b): Sections of Caecum from Rats Fed Raw Potato Starch as Sole Source of Carbohydrate (71 per cent of Diet)

(b)

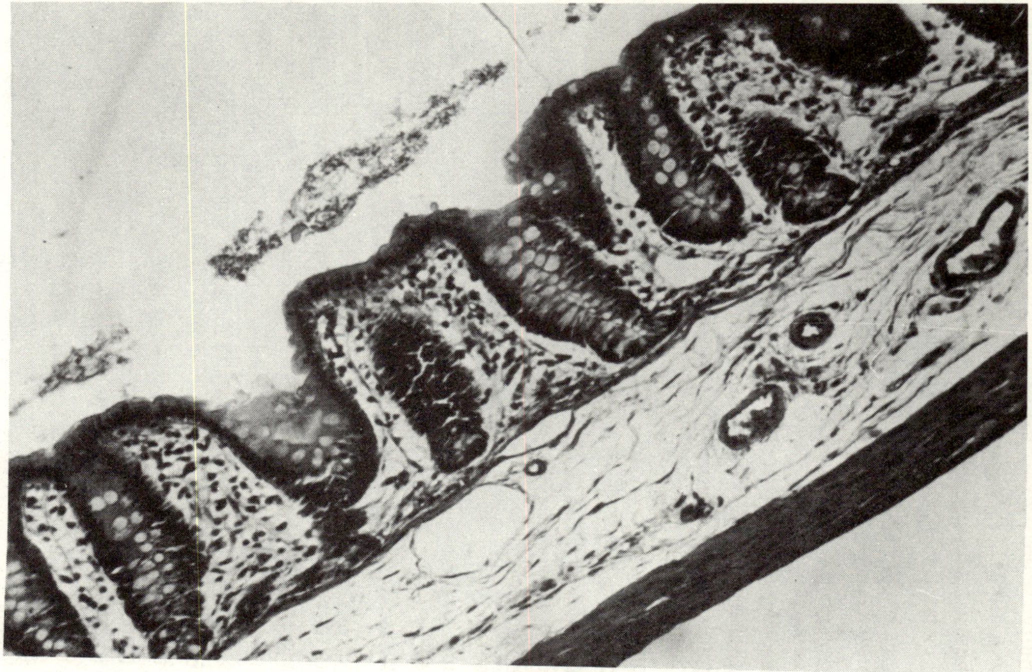

low-molecular-weight soluble fractions contained the flatus factors[261] which were fermented primarily by anaerobic clostridia in dogs.[262,263] The trisaccharide *raffinose* (*O*-α-D-galactopyranosyl-(1→6)-*O*-α-D-glucopyranosyl-(1→2)-β-D-fructofuranoside) and the tetrasaccharide, *stachyose* (*O*-α-D-galactopyranosyl-(1→6)-*O*-α-D-galactopyranosyl-(1→6)-*O*-α-D-glucopyranosyl-(1→2)-β-D-fructofuranoside) have been identified in the flatus-producing fraction of beans[264] and they are not hydrolysed by mammalian enzymes secreted in the gastrointestinal tract. The ability of raffinose and stachyose to cause flatulence when fed in otherwise non-flatus-producing diets has been demonstrated, confirming that these oligosaccharides are responsible at least in part for the flatulence associated with legumes.[265]

Fibre. Although fibre has not been shown to be an essential component of man's diet, its deficiency in diets consumed in affluent western societies has been blamed for many of the diseases and gastrointestinal disorders prevalent in those populations including appendicitis, constipation, diverticular disease, diabetes, ischaemic heart disease and colon cancer.[266]

Studies of the biological and physiological effects of fibre are complicated by the loose definition, ambiguous terminology and difficult chemical analysis of fibre.[267–269] Furthermore, fibre from different plant sources and even fibre processed in different ways has very different biological and chemical characteristics.[270] Consequently, epidemiological and experimental evidence implicating fibre in the aetiology of various diseases (discussed below) can be difficult to interpret and is often conflicting. The problems involved in obtaining such evidence were concisely, if bluntly, summarised by Mendeloff[271] – 'since no one can agree on exactly what dietary fibre is or how to measure it, it is impossible to define whether or not fibre deficiency or excess in the diet exists or can constitute a protective or noxious element in the diet'. However, this has not deterred people from trying to assess the significance of fibre in the aetiology of western diseases.

In the case of constipation and diverticular disease there is good clinical evidence that the administration of high fibre diets will relieve the conditions.[272] Although this does not mean that a deficiency of fibre in the diet was responsible for the disorders it has been shown that rats develop colonic diverticula when fed low fibre diets.[273] Burkitt[266] has proposed fibre deficiency as a major factor in the aetiology of colon cancer, since western countries have a high incidence of the disease by comparison to underdeveloped countries, the population of which consume, in general, diets high in fibre. However Drasar and Irving,[40] using data from 32 countries have shown that the incidence of colon cancer does not correlate at all with the level of fibre in the diet whether vegetable, fruit or cereal in origin. On the other hand in a recent study of the dietary habits of colon cancer patients, Graham *et al.*[274] found association of the disease with a decrease in consumption of vegetables.

The protective effect of dietary fibre on colon cancer is purported to be due to a multiplicity of actions of fibre in the gut including (a) decreased transit time, which reduces the residence time of a toxic chemical in the gut; (b) increased stool weight which dilutes the toxic chemical; (c) alteration of numbers and types of intestinal bacteria which can metabolise chemicals in the gut contents to carcinogens; (d) sequestration or absorption of the chemical by the colloid matrix of fibre which facilitates the elimination of the former.[275]

Transit Time and Stool Weight. It has been suggested that high-fibre content of the diet is responsible for the faster transit times and high faecal bulk (200–400 g/day) of rural African populations by comparison to western populations (approximately 80 h and 100 g/day respectively).[276,277] It should be noted, however, that a difference in fibre contents of the diet was only one of many cultural, dietary and economic differences between the two populations studied. In a recent study of two Scandinavian populations[197] one from rural Finland and one from Copenhagen it was found that, despite a higher intake of dietary fibre in the rural population (31 g against 17 g in Copenhagen) bowel transit times were not significantly different in the two groups, although stool weights were slightly higher in the Finnish group.

Experimental changes in the fibre content of human diets especially by addition of wheat bran have in general shown that the addition of fibre increases faecal weight and decreases transit time[39,278] although some conflicting results have been obtained[279–281] and pectin has been shown to have no effect on colonic function. There are, however, major differences in the response of colonic function to different sources of fibre, possibly related to pentose-polysaccharide content, and a wide variation in the response of individuals to standard fibre intake.[282,283]

Effect on Intestinal Bacteria. Most studies have shown that the alteration of the fibre content of the diet has little, if any, effect on the numbers of bacteria in the faeces[35,39,278] although differences in faecal bacteria are found when geographically distinct populations are studied (see p. 190).

Sequestration by Fibre. Studies of the absorptive capacity of fibre *in vitro* have concentrated on the binding of bile salts which are purported to be implicated in large bowel cancer (see p. 212). These studies showed that fibre preparations from different sources vary in their ability to bind bile salts.[284,285]

Attempts to assess the implications of the steroid binding capacity *in vivo* have led to conflicting results and this area is still the subject of considerable debate. The work has concentrated on two main subjects:

(a) The effect of fibre on the concentration of faecal steroids and on the bacterial metabolism of bile acids since if bile acids and their metabolites are involved in colon cancer, a reduction in their concentration in faeces might reduce the risk of large bowel cancer.
(b) The influence of fibre on cholesterol metabolism and on serum cholesterol since the high blood levels of the latter are thought to be involved in atherosclerosis and, in addition, cholesterol and its metabolites have been proposed as possible carcinogens or co-carcinogens for the gut (see p. 215).

Although faecal steroids are found in lower concentrations in the stools of people in Africa and India (areas of low colon cancer incidence) than in people in the UK or USA[286] this difference cannot be ascribed solely to differences in fibre content of the diet since the dietary proportions of fat and protein are very different as well. A study of the faecal steroids of two Scandinavian populations, one in rural Finland (low colon cancer incidence, high dietary fibre, protein and fat) the other in Copenhagen (high colon cancer incidence, low fibre, lower fat and protein intake) revealed no significant differences in faecal concentrations of acid steroids or in the total steroids excreted.[197]

In contrast, a comparison of the Finnish population with a New York population showed that the faecal concentrations of neutral steroids was similar in the two groups, but, due to the higher faecal output (three-fold higher), of the Finnish subjects their daily excretion of cholesterol and its metabolites was higher.[287] The bile acid concentration in the stools of the Finnish population was only about one third that in the New York group but the higher stool weight of the former results in virtually identical daily outputs of total bile acids between the two groups. This investigation suggests the increased stool bulk resulting from high fibre intake dilutes faecal bile acids.

There is experimental evidence to support the diluting effect of fibre on faecal steroids in both man and animals. Walters *et al.*[278] studying the effects of dietary supplements of wheat bran or bagasse in man found the two types of fibre decreased the faecal concentration of neutral steroids but only bran reduced the faecal concentrations of bile acids. No changes in the bacterial degradation of acid or neutral steroids were detected.

The effects of fibre on faecal steroid excretion are complicated not only by the nature of the fibre used but also by the levels of fat and cholesterol in the diet, the fibre having a greater effect on acid and neutral steroid excretion at low dietary levels of fats.[288,289]

Effect on Toxicity of Foreign Compounds. The presence of fibre in the diet has been shown to reduce the toxic effects of a variety of compounds including glucasorbic acid,[290] amaranth,[291] polyethylene glycol, Tween 20[292] and sodium cyclamate.[293] In some cases, the ameliorating effects of the fibre were extremely potent, for example, amaranth, when incorporated into a stock diet at a concentration of 5 per cent had no adverse effects on weight gain or appearance of rats over a period of 21 days. However, at the same concentration in a semi-purified diet, amaranth inhibited growth and caused death in 14 days. These toxic effects could be counteracted by addition of cellulose or alfalfa.[291] There is also evidence that the tumorigenic effects of azoxymethane[294,295] and 2-acetylaminofluorene[296] can be reduced by dietary fibre.

Minor Dietary Constituents. The gastrointestinal tract may be involved in the toxicity of minor dietary constituents in two ways: the toxicity of orally ingested constituents may be modified by metabolism in the gut or may affect normal gut physiology and function. Examples of the former involvement include cycasin, cyanogenic glycosides and certain isoflavonoid components where metabolism in the gut leads to the formation of metabolites which are more toxic than the parent compound. Examples of interferences with normal physiological processes in the gut are afforded by inhibitors of the enzymes involved in the digestive processes, such as the protease inhibitors.

Cycasin. Cycasin is a glucoside of methylazoxymethanol which occurs in cycad nuts and other parts of the cycad palms (*Cycas circinalis*). Consumption of improperly prepared products from cycads has been associated with jaundice, haemorrhage and malignant tumours in liver and kidney.[297,298] Investigations into the toxic constituents involved led to the isolation of cycasin and related glycosides.[299,300]

Cycasin is toxic and highly carcinogenic when fed to conventional rats and tumours are also produced in mice, guinea pigs, hamsters and fish (see review[301]). However, it is much less toxic and non-carcinogenic when fed to germ-free rats or when administered parenterally,[147] although the aglycone, methylazoxymethanol is toxic and carcinogenic by intraperitoneal injection.[302] It is now understood that cycasin has to be deglucosylated before its toxicity is displayed and that the β-glucosidase activity of the gut microflora effects this hydrolysis: β-glucosidases are normally absent from mammalian tissues which explains the lack of carcinogenic effects of cycasin in germ-free animals or on parenteral administration.[303] In these cases, unchanged cycasin is eliminated in the urine.

An exception to the above general observations appears to be the report that cycasin is carcinogenic after injection in neonatal rats[304] but this has been rationalised by the discovery that β-glucosidase activity is present in subcutaneous tissue of early post-natal rats.[305] Alternatively, the explanation may reside in the fact that cycasin is toxic to the offspring when fed to lactating rats due to secretion of the aglycone in milk. Thus cycasin excreted in urine by the young may be ingested by the dam licking the offspring, hydrolysed in the gut and the aglycone transmitted to the suckling pups.

Cyanogenic Glycosides. Cyanogenic glycosides, i.e. glycosides which yield HCN on hydrolysis, are widely distributed in the plant kingdom and over 1000 different species have been reported to contain cyanogens. They are glycosides of α-hydroxynitriles, of which at least 20 are known, the most important in relation to possible human intoxication being amygdalin, linamarin and dhurrin (see review by Conn[306]). Amygdalin occurs in several drupes and pomes, including almond, apricot, apple, cherry and plum, linamarin is the principal cyanogen in cassava and some pulses, and dhurrin is found in sorghum.

Hydrolysis of the glycoside by β-glucosidases liberates the aglycone, an unstable cyanohydrin, which decomposes spontaneously with the liberation of HCN. As discussed above, β-glucosidase activity is absent from mammalian tissues but is displayed by the gut microflora, and the acute toxicity of the cyanogenic glycosides is much greater after oral ingestion in conventional animals than in germ-free animals or after parenteral injection.[70]

Oestrogens. Compounds which exibit oestrogenic activity are widely distributed in plants including some consumed by man such as carrots,[307] soya beans,[308] cereals and potatoes.[309,310] A number of isoflavones have been identified from these sources and shown to possess weak oestrogenic activity. Studies on the metabolism of these compounds by rat caecal microflora *in vitro* have shown that, by a sequence of reactions involving *O*-demethylation, and reduction, daidzein and formononetin are converted into the isoflavan, equol[311] and similar results have been reported with sheep rumen contents.[312] Equol is also a metabolite of genistein, biochanin A and formononetin in the hen.[313,314] Rat caecal micro-organisms can also effect heterocyclic ring fission in isoflavones such as genistein, which is hydroxylated in the 5- position, with the formation of 4-ethylphenol,[311] and biochanin A is similarly converted into 4-ethylphenol on intraruminal administration to sheep.[315] In the latter case, *O*-demethylation in the 5-position may precede ring fission.

Enzyme Inhibitors. Inhibitors of the digestive enzymes of the gastrointestinal tract may impair the utilisation of food constituents and even produce toxic effects mimicking congenital enzyme deficiencies; there may also be secondary effects on other organs.

The most widely studied group of enzyme inhibitors are the protease inhibitors.[316] Inhibitors of trypsin and chymotrypsin are found in important dietary staples such as legumes, cereals and potatoes as well as many other vegetables and eggs. Of these, the soya-bean inhibitors of proteolytic enzymes have been most comprehensively investigated.

Soya-bean trypsin inhibitor causes a depression of growth in rats and chicks, and there is an apparent impairment of protein utilisation. Heat inactivation of the inhibitor or supplementation of the diet with methionine and cystine improves protein utilisation but differences in digestibility between unheated and raw soya meal are insufficient to account for the differences in biological value. Further, the net absorption of nitrogen and sulphur is similar in both cases in the rat. In part, this can be explained by the observation that nitrogen absorption in the distal ileum is impaired but the undigested protein is hydrolysed and absorbed in the lower gut where it has poorer utility for growth. However, there is much evidence that the growth impairment is not simple nutritional deficiency resulting from impaired protein utilisation since, in rodents, there is still growth retardation when anti-tryptic preparations are included in diets containing pre-digested protein or amino acids. Furthermore, some reports have indicated that trypsin inhibitors do not depress proteolysis in the intestine of rats and mice.

A feature of the biological effects of soya-bean inhibitor is that it induces a pancreatic hypertrophy in rats and chicks (but not in calves, dogs or adult guinea pigs). There is true hyperplasia of pancreatic tissue and increased secretion of proteolytic enzymes from the hyperactive organ, and even if the diet is supplemented with amino acids to prevent growth impairment, the pancreas remains enlarged. Thus the growth depression and increased requirement for methionine may result from endogenous losses of essential amino acids from the hyperactive pancreas which is responding in an adaptive way to the presence of inhibitor in the small intestine. The endogenous losses of amino acids appear largely in the faeces of chicks but not of rats. In the latter, it appears that the amino acids are degraded by the microflora of the lower gut and the beneficial effects of antibiotics in diets containing raw soya beans may be explained in this way.

The mechanism by which trypsin inhibitors cause pancreatic hypertrophy is not clear but presence of the inhibitor in the intestine releases a humoral factor which has been partially characterised and which stimulates pancreatic secretions. The enlargement may thus be considered a consequence of prolonged stimulation due to inhibitor in the diet. There are two hypotheses, not mutually exclusive, through which the regulation of pancreatic secretions may be explained. One postulate is that trypsin and chymotrypsin regulate pancreatic activity via a negative feedback mechanism, i.e. the presence of these enzymes in the gut suppresses pancreatic secretion. Inhibitors, by combining with the enzymes, might be expected to interfere with this feedback loop. Alternatively, it might be postulated that the inhibitor itself stimulates secretion by a positive feedback mechanism.

The differences between the chick (in which there is increased faecal loss of *S*-amino acids) and rat (in which faecal loss is minimal) has been explained as follows. In the young chick the pancreatic response is delayed and the amount of trypsin secreted is insufficient to compensate for the presence of the inhibitor and faecal losses of amino acids occur. In the rat (and older hen) the proteolytic enzymes are secreted in increased amount, sufficient to compensate for the presence of the inhibitor, and much of the *S*-amino acid present in the gut is of endogenous origin. In this case, the endogenous protein is lost due to bacterial degradation in the lower gut.

The significance of these observations for man is questionable since soya-bean trypsin inhibitor does not cause pancreatic enlargement in larger species such as calves, pigs, dogs and even adult guinea pigs. This led to the recognition that pancreatic enlargement was only seen in those animals where the normal pancreas is greater than about 0.4 per cent of body weight. In other species with a smaller relative weight of the pancreas, no enlargement occurs. Thus man, with a relative pancreas weight of 0.09–0.12 per cent of body weight, might not be affected in this way.

Haemagglutins (lectins). Lectins are proteins which possess the ability to agglutinate erythrocytes in a similar way to antibodies. The most significant dietary sources of these compounds are the legumes but lectins have also been detected in potato and cereals (see review by Jaffe[317]). In most cases, the lectins are substantially more toxic by parenteral routes of administration than on oral ingestion and in order to exhibit oral toxicity the lectin must be resistant to proteolysis in the gut. In this respect, kidney-bean lectin is particularly resistant to digestion and has most frequently been involved in human intoxication. It appears that the bean lectin combines with mucosal cells lining the intestine causing non-specific interference with absorption of nutrients and an acute gastro-enteritis. Despite the long-recognised association between consumption of beans which have been insufficiently heat treated and gastrointestinal toxicity, cases of human intoxication continue to be reported.[318]

Food Additives

The gastrointestinal tract may feature in the toxicology of food additives in three principal ways, namely: (a) the additive may be extensively metabolised in the gut, thus modifying the toxicity of the original compound; (b) the additive, or its metabolites, may undergo enterohepatic circulation (or in the case of nitrate, gastro-salivary circulation) prolonging exposure and delaying excretion; or (c) the additive may induce changes in gut morphology (sometimes of a pathological nature), function and/or microbial population. Examples of all of these features are included in this section.

Food Colours. Most of the synthetic colours permitted for use in food is the yellow, orange, red and brown range are azo compounds. The black colours, Black 7984 and Black PN are *bis*-azo compounds with two azo groups, as are some of the components of Brown FK. The metabolism of azo compounds has been studied more comprehensively than most other classes of food additive (see review by Walker[97]) and the important role of metabolism by the gut microflora is well established.

Early studies on the metabolism of Orange 1 in dogs[319] and on the azo-drug, Prontosil, in mice, rabbits and man,[95,320] indicated the importance of reductive cleavage of the azo link *in vivo*, but it was not until the comprehensive study by Radomski and Mellinger[321] on Amaranth (FD & C Red No. 2), Ponceau SX (FD & C Red No. 4) and Sunset Yellow FCF (FD & C Yellow No. 6) that the importance of the gut microflora in effecting this reduction was established. Amaranth, Ponceau SX and Sunset Yellow FCF, which are water-soluble, sulphonated azo compounds, were observed to be poorly absorbed from the gut to the extent of 2–4 per cent of the orally administered dose in rats and the extent of azo-reduction was diminished by co-administration of antibiotics. Similarly, Scheline and Longberg[135] reported low absorption of Acid Yellow given to rats *per os* and concluded that this sulphonated azo dye was almost quantitatively reduced by the gut microflora. In both of the above studies, when the colours were administered parenterally, they were excreted in urine and/or bile with the azo group intact, confirming that reduction by mammalian azo reductases was insignificant compared to the gut microflora. These observations were not confined to the rat since Jones *et al.*[322] carried out a comparative study of the metabolism of tartrazine in rat, rabbit and man, and found that the reduction product, sulphanilic acid, was excreted virtually quantitatively after oral administration, whereas unreduced dye was excreted after intraperitoneal injection.

Since these early studies, there has been adequate confirmation that the gut microflora of a range of experimental animals and man can quite generally reduce azo food colours *in vitro* and *in vivo*. Dyes which have been studied in this way include Brown FK,[323] Red 10B, Red 2G and Red 6G,[324] Lissamine Fast Yellow 2G,[325] Orange G[326] and Ponceau 4R,[327] the *bis*-azo dye, Black PN was observed to undergo sequential reduction of the two azo links both in rat and man.[328] Even polymeric azo food colours have been observed to undergo reduction by the gastrointestinal microflora.[329]

In some cases, sulphonated azo dyes which have been administered parenterally are excreted to a significant extent in bile and hence are exposed to metabolism by the gut microflora. Thus, in a comparative study with 16 sulphonated azo dyes administered intravenously,[330] Lissamine Fast Yellow 2G was shown to be excreted almost quantitatively in bile while Red 2G and Ponceau SX were excreted by this route to the extent of 64 per cent and 50 per cent of the administered dose, respectively. Other azo dyes were found at lower levels in bile. No simple relationship between molecular weight or structure and the extent of biliary excretion has been observed. Clearly, in the absence of biliary excretion studies, the extent to which parenterally administered azo compounds are metabolised by the gut microflora cannot be accurately assessed.

The reduction of azo compounds in the gastrointestinal tract has a number of important toxicological sequelae. The amines produced by azo-reduction may themselves be toxic or be converted into toxic metabolites in the tissue after absorption. This in turn may lead to quantitative differences in the toxicity of azo compounds when given orally compared with parenteral dosing. Furthermore, interspecies differences in gut metabolism may give rise to differences in toxic response. Two examples illustrate these points.

In studies of the toxicity of Brown FK it was observed that, when this dye was repeatedly administered intragastrically to rats it produced a vacuolar myopathy of cardiac and skeletal muscle which was not observed after intra-peritoneal dosing.[331] Similar lesions were also observed in rabbits and guinea pigs but not in mice. This work suggested that amine metabolites of Brown FK, produced by azo-reduction in the gut, were responsible for the myotoxic effects and this was confirmed in a subsequent study[332] in which these amines were administered parenterally, i.e. the amines formed in the gut appeared to be absorbed and effect the muscle injury. In the latter study, the amines were also shown to be myotoxic to mice, leading to the conclusion that the interspecies differences in toxicity of intra-gastrically administered Brown FK resulted from differences in metabolism in the gut. In passing, it might also be noted that the amines from reduction of Brown FK appear also to be mutagenic.[333]

A second example is that of Orange G which liberates aniline on reduction of the azo group. This, and a number of related dyes (Red 2G, Red 10B, Orange RN), produces Heinz bodies in the erythrocyte and a haemolytic anaemia when administered orally to rats but not to ferrets.[326] The haematological effects in rats are what might be expected from exposure to aniline and consequential on intestinal azo reduction. Metabolic investigations revealed that the inter-species difference resulted from the fact that azo-reduction was occurring to a considerably lesser extent in the ferret and what reduction did occur appeared to take place in the lower gut just prior to excretion with little absorption of aniline. *In vitro* studies on Orange G revealed that the intestinal microflora of the ferret was capable of reducing the dye and this suggests that gut morphology and residence time is important in determining the extent of azo-reduction. Interestingly, metabolic studies in man revealed that azo-reduction was as complete as in the rat so that, in this case, the rat was a better metabolic model for man than was the ferret.

In contrast to the extensive metabolism of azo food colours in the gut, the other classes of water-soluble colours though poorly absorbed are not metabolised to any major degree. Thus the sulphonated triphenylmethane colours are largely excreted in the faeces unchanged[334–338] and this accounts for their low oral toxicity. Rather surprisingly, unlike many polar, poorly absorbed compounds, the triphenylmethane colours do not appear to exert a cathartic effect nor lead to an increase in caecal size. Conversely, Erythrosine BS which is also poorly absorbed and not degraded, has been reported to cause caecal enlargement and an increased output of faeces with higher water content than control animals.[339] This dye also causes changes in the microbial populations of the gut which may be causally related to the observed caecal enlargement (Rowland and Mason, unpublished results).

Indigo carmine (FD & C Blue No. 2) is poorly absorbed from the gastrointestinal tract and much of an orally administered dose is excreted unchanged[340,341] but some metabolism does take place as indicated by the finding of isatin-5-sulphonic acid and 5-sulphonanthranilic acid in the urine and bile of rats dosed orally. *In vitro* studies indicated that the rat gut microflora was involved in forming these metabolites and this is a rare example of oxidative metabolism in the gut.

Food Flavours. In view of the very large number of compounds used as food flavours a comprehensive review is not possible but a few cases in which gut metabolism may significantly influence the toxicity will be cited as examples.

Coumarin is a naturally occurring component of the essential oils of various plants and found use as a food flavour until hepatotoxic effects were observed in experimental animals.[342] The many subsequent studies on coumarin toxicity have been the subject of a recent, thorough review.[343] Major inter-species differences in metabolism and toxicity have been observed, with the rat being particularly sensitive to the hepatotoxic effects. In the rat, coumarin causes bile duct proliferation accompanied by fibrosis but the relevance of this for man is doubtful in view of the metabolic differences. Thus, in rat and rabbit, *O*-hydroxyphenylacetic acid is the major urinary metabolite while, in man, 79 per cent of an orally administered dose was excreted as 7-hydroxycoumarin. The differences in metabolism result, in part, from differences in hydroxylating activity in the liver but Scheline[72,130] drew attention to the possible involvement of the rat intestinal microflora in reducing coumarin to dihydrocoumarin and scission of the heterocyclic ring to yield *O*-hydroxyphenyl propionic acid. Furthermore, coumarin has been observed to undergo biliary excretion in the rat but not to any significant extent in man.[344] Biliary excretion may increase the extent of metabolism by the gut microflora and, additionally, by concentrating toxic metabolites in the bile ducts, influence the development of the lesions seen at this site.

Many esters are used as food flavourings and the role of intestinal esterases in hydrolysing them to the corresponding acids and alcohols has been demonstrated in a number of cases.[345–347] Hydrolysis of simple esters appears to be general and essentially complete after oral ingestion so that the toxicological properties of esters are largely predictable from a knowledge of the acids and alcohols liberated.

Oxygenated monoterpenes are major components of essential oils and widely used food flavours. Metabolic data on this class of compound are sparse but, in the case of linaloöl, substantial entero-hepatic circulation was observed after oral administration to the rat.[73] In addition to hydrolysing biliary conjugates of linaloöl, the gut microflora was also implicated in the reduction of double bonds to yield dihydro- and tetrahydro- linaloöl. Enterohepatic circulation may also occur with metabolites of citral[348] and the non-oxygenated terpene *d*-limonene.[349] Reduction of the monoterpene ketone, carvone, to yield the corresponding tertiary alcohol and metabolites in which an ethylene link was also reduced has been reported *in vivo* in the rabbit[350,351] and *in vitro* studies with rat caecal microflora have indicated that these reductions may be a feature of metabolism in the gut (R. Walker, unpublished results). The toxicological significance of reduction and enterohepatic circulation of terpenoid flavours remains to be established but clearly leads to delayed excretion and prolonged exposure to metabolites.

Synthetic Sweeteners. While a number of studies have indicated that saccharin is incompletely absorbed when administered at high levels in the diet, and this sweetener does not undergo metabolism to an appreciable degree,[352] effects on the gastrointestinal tract may still be of importance toxicologically. At high dietary concentrations, saccharin produces a transient diarrhoea and an increase in faecal water content with an associated caecal enlargement in the rat,[353,354] additionally there are changes in the caecal microflora with increased counts of aerobes and qualitative changes in the anaerobes. While these changes may be of toxicological significance *per se*, it has been proposed[353] that, by secondary effects on mineral absorption and on the microbial urease, urine composition and pH is altered and that this may be important in relation to the production of bladder tumours. The urinary changes could conceivably lead to crystalluria which frequently is associated with bladder carcinoma in the rat.

At high oral dose levels, cyclamate also produces similar effects on the gastrointestinal tract with increased water retention and an increase in faecal water content. However, cyclamate may also undergo metabolism by the gut microflora to yield cyclohexylamine and this conversion has been observed in several species including man, dog, cat, rat, guinea pig and rabbit (see review[355]). Interestingly, Renwick and Williams[148] found that this reaction did not occur in naive rats given a single oral dose of cyclamate but, after prolonged exposure to cyclamate in drinking water, the gut microflora developed the ability to convert cyclamate into cyclohexylamine. These observations gave an explanation to the variable results in other metabolic studies, including experiments in man. Thus studies in humans[356–358] revealed major inter-individual differences in the ability to produce cyclohexylamine and Japanese subjects appeared better converters than US citizens. Other dietary factors as well as cyclamate exposure may lead to differences in gastrointestinal microflora and account for these observations. The significance of these observations in relation to the potential carcinogenicity of cyclamate requires further clarification but cyclohexylamine has been reported to be mutagenic in some *in vitro* assays.

One general point to emerge from these studies on cyclamate and adaptation by the gut microflora is that qualitative differences in metabolism and toxicity may result from prolonged exposure to test compounds. It is rare to find such comprehensive studies on metabolism by the gut microflora *in vivo* and there may be many more examples of such adaptations of which we are not yet aware. It would seem advisable that metabolic studies should be performed not only as is usual on naive animals but also at intervals after continuous exposure to test compounds.

Emulsifiers, Stabilisers and Thickening Agents. The sulphated polysaccharides, carrageenans, have been extensively studied in relation to pathological changes produced in the gastrointestinal tract and there are marked interspecies differences in this response to carrageenans. This topic is discussed on p. 249. Interestingly, alginates which are also seaweed polysaccharides, do not appear to cause adverse effects on the gastrointestinal tract of guinea pigs.[359]

Studies on a number of chemically modified starches used as thickening agents have indicated that they produce effects on the gastrointestinal tract in rats and pigs, particularly enlargement of the caecum.[360–362] The phenomenon of caecal enlargement is discussed in more detail on p. 246 and appears to relate to the impaired ability of pancreatic amylase to hydrolyse these materials in the small intestine. In some cases, renal pathology has also been observed in the form of increased mineralisation or nephro-calcinosis; these effects may be secondary to changes in mineral absorption resulting from the primary effects on the gut. Further work is required to clarify this point.

Modified celluloses find use as thickeners and stabilisers in a wide range of foods and generally are not absorbed to any significant extent from the gastrointestinal tract.[363] Trace amounts only of ^{14}C-labelled modified celluloses have been detected in the carcass in metabolic studies (e.g. Braun *et al.*)[364] which may have resulted from a very small degree of breakdown in the gastrointestinal tract or from persorption.

Intestinal metabolism has been shown to be extensive with Emulsifier YN, a mixture of ammonium salts of phosphatidic acids.[365,366] In the rat, mouse, guinea pig and ferret this emulsifier was rapidly hydrolysed in the gut to inorganic phosphate and lipid. Similarly, sucrose esters of fatty acids are rapidly hydrolysed to sucrose and free fatty acids in the gut[367,368] and these observations add a rationale to the low toxicity of these compounds.

Toxicity tests on a range of Spans (sorbitan esters of long chain fatty acids) showed that, at very high dietary levels, they could cause gastrointestinal irritation and diarrhoea;[369,370] and sorbitan stearate was shown to decrease the total numbers of intestinal organisms in rats at dietary levels of 25 per cent[371] but not at 10 per cent.[372] In view of the large dietary levels required to produce these effects, it seems unlikely that they are significant at normal levels of human exposure. Metabolic studies indicate that the ester link of fatty acid esters of sorbitan is hydrolysed in the gastrointestinal tract before absorption with the fatty acid entering the normal body pool and the sorbitan moeity being largely excreted unchanged in the urine.[373] The effects of Spans on other organs, particularly the liver and kidney[374,375] may thus be a feature of the sorbitan load in urine.

Ester hydrolysis in the gastrointestinal tract is also prominent in the metabolism of some other non-ionic emulsifiers. Thus polyglycerol esters are readily hydrolysed; the liberated fatty acids are utilised in normal metabolic processes but the polyglycerol moiety is not and is excreted.[376] Similarly, stearoyl propylene glycol hydrogen succinate is hydrolysed to a large extent by pancreatic enzymes in the gut prior to absorption[377] with the liberation of stearic and succinic acids.

Preservatives. In general, food preservatives control spoilage as a consequence of their anti-microbial activity and it might therefore be expected that such additives would affect the microbial population of the gut. However, systematic studies of the effects of preservatives on the gut microflora are lacking.

High oral doses of the preservative sulphur dioxide (or salts of sulphurous acid) cause irritation and ulceration of the gastric mucosa and this topic is discussed in more detail on p. 248.

The use of nitrites and nitrates as preservatives has far reaching toxicological consequences and this is discussed more fully on p. 242.

Food Contaminants

Pesticides. Much of the work on pesticide breakdown has concentrated on degradation in soil and aquatic ecosystems, while metabolism in the gastrointestinal tract is less well studied. For obvious reasons, the interaction of pesticides with ruminal micro-organisms has been investigated thoroughly. Although the types of bacteria are not the same as in the human gut or even that of laboratory animals, the ruminant studies provide some clues as to the likely fate of pesticides in human gut.

Organochlorine Pesticides

p,p-DDT[1,1,1-trichloro-2,2-bis(*p*-chlorophenyl)ethane]. After administration of *p,p*-DDT orally to rats or rabbits, *p,p*-TDE and unchanged pesticide are found in the liver.[116] In contrast, when the pesticide is given intraperitoneally, little or no *p,p*-TDE is found, observations which suggest that gut enzymes participate in gut DDT metabolism. Several genera of gut microorganisms, including bacteroides, clostridia and especially coliforms have been shown to reductively dechlorinate *p,p*-DDT to TDE *in vitro.*[378,379] Other metabolites were also found but no chlorine atoms on the aromatic ring were removed.

Dieldrin. Matsumura *et al.*[380] investigated the metabolism of dieldrin by various unidentified aquatic and intestinal bacteria. The major dieldrin metabolite produced by micro-organisms isolated from the rat gut was photodieldrin which has been reported to be more toxic than the parent compound to a variety of biological systems.[381]

Other organochlorine pesticides. Apart from DDT and dieldrin the intestinal metabolism of organochlorine pesticides has not been well studied. The majority of these pesticides, however, are in general recalcitrant to microbial degradation in anaerobic ecosystems.[378]

Pesticides Containing Nitro Groups. A wide variety of pesticides containing nitro groups is degraded by rumen contents[378] and it seems reasonable to suppose that the active nitroreductase in human gut contents can effect similar degradations. It is known that dichloran (2,6-dicholoro-4-nitroaniline) is reduced by *E. coli* to form 2,6-dichloro-*p*-phenylenediamine[382] and similarly TCNB (2,3,4,5 or 2,3,4,6-tetra-chloro-1-nitrobenzene) is reduced to tetrachloroaniline by rabbit intestinal contents.

The nitro groups on the pesticides 'Nitralin', 'Benefin', 'Dinobuton', 4,6-dinitro-*o*-cresol, 'Dinoseb' parathion and 'Trifluralin' can be reduced to amino groups by cultures of rumen microorganisms,[378] but except for 'Trifluralin'[383] their metabolism by human or rat intestinal bacteria has not been studied.

There is no evidence that the above pesticides are further metabolised by deamination. However, it is likely that the ability of ruminal organisms to *N*-dealkylate 'Trifluralin' and 'Benefin',[378] may also be shared by bacteria derived from humans.

Organophosphates. The high activity of esterases in the mammalian intestine (both of bacterial and animal origin) render the hydrolysis of organophosphates in the gut likely. In rumen fluid, tetraethylpyrophosphate and mevinphos are rapidly metabolised[384] as are parathion and malathion which are converted into paraoxon and malaoxon respectively.

Diquat and Paraquat. When given orally to rats, paraquat and diquat are extensively metabolised by intestinal bacteria, with about 70 per cent of the dose of diquat and 30 per cent of paraquat dose appearing in the faeces of the degraded form.[385]

Metals and Organometallics. A variety of transformations of toxic metals have been demonstrated in the presence of intestinal micro-organisms. Since the toxicity of such metals is profoundly influenced by their chemical form, gut flora metabolism is potentially of great significance. Methylation of inorganic mercury to form the highly toxic methylmercury occurs *in vitro* in the presence of gut contents from the rat and human faeces[136,386,387] and in rat jejunal blind loops *in vivo.*[388] Various groups of bacteria were found to be responsible for the methylation of mercury including *E. coli*, lactobacilli, bacteroides and bifidobacteria.[386]

Arsenic is also metabolised by the gut flora, sodium arsenate being reduced to arsenite and then methylated to methyl and dimethyl arsenic which are less toxic than inorganic arsenicals.[389]

Small intestine and caecal contents of the rat metabolise methylmercury to volatile products, metallic mercury and *bis*-methylmercuric sulphide, the latter being formed by the reaction of methylmercury with hydrogen sulphide generated by members of the gut flora.[390–392]

The significance of the gut flora metabolism of methylmercury to the overall metabolism, tissue distribution and toxicity of the organomercurial was explored using rats treated with antibiotics to eliminate their intestinal microflora.[47] After oral doses of methylmercury the tissue content of mercury in the kidneys, brain, blood, muscle and lung was significantly higher (by 35–40 per cent) in the antibiotic-treated rats compared with conventional rats, and furthermore the latter animals excreted over twice the amount of mercury in the faeces than did the rats without gut flora. In the kidneys of the antibiotic-treated rats a higher proportion of the mercury was present as methylmercury. Thus the gut flora would appear to reduce the tissue content of mercury after oral dosing with methylmercury, and this has now been shown to reduce the toxicity of methylmercury. Neurological signs of toxicity after methylmercury treatment were seen in four out of five antibiotic-treated rats but in only one out of five rats which possessed a gut flora,[47] indicating that the gut flora plays a protective role in methylmercury toxicity.

The metabolism of organo arsenicals has been studied in hens since several of these compounds are used as growth promoters and disease-control agents in poultry. Of the compounds studied, arsanilic acid and acetylarsanilic acid were excreted virtually unchanged after oral doses, but 4-nitrophenylarsonic acid and 4-hydroxy-3-nitrophenyarsonic acid were reduced to arsanilic acid (4-aminophenylarsonic acid) and 3-amino-4-hydroxyphenyl arsonic acid in the gut.[114,393,394] Furthermore, when incubated anaerobically with hen faeces, 4-nitrophenylarsonic acid was converted into 4-nitrophenylarsenoxide, 4-aminophenylarsenoxide and arsenilic acid.[114]

Packaging Materals

Phthalate Diesters. Diesters of phthalic acid are extensively used as plasticisers in various plastic formulations, including those used in food packaging, and the esters can comprise as much as 40 per cent of the final weight of the material.[395] The acute toxicity of the phthalate esters is low[396,397] but chronic administration to rats of some esters has led to liver enlargement, changes in hepatic enzyme activity and, in the case of di-*n*-butyl- and di-(2-ethylhexyl)-phthalate (DEHP), to testicular atrophy.[398–402]

When given orally to the rat, the diesters are excreted in the urine as derivatives of the corresponding monoesters.[403,404] The partial hydrolysis to the monoester appears to be a crucial step, not only in the metabolism of the diester but also for its toxicity, since the hepatic changes produced by DEHP can be substantially reproduced by the oral administration of mono-(2-ethylhexyl) phthalate.[400] The major site of this hydrolysis is the gut and both small intestinal contents and intestinal mucosa preparations from the rat have been shown to metabolise a wide variety of phthalate diesters.[8,405,406] In general, the diesters with short alkyl chain lengths (dimethyl-, diethyl- and dibutyl-phthalates) were hydrolysed much more rapidly than phthalates with longer side chains (di-*n*-octyl-, dicyclohexyl- and di-(2-ethylhexyl)-phthalates). Lake *et al.*[406] also showed that intestinal preparations from ferret, baboon and man could hydrolyse the diesters.

Mammalian enzymes, both extracellular lipase derived from the pancreas,[407,408] and esterases associated with mucosal cells[405,406] appear to be the major source of diester-metabolising activity, especially in the case of dimethyl-, diethyl- and dibutyl-phthalates. However, Rowland[8] showed that the gut flora contributed to the hydrolysis of DEHP in the rat.

It was also found that feeding DEHP to rats increased, by threefold, the rate of hydrolysis of DEHP by intestinal contents apparently due to increased activity in mucosal enzymes.[8]

Nitrate, Nitrite and Nitrosamines

Considerable amounts of nitrate are ingested in the diet daily, primarily from vegetables and drinking water but cured meats make an additional contribution.[409] Nitrate *per se* is relatively non-toxic but toxicological problems arise from its reduction to nitrite *in vivo*. This reduction is due almost exclusively to the action of the microflora which inhabit the alimentary tract. The extent to which this reduction occurs varies with species and, within a given species, with age and physiological condition. Ruminants, especially cattle, with a heavily populated portion of the upper gastrointestinal tract, are particularly able to reduce nitrate to nitrite and there are many well-documented cases of cattle dying from nitrite intoxication after consuming vegetation containing high levels of nitrate. In the normal adult human, the stomach is virtually sterile and nitrate is absorbed before reduction to nitrite occurs. However, the pH of the stomach of the neonatal infant is much higher and bacterial colonisation occurs which leads to a greater reduction of nitrate than in the adult.[410] As a consequence of this and the fact that fetal haemoglobin, which forms 60–80 per cent of the circulating haemoglobin at birth, is more readily oxidised by nitrite to methaemoglobin than is mature haemoglobin, many cases of infant methaemoglobinaemia have resulted from ingestion of nitrate, particularly in water.[411] Epidemiological evidence indicated that no cases of methaemoglobinaemia had been reported in the US resulting from consumption of water containing less than 45 mg nitrate/litre and this figure has been used as a maximum in WHO standards for drinking water.[412] However, clinical and sub-clinical cases have been reported in Europe where the concentration of nitrate in water was less than 45 mg/litre.[413] Interestingly, high nitrate levels in vegetables do not seem to present the same hazard (except where reduction to nitrite has occurred prior to consumption) and vegetables may contain an inhibitor of reduction by alimentary micro-organisms.

It is not only in the stomach of the neonate or achlorhydric adult that significant reduction of nitrate to nitrite can occur. It has long been known that nitrite is present in human oral saliva.[414] However, saliva taken directly from the salivary gland ducts contains nitrate but not nitrite.[415–417] Reduction of salivary nitrate is effected by a variety of micro-organisms which populate the oral cavity and may be diminished by the use of bactericidal mouth washes.[418] Recently a mammalian nitrate reductase has also been identified in the tissues of the lingua dorsa; this enzyme may play a subsidiary role in the reduction of nitrate in the oral cavity. Several studies have demonstrated that dietary nitrate is a significant source of salivary nitrite[418–424] and concentrations of nitrite in saliva depend *inter alia* on the amount of nitrate ingested. In this work, peak nitrite concentrations of several hundred mg/litre have been recorded after ingestion of foods or beverages rich in nitrate. On the bases of an average concentration of nitrite in saliva much lower than this, White[425] calculated that as much as 68 per cent of the total daily intake of nitrite in US residents arises from reduction of nitrate *in vivo*.

The major interest in the toxicological consequences of exposure to nitrate and/or nitrite relates to the possibility that such exposure may lead to the formation of carcinogenic nitrosamines and nitrosamides in the alimentary tract, particularly the stomach. The most common synthetic route to nitrosamines is via the reaction of nitrite with secondary amines under acid conditions, weakly basic amines nitrosating more readily than strongly basic ones.[426] The rate of this reaction depends on the concentration of undissociated nitrous acid and of unprotonated amine, and is therefore pH-dependent with an optimum commonly in the range 2.5-3.5, i.e. around the post-prandial pH of the stomach. It is now known that dialkylnitrosamines may also be synthesised from tertiary amines[427] and Lijinsky *et al.*[428,429] showed that a number of drugs with a tertiary amino group could readily react in this way; indeed, aminopyrine forms *N*-nitrosodimethylamine more readily than does the strongly basic dimethylamine.[430] Under appropriate conditions, primary amines may also form *N*-nitroso derivatives of secondary amines, for example the polyamines spermine and spermidine react with nitrous acid to give a mixture of nitrosamines.[431]

The nitrosation reaction is catalysed by thiocyanate[432] and this ion is formed from glucosinolates which occur in many vegetables of the *Brassica* family and others. Thiocyanate occurs normally in human saliva and it is known that smokers' saliva contains three or four times as much cyanate as that of non-smokers[432,433] possibly as a detoxication product of cyanide in tobacco smoke. Some naturally occurring polyphenols like chlorogenic acid (present in substantial amounts in coffee) and gallotannins can also catalyse nitrosation[434,435] while thiocyanate and chlorogenic acid together behave synergistically.[436]

In view of the above observations, it appears highly probable that nitrosamines are formed in the stomach of healthy, normal individuals after ingestion of dietary precursors and this has been confirmed, both directly and indirectly, in several ways, namely: (a) By incubation of precursors of nitrosamines with gastric juice (or simulated gastric contents) *in vitro*; (b) by examination of gastric contents after ingestion of precursors, including normal dietary components; (c) by the detection of nitrosamines in faeces and body fluids after oral ingestion of precursors in the diet; and (d) by carcinogenicity studies involving oral administration of precursors.

Early *in vitro* studies[437] demonstrated that nitrosamines were formed from a range of different amines on incubation with human gastric juice and Sen *et al.*[438] detected the formation of *N*-nitrosodiethylamine when diethylamine and sodium nitrite were incubated with gastric juice from rat, rabbit, cat, dog and man. Many subsequent studies have confirmed these early observations and extended them to show that normal foods contained precursor amines leading to the formation of *N*-nitrosodimethylamine, *N*-nitrosodiethylamine, *N*-nitrosopyrrolidine and *N*-nitrosopiperidine.[424,439] Furthermore, La Bar and Sander[440] demonstrated that some foods (cured meat, spinach) could serve as sources of the nitrosating agent when incubated with aminopyrine under simulated gastric conditions.

In vitro studies have also indicated that, at pH values higher than are optimal for chemical nitrosation, common microbial inhabitants of the gastrointestinal tract can both reduce nitrate to nitrite and effect nitrosation of secondary amines.[140,441] The importance of microbial reduction of nitrate to nitrite has already been stressed in relation to neonatal methaemoglobinaemia and adult achlorhydria and it also appears that bacterial colonisation of the stomach in such cases may also increase the hazards from nitrosamine formation.[442] In addition to reducing nitrate and catalysing nitrosation, the gut microflora can furnish nitrosatable amines by deamination of amino acids and from degradation of phospholipids.[443]

The results of *in vitro* studies have been amply supported by experiments carried out *in vivo*. Thus Druckrey *et al.*[444] detected low yields of *N*-nitrosodiethylamine when diethylamine and nitrite were administered to rats *per os* and Sander *et al.*[437] in similar experiments found nitrosamines derived from diphenylamine, *N*-methylaniline and *N*-methylbenzylamine in the rat. Subsequent studies have demonstrated the *in vivo* synthesis of nitrosamines from a variety of secondary and tertiary amines in several species. Examples include the nitrosation of sarcosine in mice,[445] of piperazine in dogs,[446] of piperidine in rats,[447,448] of phenometrazine in rabbits and rats,[449] of aminopyrine in rats[430] and of diphenylamine in man.[450] The significance of nitrosamine synthesis *in vivo* has been further underlined by the observation that *N*-nitrosopiperidine could be detected in the gastric contents of human volunteers who had consumed homogenised but otherwise normal food materials containing nitrite.

The work of Sander and Sief[450] showed that *N*-nitrosodiphenylamine could be detected in the stomach contents of about one third of human volunteers who had ingested sodium nitrate and diphenylamine while

Alam *et al.*[448] found *N*-nitrosopiperidine in the gastrointestinal tract of rats dosed orally with nitrate and piperidine. This demonstrated that, both in experimental animals and man, nitrate could be converted into a nitrosating agent in the stomach and, as previously indicated, this is likely to be particularly important in cases of achlorhydria when the stomach is colonised by enteric bacteria.

Further confirmation of the formation of nitrosamines *in vivo*, and most probably in the gastrointestinal tract, is afforded by the detection of several volatile nitrosamines in the faeces of normal humans.[451] *N*-nitrosodimethylamine has also been detected in human blood samples.[452,453] Higher levels of this nitrosamine were found in Japanese than in US residents which Yamamoto *et al.*[453] attributed to the higher concentrations of nitrate and secondary amines in Japanese diets.

In appropriate circumstances, sufficient nitrosation can occur in the gut to produce a carcinogenic response in experimental animals. Thus, long-term studies involving co-administration of nitrite and amines in the diet led to the production of tumours by *N*-methylbenzylamine, morpholine, aminopyrine or heptamethyleneimine in rats[454–456] and by morpholine in Syrian hamsters and in Swiss mice.[456–457] Claims that nitrite alone can cause increased tumour incidence, presumably by interaction with endogenous or unidentified dietary amines, remain to be substantiated.

From the above, it is clear that precursors of nitrosamines are widely distributed in the diet and may be converted into nitrosamines in the stomach. However, the significance of these observations in the aetiology of human cancer or of 'background' tumour incidence in experimental animals is less clear. The 'high risk' groups of the population are likely to be those ingesting high concentrations of nitrate and/or nitrite in food and drinking water, with high levels of nitrite and thiocyanate in saliva, and with hypochlorhydria allowing bacterial colonisation of the upper gastrointestinal tract.[442,458]

Pathology of the Gastrointestinal Tract in Safety Evaluation

Caecal Enlargement

Caecal enlargement in experimental rodents is a common observation in the safety evaluation of food components and additives. It has been recorded in the evaluation of materials with widely different structures including raw starches,[250,251] modified starches,[360,459] lactose,[258] sorbitol,[460] maltitol,[461] a high-molecular-weight fraction of glucose syrup,[259] caramel,[462] Yellow 2G,[463] pectin[464] and Erythrosine.[339] Similar effects have been seen with polyethylene glycol[465] and magnesium sulphate.[466] Particular attention has been paid to chemically modified starches since, as used in infant foods, they may contribute between 10 per cent and 30 per cent of the total available energy. In their overall diets, infants may receive 5 per cent or more of their total energy intake as modified starches.[467] Clearly, the toxicological and nutritional significance of caecal enlargement is of considerable importance in the safety evaluation of many foods and additives.

In interpreting caecal enlargement as either a toxic or an adaptive response, several questions arise.

(a) What is 'normal' caecal size? (Different, nutritionally adequate diets may produce significant differences in the weight of the caecum without obvious adverse effects.)
(b) What mechanisms control caecal size?
(c) What changes occur in caecal tissues associated with increased tissue mass?
(d) Are there adverse effects in other organs or functions consequent on caecal enlargement? (Walker[468]).

The question of what is 'normal' and what deviations from the norm may be considered adaptive is constantly a problem for the toxicologist. In respect of caecal size different 'control' diets can lead to differences which are statistically significant but which do not affect biological performance as judged by nutritional or toxicological criteria, and such differences may be considered adaptation to the different diets. This then raises the issue of what basis is used for judging whether dietary additives cause caecal enlargement. For example, in studies on modified starches, de Groot *et al.*[360] used gelatinised potato starch as the carbohydrate source in control diets and concluded that some modified starches caused caecal enlargement. However, Walker and El Harith,[469] also working with modified starches, used maize starch in control diets and did not observe significant caecal enlargement with food grade materials. In these studies, gelatinised potato starch was found to produce a smaller caecal weight than maize starch and if de Groot *et al.*[401] had used maize starch in their control diets, their conclusions might have been different! Clearly, small but statistically significant differences in caecal size may be of no toxicological importance in some cases but merely represent variations within what might be considered normal limits.

With regard to mechanisms controlling caecal size, a common feature of the materials which effect caecal enlargement is that they are poorly absorbed from the gut and they are osmotically active (or may be degraded to lower-molecular-weight osmotically active materials in the case of starches) and this led Leegwater *et al.*[470] to postulate that caecal size is controlled by the osmotic value of the contents and should be considered an adaptive, homeostatic mechanism. This hypothesis also accommodates the increased caecal size observed in germ-free animals in which there is an increase in colloidal materials of endogenous origin in caecal contents which are degraded by the gut microflora of conventional animals. Walker[468] concurred with this postulate while drawing attention to the fact that changes in the gut microflora, pH (and possibly E_h) might also be involved. However, at the cellular level, further work is clearly needed to elucidate the mechanism by which the hyperplasia associated with caecal enlargement is controlled. Even the gross caecal enlargement caused by raw potato starch is reversible in a relatively short time[471] which supports the view that the hyperplasia does not indicate permanent changes in the tissues and might be considered adaptation.

If changes in caecal size are considered adaptive, however, the process is not well controlled and can lead to pathological changes when the degree of enlargement is gross. Thus the remarkable increase in caecal size brought about by raw potato starch (see p. 220) lead to increased mortality due to respiratory insufficiency, and oedematous changes in the caecal mucosa and sub-mucosa were also recorded along with loss of muscle tone. At the biochemical level, significant increases in serum GPT and alkaline phosphatase levels were observed when caecal enlargement was gross[471] and there may be changes in the absorptive process in the gastrointestinal tract.[465]

Clearly some small changes in caecal size must be considered tolerable in toxicity tests, particularly where unrealistically high dietary levels of food additives are administered, but much more work is required to elucidate the limits beyond which caecal size must be considered abnormal.

Ulceration of the Gut

Effects of Sulphite on the Gastric Mucosa. The use of sulphur dioxide and sulphites as food preservatives and in medicines goes back many centuries and they are still extensively used in many foods and drinks such as dried fruit, fruit pulps, wine, beer and soft drinks.[472] Pathological changes have been seen in the stomachs of experimental animals on administration of relatively high levels of sulphite in the diet. At dietary concentrations of 0.5–2 per cent sodium metabisulphite, mild hyperplasia of gastric squamous epithelium has been observed in rats and pigs, which were also given large amounts of dietary thiamine to counteract the known anti-thiamine action of sulphite.[473,474] At higher dietary levels (4–8 per cent sulphite) in rats ulceration and papillomatous elevations were observed in the forestomach and necrosis, haemorrhagic erosions, inflammation and an atypical glandular hyperplasia were seen in the glandular stomach.[473,475] The changes were reversible within a few weeks after cessation of sulphite feeding.

Carrageenan. Carrageenans are a group of sulphated polygalactans of molecular weight 250,000–750,000 extracted from various genera of red algae within the class Rhodophyceae.[476] The native, or undegraded carrageenan (UC), can be fractionated into three main types known as λ, κ and ι carrageenans which differ slightly in sulphate content and chemical linkages. *Euchema spinosum* contains predominantly ι carrageenan which when subjected to mild acid hydrolysis yields a degraded form (DC) with a molecular weight of 30,000 or less. UC is widely used in food and drink as a texturising and suspending agent and DC is used in some countries in the treatment of gastric ulcers.[476,477]

In general carrageenan is poorly absorbed from the gut of most mammals[478] but Sharratt *et al.*[479] showed that DC, and to a lesser extent UC, can be absorbed and becomes located in macrophages of the lamina propria of the caecum in guinea pigs and rabbits and in the spleen, liver and kidney of guinea pigs.[480] The carrageenan found in the livers of guinea pigs had a molecular weight of about 10,000 regardless of the molecular weight of the carrageenan fed indicating that degradation of high-molecular-weight carrageenan occurs.[481]

It is now well established that DC, and to a much lesser extent UC (ι fractions), cause ulcerative lesions in the caecum and colon of guinea pigs and rabbits when given orally.[478,482,483]

The ulcerative lesions showed features of inflammatory infiltration as well as crypt abscesses, the infiltration consisting of lymphocytes, plasma cells and macrophages (some of which contained carrageenans).[484] The ulcerogenic process appears to proceed initially via the uptake (persorption) of carrageenan by the macrophages of the caecal lamina propria which Abraham *et al.*[485] postulate stimulates the release of lysosomal enzymes which damage the surrounding tissue eventually leading to a destruction of the overlying epithelium.

Recently, intestinal micro-organisms have been implicated in the aetiology of the ulcerative lesion in the guinea pig, since germ-free guinea pigs or guinea pigs treated with certain antibiotics did not develop ulcers after DC treatment.[486]

There is little epidemiological evidence to suggest that carrageenan can induce large bowel ulcers in man. For example, the seaweeds from which carrageenans are derived form a major part of the diet of populations in Ireland, the South Pacific Islands and Japan, but the incidence of ulcerative colitis does not appear to be elevated in those countries[487] and furthermore no instances of ulcerative colitis were found in patients given DC (about 5 g/day) as treatment for peptic ulcer.[477]

Secondary Pathological Effects. Pathological conditions in the gastrointestinal tract may be observed as secondary consequences of primary lesions at other sites. For example, conditions which affect the synthesis and secretion of bile acids by the liver can seriously impair fat digestion and absorption, leading to steatorrhoea.[488] The conditions may be congenital or a consequence of liver damage either by disease or toxic substances.

References

1. L.V. Holdeman and W.E.C. Moore, *Anaerobe Laboratory Manual*, 3rd edn. (VPI Anaerobe Laboratory, Blacksburg, Virginia, 1972).

2. B.S. Drasar, 'Cultivation of Anaerobic Intestinal Bacteria', *J. Pathol. Bacteriol., 94* (1967), pp. 417–427.

3. W.E.C. Moore, E.P. Cato and L.V. Holdeman , 'Some Current Concepts in Intestinal Bacteriology', *Am. J. clin. Nutr., 31* (1978), pp. 533–542.

4. D.C. Savage, 'Interactions Between the Host and its Microbes' in R.T.J. Clark and T. Bauchop (eds.), *Microbial Ecology of the Gut* (Academic Press, London, 1977), pp. 277–310.

5. J.P. Brown, 'Role of Gut Bacterial Flora in Nutrition and Health: a Review of Recent Advances in Bacteriological Techniques, Metabolism and Factors Affecting Flora Composition', *Crit. Rev. Fd Sci. Nutr., 8* (1977), pp. 229–336.

6. B.S. Drasar, M.J. Hill and R.E.O. Williams, 'The Significance of the Gut Flora in Safety Testing of Food Additives' in F.J.C. Roe (ed.), *Metabolic Aspects of Food Safety* (Blackwell Scientific Publications, Oxford, 1970), pp. 245–260.

7. R.A. Giannella, S.A. Broitman and N. Zamcheck, 'Gastric Acid Barrier to Ingested Microorganisms in Man: Studies *in vivo* and *in vitro*', *Gut, 13* (1972), pp. 251–256.

8. I.R. Rowland, 'Metabolism of di-(2-ethylhexyl) Phthalate by the Contents of the Alimentary Tract of the Rat', *Fd Cosmet. Toxicol., 12* (1974), pp. 293–302.

9. P.A. Sykes, K.H. Boulter and P.F. Schofield, 'Alterations in Small-Bowel Microflora in Acute Intestinal Obstruction', *J. Med. Microbiol., 9* (1976), pp. 13–22.

10. S.L. Gorbach, R. Mitra and B. Jacobs, 'Bacterial Contamination of the

Upper Small Bowel in Tropical Sprue', *Lancet, i* (1969), pp. 74–77.

11. P. Bhat, D. Shantakumari, D. Rajan, V.I. Mathan, C.R. Kapadia, C. Swarnabai and S.J. Baker, 'Bacterial Flora of the Gastrointestinal Tract in Southern Indian Control Subjects and Patients with Tropical Sprue', *Gastroenterology, 62* (1972), pp. 11–12.

12. S.L. Gorbach, 'Microflora of the Gastrointestinal Tract in Tropical Enteritis: A Current Appraisal', *Am. J. clin. Nutr., 25* (1972), pp. 1127–1130.

13. L.J. Mata, F. Jimenez, M. Cordon, R. Rosales, E. Prera, R.E. Schneider and F. Viteri, 'Gastrointestinal Flora of Children with Protein-Calorie Malnutrition', *Am. J. clin. Nutr., 25* (1972), pp. 1118–1126.

14. M. Gracey, Suharjono, Sunoto and D.E. Stone, 'Microbial Contamination of the Gut: Another Feature of Malnutrition', *Am. J. clin. Nutr., 26* (1973), pp. 1170–1174.

15. W.E.C. Moore and L.V. Holdeman, 'Human Faecal Flora. The Normal Flora of 20 Japanese–Hawaiians', *Appl. Microbiol., 27* (1976), pp. 961–979.

16. L.V. Holdeman, I.J. Good and W.E.C. Moore, 'Human Fecal Flora: Variation in Bacterial Composition Within Individuals and a Possible Effect of Emotional Stress', *Appl. Environ. Microbiol., 31* (1975), pp. 359–375.

17. A.J. Mastromarino, B.S. Reddy and E.L. Wynder, 'Faecal Profiles of Anaerobic Microflora of Large Bowel Cancer Patients and Patients with Non-hereditary Large Bowel Polyps', *Cancer Res., 38* (1978), pp. 4458–4462.

18. J. Gossling and J.M. Slack, 'Predominant Gram-Positive Bacteria in Human Faeces: Numbers, Variety and Persistence', *Infection and Immunity, 9* (1974), pp. 719–729.

19. B.G. Veilleux and I. Rowland, 'Simulation of the Rat Intestinal Ecosystem using a Two-stage Continuous Culture System', *J. gen. Microbiol., 123*, (1981), pp. 103–115.

20. C.P. Davis, D. Cleven, E. Balish and C.E. Yale, 'Bacterial Association in the Gastrointestinal Tract of Beagle Dogs', *Appl. Environ. Microbiol., 34* (1977), pp. 194–206.

21. A.W. Brinkley and G.E. Mott, 'Anaerobic Faecal Bacteria of the Baboon', *Appl. Environ. Microbiol., 36* (1978), pp. 530–532.

22. A. Lee and M. Phillips, 'Isolation and Cultivation of Spirochetes and Other Spiral-shaped Bacteria Associated with the Cecal Mucosa of Rats and Mice', *J. Appl. Environ. Microbiol., 35* (1978), pp. 610–613.

23. D.C. Savage and R.V. Blumershine, 'Surface – Surface Association in Microbial Communities Populating Epithelial Habitats in the Murine Gastrointestinal Ecosystem, Scanning Electron Microscopy', *Infect. Immunol., 10* (1974), pp. 240–250.

24. D.C. Savage, J.S. McAllister and C.P. Davis, 'Anaerobic Bacteria on the Mucosal Epithelium of the Murine Large Bowel', *Infect. Immunol., 4* (1971), pp. 492–502.

25. D.P. Nelson and L.J. Mata, 'Bacterial Flora Associated with the Human Gastrointestinal Mucosa', *Gastroenterol., 58* (1970), pp. 56–61.

26. H. Haenel, 'Human Normal and Abnormal Gastrointestinal Flora', *Am. J. clin. Nutr., 23* (1970), pp. 1433–1439.

27. C.L. Bullen, P.V. Tearle and A.T. Willis, 'Bifidobacteria in the Intestinal Tract of Infants: An *in vivo* Study', *J. Med. Microbiol., 9* (1976), pp. 325–333.

28. R.B. Ellis-Pegler, C. Crabtree and H.P. Lambert, 'The Faecal Flora of Children in the United Kingdom', *J. Hyg., Camb., 75* (1975), pp. 135–142.

29. B.S. Drasar and M. Shiner, 'Studies on the Intestinal Flora. Part II – Bacterial Flora of the Small Intestine in Patients with Gastrointestinal Disorders', *Gut, 10* (1969), pp. 812–819.

30. V.C. Aries, J.S. Crowther, B.S. Drasar, M.J. Hill and R.E.O. Williams,

Bacteria and the Aetiology of Cancer of the Large Bowel', *Gut, 10* (1969), pp. 334–335.

31. B.S. Drasar, J.S. Crowther, P. Goddard, Gabrielle Hawksworth, M.J. Hill, Susan Peach, R.E.O. Williams and A. Renwick, 'The Relation Between Diet and the Gut Flora in Man', *Proc. Nutr. Soc., 32* (1973), pp. 49–52.

32. J.S. Crowther, B.S. Drasar, M.J. Hill, R. MacLennan, D. Magnin, S. Peach and C.H. Teoh-Chan, 'Faecal Steroids and Bacteria and Large Bowel Cancer in Hong Kong by Socio-Economic Groups', *Br. J. Cancer, 34* (1976), pp. 191–197.

33. W.E.C. Moore and L.V. Holdeman, 'Discussion of the Current Bacteriological Investigations of the Relationships Between Intestinal Flora, Diet and Colon Cancer', *Cancer Res., 35* (1975), pp. 3418–3420.

34. M. Winitz, R.F. Adams, D.A. Seedman, P.N. Davis, L.G. Jayko and J.A. Hamilton, 'Studies in Metabolic Nutrition Employing Chemically Defined Diets', *Am. J. clin. Nutr., 23* (1970), pp. 546–559.

35. H.R. Atterbery, V.L. Sutter and S.M. Finegold, 'Effect of a Partially Chemically Defined Diet on Normal Human Faecal Flora', *Am. J. clin. Nutr., 25* (1972), pp. 1391–1398.

36. R.S. Speck, D.H. Calloway and W.K. Hadley, 'Human Faecal Flora Under Controlled Diet Intake', *Am. J. clin. Nutr., 23* (1970), pp. 1488–1494.

37. B.S. Drasar and M.J. Hill, *Human Intestinal Flora* (Academic Press, London, 1974).

38. V.C. Aries, J.S. Crowther, B.S. Drasar, M.J. Hill and F.R. Ellis, 'The Effect of a Strict Vegetarian Diet on the Faecal Flora and Faecal Steroid Concentration', *J. Pathol., 103* (1971), pp. 54–56.

39. B.S. Drasar, D.J.A. Jenkins and J.H. Cummings, 'The Influence of a Diet Rich in Wheat Fibre on the Human Faecal Flora', *J. Med. Microbiol., 9* (1976), pp. 423–431.

40. B.S. Drasar and D. Irving, 'Environmental Factors and Cancer of the Colon and Breast', *Br. J. Cancer, 27* (1973), pp. 167–172.

41. B.S. Reddy, J.H. Weisburger and E.L. Wynder, 'Effects of High Risk and Low Risk Diets for Colon Carcinogenesis on Faecal Microflora and Steroids in Man', *J. Nutr., 105* (1975), pp. 878–884.

42. B.S. Reddy, J.H. Weisburger and E.L. Wynder, 'Faecal Bacterial β-glucuronidase: Control by Diet', *Science, 183* (1974), pp. 416–417.

43. D.J. Hentges, B.R. Maier, G.C. Burton, M.A. Flynn and R.K. Tsutakawa, 'Effects of a High Beef Diet on the Fecal Bacterial Flora of Humans', *Cancer Res., 37* (1977), pp. 568–571.

44. J.H. Cummings, H.S. Wiggins, D.J.A. Jenkins, H. Houston, T. Jivraj, B.S. Drasar and M.J. Hill, 'Influence of Diets High and Low in Animal Fat on Bowel Habit, Gastrointestinal Transit Time, Fecal Microflora, Bile Acid and Fat Excretion', *J. clin. Invest., 61* (1978), pp. 953–963.

45. K.T. Chung, G.E. Full and J. Silverman, 'Dietary Effects on the Composition of Fecal Flora of Rats', *Appl. Environ. Microbiol. 33* (1977), pp. 654–659.

46. B. Goldin and S.L. Gorbach, 'Alterations in Fecal Microflora Enzymes Related to Diet, Age, Lactobacillus Supplements and Dimethylhydrazine', *Cancer, 40* (1977), pp. 2421–2426.

47. I.R. Rowland, M.J. Davies and J.G. Evans, 'Tissue Content of Mercury in Rats Given Methylmercuric Chloride Orally. Influence of Intestinal Flora', *Arch. Environ. Hlth, 35* (1980), pp. 155–160.

48. B.S. Wostman and E. Bruckner-Kardoss, 'Oxidation-Reduction Potentials in Cecal Contents of Germfree and Conventional Rats', *Proc. Soc. Exp. Biol. Med., 121* (1966), pp. 1111–1114.

49. D.J. Ford, 'The Effect of the Microflora on Gastrointestinal pH in the Chick', *British Poultry Sci.,* (1974), pp. 131–140.

50. M.E. Coates and R. Fuller, 'The Gnotobiotic Animal in the Study of Gut Microbiology' in R.T.J. Clark and T. Bauchop (eds.), *Microbial Ecology of the Gut* (Academic Press, London, 1977), pp. 311–346.

51. T.L. Luckey, *Germfree Life and Gnotobiology* (Academic Press, New York, 1963).

52. M. Lev and A.F. Milford, 'Apparatus for Metabolic Studies with Anaerobes', *Appl. Microbiol., 21* (1971), pp. 555–556.

53. R.J. Allan, A.A. Tesoriero and J.J. Roxon, 'An Incubation Flask to Study Metabolism of Foreign Compounds by Intestinal Microflora', *Xenobiotica, 5* (1975), pp. 407–411.

54. P.N. Hobson, 'Continuous Culture of Rumen Bacteria: Apparatus', *J. Gen. Microbiol., 38* (1965), pp. 161–166.

55. P. Goddard, F. Fernandez, B. West, M.J. Hill and P. Barnes, 'The Nuclear Dehydrogenation of Steroids by Intestinal Bacteria', *J. Med. Microbiol., 8* (1975), pp. 429–435.

56. E.C. McCoy, W.T. Speck and H.S. Rosenkranz, 'Activation of a Procarcinogen to a Mutagen by Cell Free Extracts of Anaerobic Bacteria', *Mutat. Res., 46* (1977), pp. 261–264.

57. J.S. Brown and P.S. Dietrich, 'Mutagenicity of Anthraquinone and Benzanthrone Derivatives in the *Salmonella* Microsome Test: Activation of Anthroquinone Glycosides by Enzymic Extracts of Rat Cecal Bacteria', *Mutat. Res., 66* (1979), pp. 9–24.

58. M.A. Harris, C.A. Reddy and G.R. Carter, 'Anaerobic Bacteria from the Large Intestine of Mice', *Appl. Environ. Microbiol., 31* (1976), pp. 907–912.

59. J.H. Gordon and R. Dubos, 'The Anaerobic Bacterial Flora of the Mouse Cecum', *J. Exp. Med., 132* (1970), pp. 251–260.

60. H. Williams Smith, 'The Development of the Bacterial Flora of the Faeces of Animals and Man: The Changes that Occur during Ageing', *J. Appl. Bacteriol., 24* (1961), pp. 235–241.

61. W.L. Beeken and R.E. Kanich, 'Microbial Flora of the Upper Small Bowel in Crohn's Disease', *Gastroenterology, 65* (1973), pp. 390–397.

62. J.D.A. Gray and M. Shiner, 'Influence of Gastric pH on Gastric and Jejunal Flora', *Gut, 8* (1967), pp. 574–581.

63. B.S. Drasar, M. Shiner and G.M. Mcleod, 'Studies on the Intestinal Flora of the Gastrointestinal Tract in Healthy and Achlorhydria Persons', *Gastroenterology, 56* (1969), pp. 71–79.

64. A.M. Tomkins, B.S. Drasar and W.P.T. James, 'Bacterial Colonisation of Jejunal Mucosa in Acute Tropical Sprue', *Lancet, i* (1975), p. 59.

65. J.A.A. Van Der Wiel Korstanje and K.C. Winkler, 'The Faecal Flora in Ulcerative Colitis', *J. Med. Microbiol., 8* (1975), pp. 491–501.

66. S.L. Gorbach, J.G. Banwell, B.D. Chatterjee, B. Jacobs and R.B. Sack, 'Acute Undifferentiated Human Diarrhoea in the Tropics. I. Alterations in Intestinal Microflora', *J. clin. Invest., 50* (1971), pp. 881–889.

67. R. Cohen, M.H. Kalser, I. Arteaga, E. Yann, D. Frazier, C.A. Leite, D.G. Ahearn and F. Roth, 'Microbial Intestinal Flora in Acute Diarrhoeal Disease', *J. Am. Med. Assoc., 201* (1967), pp. 835–840.

68. R.R. Scheline, 'Toxicological Implications of Drug Metabolism by Intestinal Bacteria', *Proc. Eur. Soc. Study of Drug Toxicity* (Excerpta Medica, Amsterdam, 1972), vol. XIII, pp. 35–43.

69. R. Scheline, 'Metabolism of Foreign Compounds by Gastrointestinal Microorganisms', *Pharmacol. Rev., 25* (1973), pp. 451–523.

70. R.T. Williams, in Discussion of 'The Significance of the Gut Flora in Safety Testing of Food Additives' by Drasar, Hill and Williams, in F.J.C. Roe (ed.), *Metabolic Aspects of Food Safety* (Blackwell, London, 1970), pp. 255–260.

71. A. Taurog, F.N. Briggs and I.L. Chaikoff, 'I^{131}-labelled *L*-thyroxine; Nature of Excretion Products in Bile', *J. Biol. Chem., 194* (1952), pp. 655–668.

72. R.R. Scheline, 'Metabolism of Phenolic Acids by the Rat Intestinal Microflora', *Acta Pharmacol. Toxicol., 26* (1968), pp. 189–205.

73. D.V. Parke, Kh.M.Q. Rahman and R. Walker, 'Absorption, Distribution and Excretion of Linalool in the Rat', *Biochem. Soc. Trans., 2* (1974), pp. 612–615.

74. G. Levy, L. Weintraub, T. Matsuzawa and S. Oles, 'Absorption, Metabolism and Excretion of Salicylic Phenolic Glucuronide in Rats', *J. Pharmacol. Sci., 5* (1966), pp. 1319–1320.

75. L.A. Woods, 'Distribution and Fate of Morphine in Non-Tolerant and Tolerant Dogs and Rats', *J. Pharmacol. Exp. Ther., 112* (1954), pp. 158–175.

76. D.J. Hanahan, E.G. Daskalakis, T. Edwards and H.J. Bauden, 'Reduction of Water-soluble Azo Dyes by Intestinal Bacteria. Metabolic Patterns of ^{14}C-diethylstilboestrol', *Endocrinology, 53* (1953), pp. 163–170.

77. A.J. Glazko, W.A. Dill and L.M. Wolf, 'Observations on the Metabolic Disposition of Chloramphenicol (Chloromycetin) in the Rat', *J. Pharmacol. Exp. Ther., 104* (1952), pp. 452–458.

78. C.A. Marsh, F. Alexander and G.A. Levvy, 'Glucuronide Decomposition in the Digestive Tract', *Nature (Lond.), 170* (1952), pp. 163–164.

79. J.H. Weisburger, P.H. Grantham, R.E. Horton and E.K. Weisburger, 'Metabolism of the Carcinogen *N*-hydroxy-*N*-2-fluorenylacetamide in Germ-free Rats', *Biochem. Pharmacol., 19* (1970), pp. 151–162.

80. J.R. Williams, P.H. Grantham, H.H. Marsh, J.H. Weisburger and E.K. Weisburger, 'Participation of Liver Fractions and of Intestinal Bacteria in the Metabolism of *N*-hydroxy-*N*-2-fluorenylacetamide in the Rat', *Biochem. Pharmacol., 19* (1970), pp. 173–188.

81. L. Weinstein, 'The Sulphonamides' in L.S. Goodman and A. Gilman (eds.), *The Pharmacological Basis of Therapeutics* (Macmillan, London, 1970), pp. 1177–1203.

82. D.S. Zaharko, H. Bruckner and V.T. Oliverio, 'Antibiotics Alter Methotrexate Metabolism and Excretion', *Science, 166* (1969), pp. 887–888.

83. D.S. Zaharko and V.T. Oliverio, 'Reinvestigation of Methotrexate Metabolism in Rodents', *Biochem. Pharmacol., 19* (1970), pp. 2923–2925.

84. D.M. Valerino, D.G. Johns, D.S. Zaharko and V.T. Oliverio, 'Studies on the Metabolism of Methotrexate by Intestinal Flora I. Identification and Study of Biological Properties of the Metabolite 4-amino-4-deoxy-N^{10}-Methylpteroic Acid', *Biochem. Pharmacol., 21* (1972), pp. 821–831.

85. R.E. Ranney and S.J. Eich, 'The Metabolism of Sodium Amylopectin Sulfate in the Rat', *Toxicol. Appl. Pharmacol., 15* (1969), pp. 593–603.

86. D.J. Hearse, G.M. Powell, A.H. Olaveson and K.S. Dodgson, 'The Influence of Some Physicochemical Factors on the Biliary Excretion of a Series of Structurally Related Aryl Sulphate Esters', *Biochem. Pharmacol., 18* (1969), pp. 181–195.

87. G. Pala, G. Coppi and E. Crescenzi, 'On the Laxative Properties of Sulfuric Esters of Phenols, with Particular Reference to 4,4′-(2-picolylidene)-bis-phenylsulfuric acid Disodium Salt', *Arch. Int. Pharmacodyn. Ther., 164* (1966), pp. 356–369.

88. F.J. DiCarlo, C.B. Coutinho and M.C. Crew, 'Sites of Absorption of Pentaerythritol Tetranitrate', *Arch. Int. Pharmacodyn. Ther., 167* (1967), pp. 163–170.

89. G.W. Lanigan and L.W. Smith, 'Metabolism of Pyrrolizidine Alkaloids in the Ovine Rumen. I. Formation of 7α-hydroxy-1α-methyl-8α-pyrrolizidine from

Heliotrine and Lasiocarpine', *Aust. J. Agr. Res., 21* (1970), pp. 493–500.
90. P. Iveson, W.E. Lindup, D.V. Parke and R.T. Williams, 'The Metabolism of Carbenoxolone in the Rat', *Xenobiotica, 1* (1971), pp. 79–95.
91. D.V. Parke, T.C. Hunt and P. Iveson, 'The Fate of [^{14}C] Carbenoxolone in Patients with Gastric Ulcer', *Clin. Sci., 43* (1972), pp. 393–400.
92. H.G. Bray, Z. Hybs, S.P. James and W.V. Thorpe, 'The Metabolism of 2,3,5,6- and 2,3,4,5- Tetrachloronitro-benzenes in the Rabbit and the Reduction of Aromatic Nitro compounds in the Intestine', *Biochem. J., 53* (1953), pp. 266–273.
93. E.J. Kuchar, F.O. Geenty, W.P. Griffith and R.J. Thomas, 'Analytical Studies of Metabolism of Terraclor in Beagle Dogs, Rats and Plants', *J. Agr. Food Chem., 17* (1969), pp. 1237–1240.
94. J.W. Cook, '*In vitro* Destruction of Some Organo-phosphate Pesticides by Bovine Rumen Fluid', *J. Agr. Fd Chem., 5* (1957), pp. 859–863.
95. J. Tréfouël, J. Tréfouël, F. Nitti and D. Bovet, 'Activité du *p*-aminophénylsulfamide sur les infections streptococcignes. Experimentales de la souris et du lapin', *C. & Séanc. Soc. Biol., 120* (1935), p. 756.
96. W.W. Spink, F.W. Hurd and J. Jermsta, '*In vitro* Conversion of Prontosil-soluble to Sulphanilamide by Various Types of Microorganism', *Proc. Soc. exp. Biol. Med., 43* (1940), p. 172.
97. R. Walker, 'The Metabolism of Azo Compounds: A Review of the Literature', *Fd Cosmet. Toxicol., 8* (1970), pp. 659–676.
98. R. Gingell, J.W. Bridges and R.T. Williams, 'The Role of the Gut Flora in the Metabolism of Prontosil and Neoprontosil in the Rat', *Xenobiotica, 1* (1971), pp. 143–156.
99. T. Okabayashi and A. Yashimoto, 'Reduction of 4-nitroquinoline 1-oxide by Micro-organisms', *Chem. Pharm. Bull. (Japan), 10* (1962), pp. 1221–1226.
100. A.T. Dick, A.T. Dann, L.B. Bull and C.C.J. Culvenor, 'Vitamin B_{12} and the Detoxification of Hepatotoxic Pyrrolizidine Alkaloids in Rumen Liquor', *Nature (Lond.), 197* (1963), pp. 207–208.
101. A.H. Beckett, J.W. Gorrod and P. Jenner, 'Absorption of (-)-nicotine-1-*N*-oxide in Man and its Reduction in the Gastrointestinal Tract', *J. Pharm. Pharmacol., 22* (1970), pp. 722–723.
102. P.N. Grantham, R.E. Horton, E.K. Weisburger and J.H. Weisburger, 'Metabolism of the Carcinogen *N*-2-fluorenyl-acetamide in Germ-free and Conventional Rats', *Biochem. Pharmacol., 19* (1970), pp. 163–171.
103. R.R. Scheline, 'The Metabolism of Some Aromatic Aldehydes and Alcohols by the Rat Intestinal Microflora', *Xenobiotica, 2* (1972), pp. 227–236.
104. L.A. Griffiths and G.E. Smith, 'Metabolism of Myricetin and Related Compounds in the Rat'. Metabolite Formation *in vivo* and by the Intestinal Microflora *in vitro*', *Biochem. J., 130* (1972), pp. 141–151.
105. T. Meyer and R.R. Scheline, '3,4,5-trimethoxy-cinnamic Acid and Related Compounds. I. Metabolism by the Rat Intestinal Microflora', *Xenobiotica, 2* (1972), pp. 383–390.
106. T. Meyer and R.R. Scheline, '3,4,5-trimethoxy-cinnamic Acid and Related Compounds. II. Metabolism in the Rat', *Xenobiotica, 2* (1972), pp. 391–398.
107. T.W. Scott, P.F.V. Ward and R.M.C. Dawson, 'Formation of Aromatic Fatty Acids from Amino Acids by Rumen Microorganisms', *Biochem. J., 87* (1963), pp. 3–4.
108. T.W. Scott, P.F.V. Ward and R.M.C. Dawson, 'The Formation and Metabolism of Phenyl-substituted Fatty Acids in the Ruminant', *Biochem. J., 90* (1964), pp. 12–24.
109. M. Kaihara and J.M. Price, 'The Metabolism of Quinaldic Acid,

Kynurenic Acid and Xanthurenic Acid in the Rabbit', *J. Biol. Chem., 237* (1962), pp. 1727–1729.

110. M. Kaihara and J.M. Price, 'The Effect of Feeding Neomycin on the Dehydroxylation of Xanthurenic Acid to 8-hydroxy-quinaldic Acid by the Rabbit', *J. Biol. Chem., 238* (1963), pp. 4082–4084.

111. G.M. Lower and G.T. Bryan, 'The Metabolism of the 8-methyl ether of Xanthurenic Acid in the Mouse', *Cancer Res., 29* (1969), pp. 1013–1018.

112. G.M. Lower and G.T. Bryan, 'The Metabolism of the 8-methyl ether of Xanthurenic Acid in the Rabbit', *J. Biol. Chem., 244* (1969), pp. 2567–2571.

113. L.P. Strand and R.R. Scheline, 'The Metabolism of Vanillin and Iso Vanillin in the Rat', *Xenobiotica, 5* (1975), pp. 49–63.

114. J.P. Moody and R.T. Williams, 'The Fate of 4-nitro-phenylarsonic Acids in Hens', *Fd Cosmet. Toxicol., 2* (1964), pp. 695–706.

115. J.E. Peterson and W.H. Robinson, 'Metabolic Products of *p,p'*-DDT in the Rat', *Toxicol. Appl. Pharmacol., 6* (1964), pp. 321–327.

116. J.L. Mendel and M.S. Walton, 'Conversion of *p,p'*-DDT to *p,p'*-DDD by Intestinal Flora of the Rat', *Science, 151* (1966), pp. 1527–1528.

117. J.H.V. Stenersen, 'DDT-metabolism in Resistant and Susceptible Stable-flies and in Bacteria', *Nature (Lond.), 208* (1965), p. 200.

118. W.R. Jondorf, D.V. Parke and R.T. Williams, 'Studies in Detoxication. 76. The Metabolism of Halogenobenzenes. 1,2,3,4-, 1,2,3,5- and 1,2,4,5-tetra-chlorobenzenes', *Biochem. J., 69* (1958), pp. 181–189.

119. D.V. Parke and R.T. Williams, 'Studies in Detoxication. 81. The Metabolism of Halogenobenzenes: (a) Penta- and Hexachlorobenzenes (b) Further Observations on 1,3,5-trichlorobenzene', *Biochem J., 74* (1960), pp. 5–9.

120. T.L. Perry, M. Hestrin, L. MacDougall and S. Hansen, 'Urinary Amines of Intestinal Bacterial Origin', *Clin. Chim. Acta, 14* (1966), pp. 116–123.

121. O.M. Bakke, 'Urinary Simple Phenols in Rats Fed Diets Containing Different Amounts of Casein and 10% Tyrosine', *J. Nutr., 98* (1969), pp. 217–221.

122. O.M. Bakke, 'Minor Urinary Metabolites of *L*-dopa in the Rat', *Acta Pharmacol. Toxicol., 29* (1971), pp. 106–107.

123. J. Bergmark, A. Carlsson, A-K. Granerus, R. Jagenburg, T. Magnusson and A. Svanborg, 'Decarboxylation of Orally Administered *L*-dopa in the Human Digestive Tract', *Naunyn-Schmiedebergs Arch. Pharmakol. Exp. Pathol., 272* (1972), pp. 437–440.

124. G. Curzon and R.T.C. Pratt, 'Origin of Urinary Resorcinol Sulphate', *Nature (Lond.), 204* (1964), pp. 383–384.

125. R.R. Scheline, 'The Metabolism of Drugs and other Organic Compounds by the Intestinal Microflora', *Acta Pharmacol. Toxicol., 26* (1968), pp. 332–342.

126. R. Minder, F. Schnetzer and M.H. Bickel, 'Hepatic and Extrahepatic Metabolism of the Psychotropic Drugs, Chlorpromazine, Imipramine and Imipramine-*N*-oxide', *Naunyn-Schmiedebergs Arch. Pharmakol. Exp. Pathol., 268* (1971), pp. 334–347.

127. J. Closon, G. Salvatore, R. Michel and J. Roche, 'Dégradation de l'ester sulfurique de la 3,5,3'-triiodo-*L*-thyronine par les bactéries intestinales du rat', *C. r. Séances Soc. Biol., 153* (1959), pp. 1120–1125.

128. B.A. Koechlin, F. Rubio, S. Palmer, T. Gabriel and R. Duschinsky, 'The Metabolism of 5-fluoro-cytosine-2^{14}C and of Cytosine ^{14}C in the Rat and the Disposition of 5-fluoro-cytosine-2^{14}C in Man', *Biochem. Pharmacol., 15* (1966), pp. 435–466.

129. J.J. Roxon, A.J. Ryan and S.E. Wright, 'Enzymatic Reduction of Tartrazine by *Proteus vulgaris* from Rats', *Fd Cosmet. Toxicol., 5* (1967), pp. 645–656.

130. R.R. Scheline, 'Studies on the Role of the Intestinal Microflora in the Metabolism of Coumarin in Rats', *Acta Pharmacol. Toxicol., 26* (1968), pp. 325–331.

131. S.R. Indahl and R.R. Scheline, 'The Metabolism of Umbelliferone and Herniarin in Rats and by Rat Intestinal Microflora', *Xenobiotica, 1* (1971), pp. 13–24.

132. Y. Nakagawa, M.R. Shetlar and S.H. Wender, 'Urinary Products from Quercetin in Neomycin-treated Rats', *Biochim. Biophys. Acta, 97* (1965), pp. 233–241.

133. R.S. Rosenfeld, I. Paul and T. Yamuchi, 'Sterol Esterification in Feces', *Arch. Biochem. Biophys., 122* (1967), pp. 653–657.

134. O. Sjaastad, 'Fate of Histamine and *N*-acetylhistamine Administered into the Human Gut', *Acta Pharmacol. Toxicol., 24* (1966), pp. 189–202.

135. R.R. Scheline and B. Longberg, 'The Absorption, Metabolism and Excretion of the Sulphonated Azo Dye, Acid Yellow, by Rats', *Acta Pharmacol. Toxicol., 23* (1965), pp. 1–14.

136. I. Rowland, M. Davies and P. Grasso, 'Biosynthesis of Methylmercury Compounds by the Intestinal Flora of the Rat', *Arch. Environ. Hlth, 32* (1977), pp. 24–28.

137. P.J. Thomas, '*In vitro* Conversion of Oleic Acid to Hydroxystearic Acid by Intestinal Bacteria', *Clin. Res., 18* (1970), p. 609.

138. P.J. Thomas, 'Identification of some Enteric Bacteria which Convert Oleic Acid to Hydroxystearic Acid *in vitro*', *Gastroenterology, 62* (1978), pp. 430–435.

139. P. Klubes and W.R. Jondorf, 'Dimethylnitrosamine Formation From Sodium Nitrite and Dimethylamine by Bacterial Flora of Rat Intestine', *Res. Commun. Chem. Pathol. Pharmacol., 2* (1971), pp. 24–34.

140. G. Hawksworth and M.J. Hill, 'Bacteria and the *N*-Nitrosation of Secondary Amines', *Br. J. Cancer, 25* (1971), pp. 520–526.

141. P. Kemp and R.M.C. Dawson, 'Isomerisation of Linolenic Acid by Rumen Micro-organisms', *Biochem. J., 109* (1968), pp. 477–478.

142. R.H. Adamson, J.W. Bridges, M.E. Evans and R.T. Williams, 'Species Differences in the Aromatisation of Quinic Acid *in vivo* and the Role of Gut Bacteria', *Biochem. J., 116* (1970), pp. 437–443.

143. P. Goddard and M.J. Hill, 'The Dehydrogenation of the Steroid Nucleus by Human-gut Bacteria', *Biochem. Soc. Trans., 1* (1973), pp. 1113–1115.

144. R.T. Williams, P. Millburn and R.L. Smith, 'The Influence of Entero-hepatic Circulation on the Toxicity of Drugs', *Ann. N.Y. Acad. Sci., 123* (1965), p. 110.

145. A.G. Clark, L.J. Fischer, P. Millburn, R.L. Smith and R.T. Williams, 'The Role of Gut Flora in the Enterohepatic Circulation of Stilboestrol in the Rat', *Biochem. J., 112* (1969), p. 17.

146. L.G. Ladomery, A.J. Ryan and S.E. Wright, 'Excretion of ^{14}C Butylated Hydroxytoluene (BHT) in the Rat', *Fd Cosmet. Toxicol., 3* (1967), p. 547.

147. M. Spalz, D.W.E. Smith, E.G. McDaniel and G.L. Laqueur, 'Role of Intestinal Microorganisms in Determining Cycasin Toxicity', *Proc. Soc. exp. Biol. Med. (NY), 124* (1967), p. 691.

148. A.G. Renwick and R.T. Williams, 'Gut Bacteria and the Metabolism of Cyclamate in the Rat', *Biochem. J., 114* (1969), p. 78.

149. M.J. Hill, B.S. Drasar, V. Aries, J.S. Crowther, G. Hawksworth and R.E.O. Williams, 'Bacteria and Aetiology of Cancer of the Large Bowel', *Lancet*, pp. 95–100.

150. K. Hartiala, 'Metabolism of Hormones, Drugs and Other Substances by the Gut', *Physiological Rev., 53* (1973), pp. 496–534.

151. R.S. Chhabra and J.R. Fouts, 'Biochemical Properties of Some Microsomal Xenobiotic-metabolising Enzymes in Rabbit Small Intestine', *Drug Metabolism & Disposition, 4* (1976), pp. 208–214.

152. R.S. Chhabra and J.M. Tredger, 'Interactions of Drugs and Intestinal Mucosal Endoplasmic Reticulum', *Nutrition & Drug Interrelations*, (1978), pp. 253–277.

153. S.J. Stohs, R.C. Grafström, D.M. Burke and S. Orrenius, 'Xenobiotic Metabolism and Enzyme Induction in Isolated Rat Intestinal Microsomes', *Drug Metabolism & Disposition, 4* (1976), pp. 517–521.

154. M.H. Floch, S. Van Noorden and H.M. Spiro, 'Histochemical Localisation of Gastric and Small Bowel Mucosal Enzymes of Man, Monkey and Chimpanzee', *Gastroenterology, 52* (1967), pp. 230–238.

155. B. Friedman, D.S. Strachan and M.M. Dewey, 'Histochemical and Biochemical Analysis of the Non-specific Esterases of the Small Intestine of the Rat', *J. Histochem. Cytochem., 14* (1966), pp. 560–566.

156. D.W. Piper and B.H. Fenton, 'Beta-glucuronidase of Gastric, Mucosal and Carcinomatous Origin', *Clin. Chem. Acta, 24* (1969), pp. 215–219.

157. R.P. Spencer, K.R. Brody and B.M. Lutters, 'Some Effects of Ethanol on the Gastrointestinal Tract', *Am. J. Dig. Dis., 9* (1964), pp. 599–604.

158. W.-F. Fang and H.W. Strobel, 'The Drug and Carcinogen Metabolism System of Rat Colon Microsomes', *Arch. Biochem. Biophys., 186* (1978), pp. 128–138.

159. J. Kubistova, 'Parathion Metabolism in Female Rats', *Arch. Int. Pharmacodyn. Ther., 118* (1959), pp. 308–316.

160. P.D. Lotlikar, E.C. Miller, J.A. Miller and A. Margreth, 'Enzymic Reduction of the *N*-hydroxy Derivatives of 2-acetylaminofluorene and Related Carcinogens by Tissue Preparations', *Cancer Res., 25* (1965), pp. 1743–1752.

161. A. Aitio, 'UDP-glucuronyl Transferase Activity in Various Rat Tissues', *Int. J. Biochem., 5* (1974), pp. 325–330.

162. M.O. Pulkkinen, 'Sulphate Conjugation during Development in Human, Rat and Guinea Pig', *Acta Physiol. Scand., 66* (1966), pp. 115–119.

163. L.M. Pinkus, J.N. Ketley and W.B. Jakoby, 'The Gluthathione *S*-transferases as a possible Detoxification System of Rat Intestinal Epithelium', *Biochem. Pharmacol., 26* (1977), pp. 2359–2363.

164. C. Davidson, E.F. Zimmerman and P.K. Smith, 'On the Metabolism and Toxicity of Methyl Salicylate', *J. Pharm. Expt. Ther., 132* (1961), pp. 207–211.

165. J.R. Poley, 'Fat Digestion and Absorption in Lipase and Bile Acid Deficiency' in K. Rommel and R. Bohmer (eds.), *Lipid Absorption: Biochemical and Clinical Aspects* (Univ. Park Press, Baltimore, 1976), pp. 151–202.

166. V.L.W. Go, J.R. Poley, A.F. Hofmann and W.H.J. Summerskill, 'Disturbances in Fat Digestion Induced by Acidic Jejunal pH due to Gastric Hypersecretion in Man', *Gastroenterology, 58* (1970), pp. 638–646.

167. B. Hadorn and J.C. Haworth, 'Intestinal Enterokinase Deficiency Clinical and Biochemical Findings in Six Patients' in B. Borgström, A. Dahlqvist and L. Hambracus (eds.), *Intestinal Enzyme Deficiencies and their Nutritional Implications* (Swedish Nutrition Foundation, 1973), pp. 107–113.

168. J.C. Haworth, B. Hadorn, B. Gourley, A, Prasad and V. Troesch, 'Intestinal Enterokinase Deficiency, Occurrence in Two Siblings and Age Dependency of Clinical Expression', *Arch. Dis. Child., 50* (1973), pp. 277–282.

169. G.C. Cook, 'Incidence and Clinical Features of Specific Hypolactasia in Adult Man' in B. Borgström, A. Dahlqvist and L. Hambracus (eds.), *Intestinal Enzyme Deficiencies and their Nutritional Implications* (Swedish Nutrition Foundation, 1973), pp. 52–73.

170. A. Dahlqvist and N.-G. Asp, 'Specific Disaccharidase Deficiency in

Adults', *Trans. Biochem. Soc., 3* (1975), pp. 227–232.

171. G.C. Cook, 'Defects of Sugar Absorption. Some Observations on Racial Lactase Deficiency', *Proc. R. Soc. Med., 61* (1968), pp. 1102–4.

172. N.S. Rosensweig, 'The Influence of Dietary Carbohydrates on Intestinal Disaccharidase Activity in Man' in B. Borgström, A. Dahlqvist and L. Hambracus (eds.), *Intestinal Enzymic Deficiencies and their Nutritional Implications* (Swedish Nutrition Foundation, 1973), pp. 21-31.

173. E. Gudmund-Høyer, 'Clinical Significance and Some Nutritional Aspects of Disaccharide Malabsorption' in B. Borgström, A. Dahlqvist and L. Hambracus (eds.), *Intestinal Enzyme Deficiencies and their Nutritional Implications* (Swedish Nutrition Foundation, 1973), pp. 37–44.

174. P. Coello-Ramirez and F. Lifshitz, 'Enteric Microflora and Carbohydrate Tolerance in Infants with Diarrhoea', *Pediatrics, 49* (1972), pp. 233–242.

175. A. Jones, P.R. Flanagan and G.C. Forstner, 'Pathogenesis of Mucosal Injury in the Blind Loop Syndrome. Brush Border Enzyme Activity and Glycoprotein Degradation', *J. Clin. Invest., 60* (1977), pp. 1321–1330.

176. W. Perlow, E. Baraona and C.S. Lieber, 'Symptomatic Intestinal Disaccharidase Deficiency in Alcoholics', *Gastroenterology, 72* (1977), pp. 680–684.

177. G.C. Graham and D.M. Paige, 'Nutritional Implications of Low Intestinal Lactase Activity in Children' in B. Borgström, A. Dahlqvist and L. Hambracus (eds.), *Intestinal Enzyme Deficiencies and their Nutritional Implications* (Swedish Nutrition Foundation, 1973), pp. 45–51.

178. C.E. Stirling, A.J. Schneider, M.-D. Wong and W.B. Kinter, 'Quantitative Radioautography of Sugar Transport in Intestinal Biopsies from Normal Humans and a Patient with Glucose-Galactose Malabsorption', *J. Clin. Invest., 51* (1972), pp. 438–451.

179. E.E. Wollaeyer, W.O. Lundberg and J.R. Chipault, 'Fecal Plasma Lipids', *Gastroenterology, 24* (1952), pp. 422–436.

180. C.R. Kepler and S.B. Tove, 'Biohydrogenation of Unsaturated Fatty Acids III. Purification and Properties of a Linoleate. $\triangle$12-*cis*, $\triangle$11-trans-isomerase from *Butyrivibro fibrisolvens*', *J. Biol. Chem., 242* (1967), pp. 5686–5692.

181. S.C. Mills, T.W. Scott, G.R. Russell and R.M. Smith, 'Hydrogenation of C_{18} Unsaturated Fatty Acids by Pure Cultures of a Rumen *Micrococcus*', *Aust. J. Biol. Sci., 23* (1970), pp. 1109–1113.

182. P. Kemp and R.W. White, 'The Biohydrogenation of Linolenic and Linoleic Acids by Bacteria Isolated from an Ovine Rumen', *Biochem. J., 106* (1968), p. 55 P.

183. W.C. Watson, 'Intestinal Hydrogenation of Dietary Fatty Acids', *Clin. Chim. Acta, 12* (1965), pp. 340–342.

184. A.T. James, J.P.W. Webb and T.D. Kellock, 'The Occurrence of Unusual Fatty Acids in Faecal Lipids from Human Beings with Normal and Abnormal Fat Absorption', *Biochem. J., 78* (1961), pp. 333–339.

185. E.L. Wynder and S. Takao, 'Environmental Factors of Cancer of the Colon and Rectum', *Cancer, 20* (1967), pp. 1520–1561.

186. W. Haenszel, 'Cancer Mortality among the Foreign-born in the United States', *J. Natl Cancer Inst., 26* (1961), pp. 37–132.

187. J. Staszewski, M.G. McCall and N.S. Stenhouse, 'Cancer Mortality in 1962-66 among Polish Migrants to Australia', *Br. J. Cancer, 25* (1971), pp. 599–610.

188. M.J. Hill, 'The Effect of some Factors on the Faecal Concentration of Acid Steroids, Neutral Steroids and Urobilins', *J. Path., 104* (1974), pp. 239–245.

189. J.S. Crowther, B.S. Drasar, P. Goddard, M.J. Hill and K. Johnson, 'The Effect of a Chemically Defined Diet on the Faecal Flora and Faecal Steroid Concentration', *Gut, 14* (1973), pp. 790–793.

190. W.E. Connor, D.T. Witiak, D.B. Stone and M.L. Armstrong, 'Cholesterol Balance and Faecal Neutral Steroids and Bile Acid Excretion in Normal Men Fed Dietary Fats of Different Fatty Acid Composition', *J. Clin. Invest., 48* (1965), pp. 1363–1375.

191. E. Quintao, S.M. Grundy and E.H. Ahrens, 'Effect of Dietary Cholesterol on the Regulation of Total Body Cholesterol in Man', *J. Lipid Res., 12* (1971), pp. 233–247.

192. J.D. Wilson and C.A. Lindsey, 'Studies on the Influence of Dietary Cholesterol Metabolism in the Isotopic Steady State in Man', *J. Clin. Invest., 44* (1965), pp. 1805–1814.

193. M.J. Hill, B.S. Drasar, R.E.O. Williams, T.W. Meade, A.G. Cox, J.E.P. Simpson and B.C. Morson, 'Faecal Bile-Acids and Clostridia in Patients with Cancer of the Large Bowel', *Lancet, i* (1975), pp. 535–538.

194. B.R. Goldin and S.L. Gorbach, 'The Relationship Between Diet and Rat Faecal Bacterial Enzymes Implicated in Colon Cancer', *J. Natl Cancer Inst., 57* (1976), pp. 371–375.

195. B.S. Reddy and E.L. Wynder, 'Large-Bowel Carcinogenesis Fecal Constituents of Populations with Diverse Incidence Rates of Colon Cancer', *J. Natl Cancer Inst., 50* (1972), pp. 1437–1442.

196. H.F. Mower, R.M. Ray, R. Shoff, G.N. Stemmermann, A. Nomura, G.A. Glober, S. Kamiyama, A. Shimada and H. Yamakawa, 'Fecal Bile Acids in Two Japanese Populations with Different Colon Cancer Risks', *Cancer Res., 39* (1979), pp. 328–331.

197. IARC Intestinal Microecology Group, 'Dietary Fibre, Transit-Time, Faecal Bacteria, Steroids and Colon Cancer in Two Scandinavian Populations', *Lancet, i* (1977), pp. 207–211.

198. E.L. Wynder and B.S. Reddy, 'The Epidemiology of Cancer of the Large Bowel', *Am. J. Dig. Dis., 19* (1974), pp. 937–946.

199. T. Narisawa, N.E. Magadia, J.H. Weisburger and E.L. Wynder, 'Promoting Effect of Bile Acids on Colon Carcinogenesis after Intrarectal Instillation of *N*-methyl-*N*′-nitro-*N*-nitrosoguanidine in Rats', *J. Natl Cancer Inst., 53* (1974), pp. 1093–1097.

200. S.J. Silverman and A.W. Andrews, 'Bile Acids: Comutagenic Activity in the *Salmonella*-Mammalian Microsome Mutagenicity Test: Brief Communication', *J. Natl Cancer Inst., 59* (1977), pp. 1557–1559.

201. M.I. Kelsey and R.J. Pienta, 'Transformation of Hamster Embryo Cells by Cholesterol-α-epoxide and Lithocholic Acid', *Cancer Lett., 6* (1979), pp. 143–149.

202. V. Aries and M.J. Hill, 'Degradation of Steroids by Intestinal Bacteria II. Enzymes Catalysing the Oxidoreduction of the 3α-, 7α- and 12α-hydroxy Groups in Cholic Acid and the Dehydroxylation of the 7-hydroxyl Group', *Biochim. Biophys. Acta, 202* (1970), pp. 535–543.

203. M.J. Hill and V.C. Aries, 'Faecal Steroid Composition and its Relationship to Cancer of the Large Bowel', *J. Pathol., 104* (1971), pp. 129–139.

204. F. Fernandez and M. Hill, 'A Faecal Hydrogen Acceptor for Clostridial 3-Oxo Steroid Δ4-Dehydrogenase', *Biochim. Soc. Trans., 6* (1978), pp. 376–377.

205. P. Lombardi, B. Goldin, E. Boutin and S.L. Gorbach, 'Metabolism of Androgens and Estrogens by Human Faecal Microorganisms', *J. Steroid Biochem., 9* (1978), pp. 795–801.

206. M.J. Hill, 'The Role of Colon Anaerobes in the Metabolism of Bile Acids and Steroids, and its Relation to Colon Cancer', *Cancer, 36* (1975), pp. 2387–2400.

207. S.M. Finegold, H.R. Attebery and V.L. Sutter, 'Effect of Diet on Human Fecal Flora: Comparison of Japanese and American Diets', *Am. J. Clin. Nutr., 27*

(1974), pp. 1456–1469.
208. S.M. Finegold, D.J. Flora, H.R. Attebery and V.L. Sutter, 'Fecal Bacteriology of Colonic Polyp Patients and Control Patients', *Cancer Res., 35* (1975), pp. 3407–3417.
209. B.S. Reddy, A. Mastromarino, C. Gustafson, M. Lipkin and E.L. Wynder, 'Fecal Bile Acids and Neutral Sterols in Patients with Familial Polyposis', *Cancer, 38* (1976), pp. 1694–1698.
210. B.S. Reddy and E.L. Wynder, 'Metabolic Epidemiology of Colon Cancer. Fecal Bile Acids and Neutral Sterols in Colon Cancer Patients and Patients with Adenomatous Polyps', *Cancer, 39* (1977), pp. 2533–2539.
211. B.S. Reddy, C.W. Martin and E.L. Wynder, 'Fecal Bile Acids and Cholesterol Metabolites of Patients with Ulcerative Colitis, a High Risk Group for Development of Colon Cancer', *Cancer Res., 37* (1977), pp. 1697–1701.
212. A.L. Watne, H.-Y.L. Hai, T. Mance and S. Core, 'Fecal Steroids and Bacterial Flora in Patients with Polyposis Coli', *Am. J. Surg., 131* (1976), pp. 42–46.
213. E. Bone, B.S. Drasar and M.J. Hill, 'Gut Bacteria and their Metabolic Activities in Familial Polyposis', *Lancet, i* (1975), pp. 1117–1120.
214. M. Moskovitz, C. White and M.H. Floch, 'Bile Acid and Neutral Sterol Excretion in Carcinoma of the Colon, Other Cancers and Control Subjects', *Gastroenterology, 74* (1978), p. 107.
215. T.D. Wilkins and A.S. Hackman, 'Two Patterns of Neutral Steroid Conversion in Faeces of Normal North Americans', *Cancer Res., 34* (1974), pp. 2250–2254.
216. A.A. Salyers, J.F. Sperry, T.D. Wilkins, A.R.P. Walter and N.J. Richardson, 'Neutral Steroid Concentration in the Faeces of N. American White and S. African Black Populations at Different Risks for Cancer of the Colon', *S. African Med. J., 51* (1977), pp. 823–827.
217. P. Cruse, M. Lewin and C.G. Clark, 'Dietary Cholesterol is Co-carcinogenic for Human Colon Cancer', *Lancet, i* (1979), pp. 752–755.
218. M.J. Hill, B.C. Morson and H.J.R. Bussey, 'Aetiology of Adenoma-carcinoma Sequence in Large Bowel', *Lancet, i* (1978), pp. 245–247.
219. J.P. Cruse, M.R. Lewin, G.P. Ferulano and C.G. Clark, 'Co-carcinogenic Effects of Dietary Cholesterol in Experimental Colon Cancer', *Nature (Lond.), 276* (1978), pp. 822–825.
220. J. Melnykowycz and K.R. Johansson, 'Formation of Amines by Intestinal Microorganisms and the Influence of Chlortetracycline', *J. Expl Med., 101* (1955), pp. 507–517.
221. E.A. Phear and B. Ruebner, 'The *In Vitro* Production of Ammonium and Amines by Intestinal Bacteria in Relation to Nitrogen Toxicity as a Factor in Hepatic Coma', *Br. J. Exp. Path., 37* (1956), pp. 253–262.
222. Anon, 'Headache, Tyramine, Serotonin and Migraine', *Nutr. Rev., 26* (1968), pp. 40–44.
223. S.L. Rice, R.R. Eitenmiller and R.E. Koehler, 'Biologically Active Amines in Food. A. Review', *J. Milk Fd Technol., 39* (1976), pp. 353–358.
224. W.T. Irvine, H. Duthie and N.G. Watson, 'Urinary Output of Free Histamine After a Meat Meal', *Lancet, i* (1959), pp. 1061–1064.
225. O.M. Bakke, 'Studies on the Degradation of Tyrosine by Rat Caecal Contents', *Scand. J. Gastroenterol., 4* (1969), pp. 603–608.
226. R.K. Boutwell in B.J. Finkle and V.C. Runeckles (eds.), *Phenolic Compounds and Metabolic Regulation* (Appleton-Century-Crofts, N.Y., 1967).
227. A. Angel and K.J. Rogers, 'Convulsant Action of Polyphenols', *Nature (Lond.), 217* (1968), pp. 84–85.
228. W.F. Rogers, M.P. Burdick and G.R. Burnett, 'The Effect of Antibiotics

on the Excretion of Phenolic Compounds', *J. Lab. Clin. Med., 45* (1955), pp. 87–96.

229. R.D. DeMoss and K. Moser, 'Tryptophanase in Diverse Bacterial Species', *J. Bact., 98* (1969), pp. 167–171.

230. W.F. Dunning and M.R. Curtis, 'The Role of Indole in Incidence of 2-Acetylaminofluorene-induced Bladder Cancer in Rats', *Proc. Soc. Biol. Med., 99* (1958), pp. 91–99.

231. H.J. Dyer and H.P. Morris, 'An Effect of *N*-2-fluorenylacetamide on the Metabolism of Tryptophan in Rats', *J. Natl Cancer Inst., 26* (1961), pp. 315–329.

232. G. Bryan, 'The Role of Urinary Tryptophan Metabolite in the Etiology of Bladder Cancer', *Am. J. Clin. Nutr., 24* (1971), pp. 841–847.

233. H. Ehrhart and W. Stich, 'Untersuchungen uber Experimentelle Leukamien: Die Indol. Leukamie bei der weissen Mans', *Klin. Wochenschr., 35* (1957), pp. 504–511.

234. M.O. Rauschenbakn, E.I. Zharova and T.G. Protasova, 'Blastomogenic Properties of Certain Metabolites of Tryptophan', *Acta Un. Int. Contra Cancrum., 19* (1963), p. 660.

235. J.P. Bowden, K.-T. Chung and A.W. Andrews, 'Mutagenic Activity of Tryptophan Metabolites Produced by Rat Intestinal Microflora', *N. Natl Cancer Inst., 57* (1966), pp. 921–924.

236. C.A. Benassi, F.M. Veronese and A. De Antoni, 'On the Determination of Small Amounts of Tryptophan Metabolites and their Occurrence in Normal Human Urine', *Clin. Chim. Acta, 17* (1967), pp. 383–391.

237. F.A.G. Teulings, W. Fokkens, J.G.A.H. Kaalen and B. Van Der Werf-Messing, 'The Concentration of Free and Conjugated 3-Hydroxyanthranilic Acid in the Urine or Bladder Tumour Patients Before and After Therapy, Measured with an Enzymatic Method', *Br. J. Cancer, 27* (1973), pp. 316–322.

238. J.W. Berg and M.A. Howell, 'The Geographic Pathology of Bowel Cancer', *Cancer, 34* (1974), pp. 807–814.

239. S. Graham, H. Dayal, M. Swanson, A. Mittelman and G. Wilkinson, 'Diet in the Epidemiology of Cancer of the Colon and Rectum', *J. Natl Cancer Inst., 61* (1978), pp. 709–714.

240. D.D. Kasarda, 'Coeliac Disease. Malabsorption of Nutrients Induced by a Toxic Factor in Gluten', *Bakers Digest, 46* (1972), p. 25.

241. A. Ferguson, 'Coeliac Disease (Gluten Hypersensitivity)', *J. Human Nutr., 30* (1976), pp. 193–201.

242. O.D. Lowlessar, 'Dietary Gluten Sensitivity Updated', *J. Am. Diet Assoc., 60* (1972), pp. 475–477.

243. M.L. Roselund, 'What is Celiac Disease?', *Clin Pediatrics, 9* (1970), p. 695.

244. C.M. Anderson, A.C. Frazer, J.M. French, J.W. Gerrard, H.G. Sammons and J.M. Smellie, 'Coeliac Disease: Gastro-Intestinal Studies and the Effect of Dietary Wheat Flour', *Lancet, i* (1952), pp. 836–842.

245. M. Messer and C.M. Anderson, 'Pancreatic Carboxypeptidases A and B in Coeliac Disease', *Clin. Chim. Acta, 6* (1961), pp. 276–280.

246. M. Messer, C.M. Anderson and R.R.W. Townley, 'Peptidase Activity of Biopsies of the Duodenal Mucosa of Children with and without Celiac Disease', *Clin. Chim. Acta, 6* (1961), pp. 768–775.

247. H.J. Cornell and C.J. Rolles, 'Further Evidence of a Primary Mucosal Defect in Coeliac Disease', *Gut, 19* (1978), pp. 253–259.

248. C.F. Langworthy and H.J. Deuel, 'Digestibility of Raw Corn, Potato and Wheat Starches', *J. Biol. Chem., 42* (1920), pp. 27–40.

249. J.W. Beazell, C.R. Schmidt and A.C. Ivy, 'On the Digestibility of Raw

Potato Starch in Man', *J. Nutr., 17* (1939), pp. 77–83.

250. B. Jelinek, M.C. Katayama and A.E. Harper, 'The Inadequacy of Unmodified Potato Starch as Dietary Carbohydrate for the Albino Rat', *Can. J. med. Sci., 30* (1952), p. 447.

251. E.A. El Harith, J.W.T. Dickerson and R. Walker, 'Potato Starch and Caecal Hypertrophy in the Rat', *Fd Cosmet. Toxicol., 14* (1976), pp. 115–121

252. C.T. Whittemore, I.W. Moffat and A.G. Taylor, 'Influence of Cooking upon the Nutritive Value of Potato and Maize in Diets for Growing Pigs', *J. Sci. Fd Agric., 26* (1975), pp. 1567–1576.

253. E.A. El Harith, R. Walker, G.G. Birch and G. Sukan, 'Some Factors Influencing Caecal Enlargement Induced by the Raw Potato Starch in the Rat', *Fd Chem., 2* (1977), pp. 279–289.

254. R. Walker and E.A. El Harith, 'Nutritional and Toxicological Properties of some Raw and Modified Starches', *Ann Nutr. Alim., 32* (1978), pp. 671–679.

255. A.O. Ketiku and V.A. Oyenuga, 'Changes in the Carbohydrate Constituents of Cassava Root-Tuber (Manihot Utillissima Pohl) during Growth', *J. Sci. Fd Agric., 23* (1972), pp. 1451–1456.

256. R. Walker, 'Some Observations on the Phenomenon of Caecal Enlargement in the Rat' in C.L. Galli, R. Paoletti and G. Vettorazzi (eds.), *Chemical Toxicology of Food* (Elsevier/North Holland, Amsterdam, 1978), pp. 339–348.

257. E.A. El Harith, J.W.T. Dickerson and R. Walker, 'On the Nutritive Value of Various Starches for the Albino Rat', *J. Sci. Fd Agric., 27* (1976), pp. 521–526.

258. G. Reussner, J. Andros and R. Thiessen, 'Studies on the Utilization of Various Starches and Sugars in the Rat', *J. Nutr., 80* (1963), p. 291.

259. G.G. Birch and I.J. Etheridge, 'Chemical and Physiological Properties of Glucose Syrup Components', *Starke., 25* (1973), p. 235.

260. G.G. Birch, I.J. Etheridge and L.F. Green, 'Short-term Effects of Feeding Rats with Glucose Syrup Fractions and Dextrose', *Br. J. Nutr., 29* (1973), p. 87.

261. F.R. Steggerda, E.A. Richards and J.J. Rackis, 'Effects of Various Soybean Products on Flatulence in the Adult Man', *Proc. Soc. Expl Biol. Med., 121* (1966), pp. 1235–1239.

262. E.A. Richards and F.R. Steggerda, 'Production and Inhibition of Gas in Various Regions in the Intestine of the Dog', *Proc. Soc. Expl Biol. Med., 122*, (1966).

263. E.A. Richards, F.R. Steggerda and A. Murata, 'Relationship of Bean Substrate and Certain Intestinal Bacteria to Gas Production in the Dog', *Gastroenterol., 55* (1968), p. 502.

264. G.O. Aspinwall, K. Hunt and I.N. Morrison, 'Polysaccharides of Soybeans. III. Extraction and Fractionation of Polysaccharides from Cotyledon Meal', *J. Chem. Soc.*, (1967), pp. 1065–1070.

265. D.H. Calloway and E.L. Murphy, 'The Use of Expired Air to Measure Intestinal Gas Formation', *Ann. N.Y. Acad. Sci., 150* (1968), p. 82.

266. D.P. Burkitt, 'Disease of the Alimentary Tract and Western Diets', *Path. Microbiol., 39* (1973), pp. 177–186.

267. G.A. Spiller and R.J. Amen, 'Plant Fibers in Nutrition: Need for Better Nomenclature', *Am. J. Clin. Nutr., 28* (1975), pp. 675–676.

268. D.A.T. Southgate, 'Determination of Carbohydrates in Foods. II – Unavailable Carbohydrates', *J. Sci. Fd Agric., 20* (1969), pp. 331–335.

269. R.J. Van Soest and R.W. McQueen, 'The Chemistry and Estimation of Fiber', *Proc. Nutr. Soc., 32* (1973), pp. 123–130.

270. P.J. Van Soest, 'The Uniformity and Nutritive Availability of Cellulose', *Fed. Proc., 32* (1973), pp. 1804–1808.

271. A.I. Mendeloff, 'Dietary Fiber', *Nutr. Rev., 33* (1975), pp. 321–326.

272. N.S. Painter, A.Z. Almeida and K.W. Colebourne, 'Unprocessed Bran in Treatment of Diverticular Disease of the Colon', *Br. Med. J., 2* (1972), pp. 137–140.

273. A.J. Carlson and F. Moelzel, 'Relation of Diet to Diverticulosis of the Colon in Rats', *Gastroenterology, 12* (1949), pp. 108–115.

274. S. Graham, H. Dayal, M. Swanson, A. Mittelman and G. Wilkinson, 'Diet in the Epidemiology of Cancer of the Colon and Rectum', *J. Natl Cancer Inst., 61* (1978), pp. 709–714.

275. M.A. Eastwood and D. Hamilton, 'Studies on the Adsorption of Bile Salts to Non-Absorbed Components of Diet', *Biochim. Biophys. Acta, 152* (1968), pp. 165–173.

276. D.P. Burkitt, 'Epidemiology of Cancer of the Colon and Rectum', *Cancer, 28* (1971), pp. 3–13.

277. D.P. Burkitt, A.R.P. Walker and N.S. Painter, 'Effect of Dietary Fibre on Stools and Transit-times, and its Role in the Causation of Disease', *Lancet, ii* (1972), pp. 1408–1409.

278. R.L. Walters, I. McLean Baird, P.S. Davies, M.J. Hill, B.S. Drasar, D.A.T. Southgate, J. Green and B. Morgan, 'Effects of Two Types of Dietary Fibre on Faecal Steroid and Lipid Excretion', *Br. Med. J., 2* (1975), pp. 536–538.

279. J.M. Findlay, A.N. Smith, W.D. Mitchell, A.J.B. Anderson and M.A. Eastwood, 'Effects of Unprocessed Bran on Colon Function in Normal Subjects and in Diverticular Disease', *Lancet, i* (1974), pp. 146–149.

280. M.A. Eastwood, J.R. Kirkpatrick, W.D. Mitchell, A. Bone and T. Hamilton, 'Effects of Dietary Supplements of Wheat Bran and Cellulose on Faeces and Bowel Function', *Br. Med. J., 4* (1978), pp. 392–394.

281. M. Stasse-Wolthuis, H.F.F. Albers, J.G.C. Van Jeveren, J.W. de Jong, J.G.A.J. Hautvast, R.J.J. Hermus, M.B. Katan, W.G. Brydon and M.A. Eastwood, 'Influence of Dietary Fiber from Vegetables and Fruit, Bran or Citrus Pectin on Serum Lipids, Fecal Lipids, and Colonic Function', *Am. J. Clin. Nutr., 33* (1980), pp. 1745–56.

282. J.H. Cummings, D.A.T. Southgate, W. Branch, H. Houston, D.J.A. Jenkins and W.P.T. James, 'Colonic Response to Dietary Fibre from Carrot, Cabbage, Apple, Bran and Guar Gum', *Lancet, i* (1978), pp. 5–8.

283. O.K. Payler, E.W. Pomare, K.W. Heaton and R.F. Harvey, 'The Effect of Wheat Bran on Intestinal Transit', *Gut, 16* (1978), pp. 209–213.

284. D. Kritchevsky and J.A. Story, 'Binding of Bile Salts *in vitro* by Non-nutritive Fiber', *J. Nutr., 104* (1974), pp. 458–461.

285. M.A. Eastwood, R. Anderson, W.D. Mitchell, J. Roberton and S. Pocock, 'A Method to Measure the Adsorption of Bile Salts to Vegetable Fibre of Differing Water Holding Capacity', *J. Nutr., 106* (1976), pp. 1429–1432.

286. M.J. Hill, J.S. Crowther, B.S. Drasar, G. Hawksworth, V. Aries and R.E.O. Williams, 'Bacteria and Aetiology of Cancer of Large Bowel', *Lancet, i* (1971), pp. 95–100.

287. B.S. Reddy, A.R. Hedges, K. Laakso and E.L. Wynder, 'Metabolic Epidemiology of Large Bowel Cancer. Fecal Bulk and Constituents of High-Risk North American and Low-Risk Finnish Population', *Cancer, 42* (1978), pp. 2832–2838.

288. A. Antonis and I. Bersohn, 'The Influence of Diet on Fecal Lipids in South African White and Bantu Prisoners', *Am. J. Clin. Nutr., 11* (1962), pp. 142–155.

289. G.A. Spiller, 'Interactions of Plant Fibers and Levels of Dietary Cholesterol on Fecal Neutral Sterols Excretion', *J. Nutr., 7* (1976), XXVI.

290. B.H. Ershoff, 'Beneficial Effects of Alfalfa and Other Succulent Plants on Glucascorbic Acid Toxicity in the Rat', *Proc. Soc. Exp. Biol. Med., 85* (1957),

pp. 656–659.
291. B. Ershoff and E.W. Thurston, 'Effect of Diet on Amaranth (FD&C Red No. 2) Toxicity in the Rat', *J. Nutr., 104* (1974), pp. 937–942.
292. T. Kimura, H. Furuta, Y. Matsumoto and A. Yoshida, 'Ameliorating Effect of Dietary Fiber on Toxicities of Chemicals Added to a Diet in the Rat', *J. Nutr., 110* (1980), pp. 513–521.
293. B.H. Ershoff, 'Comparative Effects of a Purified and Stock Ration on Sodium Cyclamate Toxicity in Rats', *Proc. Soc. Exp. Biol. Med., 141* (1972), pp. 857–862.
294. J.M. Ward, R.S. Yammamoto and J.H. Weisburger, 'Cellulose Dietary Bulk and Azoxymethane-induced Intestinal Cancer', *J. Natl Cancer Inst., 51* (1973), pp. 713–715.
295. N.D. Nigro, A.W. Bull, B.A. Klopfer, M.S. Pak and R.L. Campbell, 'Effect of Dietary Fiber on Azoxymethane-induced Intestinal Carcinogenesis in Rats', *J. Natl Cancer Inst., 62* (1979), pp. 1097–1102.
296. R.W. Engel and D.H. Copeland, 'Protective Action of Stock Diets Against the Cancer-inducing Action of 2-acetylaminofluorene in Rats', *Cancer Res., 12* (1952), pp. 211–215.
297. M.G. Whiting, 'Toxicity of Cycads', *Econ. Bot., 17* (1963), p. 271.
298. G.L. Laqueur and M. Spatz, 'Toxicology of Cycasin', *Cancer Res., 28* (1968), p. 2262.
299. K. Nishida, A. Kobayashi and T. Nagahama, 'Cycasin, a New Toxic Glycoside of *Cycas revoluta* Thumg., 1. Isolation and Structure of Cycasin', *Bull. Agric. Chem. Soc. Jap., 19* (1955), p. 77.
300. N.V. Riggs, 'Glucosylazoxymethane, a Constituent of the Seeds of *Cycas circinalis* L.', *Chemy. Ind.*, (1956), p. 926.
301. Anon, 'Cycad: The Fruits of Conversion', *Fd Cosmet. Toxicol., 10* (1972), p. 246.
302. G.L. Laqueur and H. Matsumoto, 'Neoplasms in Female Fisher Rats Following Intraperitoneal Injection of Methylazoxy-methanol', *J. Natl Cancer Inst., 37* (1966), pp. 217–232.
303. K. Nishida, A. Kobayashi and T. Nagahama, 'Studies on Cycasin, a New Toxic Glycoside of *Cycas revoluta* Thunb, Part IV. Pharmacology of Cycasin', *Seikagaku, 28* (1956), pp. 218–23.
304. P.B. Magee, *IVth Conference on the Toxicity of Cycads*, Bethesda, Maryland, 1965.
305. G.L. Laqueur, *Vth Conference on Cycad Toxicity*, Univ. Miami, Florida, 1967.
306. E.E. Conn, 'Cyanogenic Glycosides', *J. Agric. Fd Chem., 17* (1969), pp. 519–526.
307. R. Ferrando, M.M. Guilleux and A. Guerilott-Vinet, 'Oestrogen Content of Plants as a Function of Conditions of Culture', *Nature, Lond., 192* (1961), p. 1205.
308. M.W. Carter, G. Matrone and W.G. Smart, 'Effect of Genistin on Reproduction of the Mouse', *J. Nutr., 55* (1955), pp. 639–645.
309. R.B. Bradbury and D.E. White, 'The Chemistry of Subterranean Clover. I. Isolation of Formononetin and Genistein', *J. Chem. Soc.* (1951), pp. 3447–3449.
310. A.N. Booth, E.M. Bickoff and G.O. Kohler, 'Estogen-like Activity in Vegetable Oil and Mill By-products', *Science, 131* (1960), pp. 1807-1808.
311. L.A. Griffiths and G.E. Smith, 'Metabolism of Apigenin and Related Compounds in the Rat. Metabolite Formation *In vivo* and by the Intestinal Microflora *In Vitro*', *Biochem. J., 128* (1972), pp. 901–911.
312. A. Nilsson, J.L. Hill and H.L. Davies, 'An *In Vitro* study of Formononetin

and Biochanin A Metabolism in Rumen Fluid from Sheep', *Biochim. Biophys Acta, 148* (1967), pp. 92–98.

313. M.N. Cayen, A.L. Carter and R.H. Common, 'The Conversion of Genistein to Equol in the Fowl', *Biochim. Biophys. Acta, 86* (1964), pp. 56–64.

314. M.N. Cayen, G. Tang and R.H. Common, 'Urinary Conversion Products of Biochanin A and Formononetin in the Fowl', *Biochim. Biophys. Acta, 111* (1965), pp. 349–357.

315. T.J. Batterham, N.K. Hart, J.A. Lamberton and A.W.H. Braden, 'Metabolism of Oestrogenic Isoflavones in Sheep', *Nature, Lond., 206* (1965), p. 509.

316. I.E. Liener, *Toxic Constituents of Plant Foodstuffs*, Chapter 2 (Academic Press, London & New York, 1980).

317. W.G. Jaffe, 'Haemagluttinins (Lectins)', in I.E. Liener (ed.), *Toxic Constituents of Plant Foodstuffs* 2nd edn (Academic Press, London & New York, 1980), Chapter 3.

318. N.D. Noah, A.E. Bender, G.B. Reaidi and R.J. Gilbert, 'Food Poisoning from Raw Red Kidney Beans', *Br. Med. J., 281* (1980), pp. 236–237.

319. P. Sisley and C. Porscher, 'Du Sort Des Matières Colorantes Dans l'organisme Animal', *C. r. Hebd. Séanc. Acad. Sci., Paris, 152* (1911), p. 1062.

320. A.T. Fuller, 'Is *p*-aminobenzenesulphonamide the Active Agent in Prontosil Therapy?', *Lancet, i* (1937), p. 194.

321. J.L. Radomski and T.J. Mellinger, 'The Absorption, Fate and Excretion in Rats of the Water-soluble Azo Dyes, FD&C Red No. 2, FD&C Red No. 4 and FD&C Yellow No. 6', *J. Pharm. exp. Ther., 136* (1962), pp. 259–266.

322. R. Jones, A.J. Ryan and S.E. Wright, 'The Metabolism and Excretion of Tartrazine in the Rat, Rabbit and Man', *Fd Cosmet. Toxicol., 2* (1964), pp. 447–452.

323. H. Fore, R. Walker and L. Golberg, 'Studies on Brown FK II. Degradative Changes Undergone *In Vitro* and *In Vivo*', *Fd Cosmet. Toxicol., 5* (1967), pp. 459–473.

324. R. Walker and A.J. Ryan, 'Some Molecular Parameters Influencing Rate of Reduction of Azo Compounds by Intestinal Microflora', *Proceedings of Symposium on the Biological Oxidation of Nitrogen in Organic Molecules*, (Taylor & Francis, London, 1971), pp. 171–174.

325. A.J. Ryan, J.J. Roxon and A. Sivayavirojana, 'Bacterial Azo Reduction: a Metabolic Reaction in Mammals', *Nature, Lond., 219* (1968), pp. 854–855.

326. I.F. Gaunt, 'Studies on the Relationship Between Heinz Bodies and Haemolysis in Laboratory Animals and the Evaluation of Heinz Body Production in Toxicological Investigations', Ph.D. Thesis, University of London.

327. T. Watabe, N. Ozawa, Kobayashi and H. Kurata, 'Reduction of Sulphonated Water-soluble Azo Dyes by Microorganisms from Human Faeces', *Fd Cosmet. Toxicol., 18* (1980), pp. 349–352.

328. A.J. Ryan and P.G. Welling, 'The Metabolism and Excretion of Black PN in the Rat and Man', *Fd Cosmet. Toxicol., 8* (1970), pp. 487–497.

329. T. Honohan, F.E. Enderlin, B.A. Ryerson and T.M. Parkinson, 'Intestinal Absorption of Polymeric Derivatives of the Food Dyes Sunset Yellow and Tartrazine in Rats', *Xenobiotica, 7* (1977), pp. 765–774.

330. A.J. Ryan and S.E. Wright, 'The Excretion of Some Azo Dyes in Rat Bile', *J. Pharm. Pharmacol., 13* (1961), pp. 492–495.

331. P. Grasso and L. Golberg, 'Problems Confronted and Lessons Learnt in the Safety Evaluation of Brown FK', *Fd Cosmet. Toxicol., 6* (1968), pp. 737–747.

332. R. Walker, P. Grasso and I.F. Gaunt, 'Myotoxicity of Amine Metabolites from Brown FK', *Fd Cosmet. Toxicol., 8* (1970), pp. 539–542.

333. S. Vennitt and C.T. Bushell, 'Mutagenicity of the Food Colour Brown FK and Constituents in *Salmonella typhimurium*', *Mutat. Res., 40* (1976), p. 309.

334. S.M. Hess and O.G. Fitzhugh, 'Absorption and Excretion of Certain Triphenylmethane Colours in Rats and Dogs', *J. Pharmacol. exp. Ther., 114* (1955), p. 38.

335. I.F. Gaunt, J. Hardy, I.S. Kiss and S.D. Gangolli, 'Short-term Toxicity of Violet 6B (FD&C Violet No. 1) in the Rat', *Fd Cosmet. Toxicol., 12* (1974), pp. 11–19.

336. P. Grasso, J. Hardy, I.F. Gaunt, P.L. Mason and A. Lloyd, 'Long-term Toxicity of Violet 6B (FD&C Violet No. 1) in Mice', *Fd Cosmet. Toxicol., 12* (1974), pp. 21–31.

337. J.P. Brown, A. Dorsky, F.E. Enderlin, R.L. Hale, V.A. Wright and T.M. Parkinson, 'Synthesis of ^{14}C-labelled FD&C Blue No. 1 (Brilliant Blue FCF) and its Intestinal Absorption and Metabolic Fate in Rats', *Fd Cosmet. Toxicol., 18* (1980), pp. 1–5.

338. J.C. Phillips, D. Mendis, C.T. Eason and S.D. Gangolli, 'The Metabolic Disposition of ^{14}C-labelled Green S and Brilliant Blue FCF in the Rat, Mouse and Guinea Pig', *Fd Cosmet. Toxicol., 18* (1980), pp. 7–13.

339. K.R. Butterworth, I.F. Gaunt, P. Grasso and S.D. Gangolli, 'Acute and Short-term Toxicity Studies on Erythrosine BS in Rodents', *Fd Cosmet. Toxicol., 14* (1976), pp. 525–531.

340. E.J. Ethco and J.M. Webb, 'The Fate of FD&C Blue No. 2 in Rats', *J. Pharmacol. exp. Ther., 154* (1966), pp. 384–389.

341. I.F. Gaunt, I.S. Kiss, P. Grasso and S.D. Gangolli, 'Short-term Toxicity Study on Indigo Carmine in the Pig', *Fd Cosmet. Toxicol., 7* (1969), pp. 17–24.

342. L.W. Hazleton, T.W. Tusing, B.R. Zeitlin, R. Thiessen and M.K. Murer, 'Toxicity of Coumarin', *J. Pharmacol. exp. Ther., 118* (1956), pp. 348–358.

343. A.J. Cohen, 'Critical Review of the Toxicology of Coumarin with Special Reference to Interspecies Differences in Metabolism and Hepatotoxic Response and their Significance to Man', *Fd Cosmet. Toxicol., 17* (1979), pp. 277–289.

344. W.H. Shilling, R.F. Crampton and R.C. Longland, 'Metabolism of Coumarin in Man', *Nature, Lond., 221* (1969), p. 664.

345. R.C. Longland, W.H. Shilling and S.D. Gangolli, 'The Hydrolysis of Flavouring Esters by Artificial Gastro-intestinal Juices and Rat Tissue Preparations', *Toxicology, 8* (1977), pp. 197–204.

346. FEMA, Scientific Literature Review of Aliphatic Primary Alcohols, Aldehydes, Esters and Acids in Flavor Usage. (National Information Services/ Food & Drug Administration: Washington DC, 1974).

347. F. Grundschober, 'Toxicological Assessment of Flavouring Esters', *Toxicology, 8* (1977), pp. 387–390.

348. J.C. Phillips, J. Kingsnorth, S.D. Gangolli and I.F. Gaunt, 'Studies on the Absorption, Distribution and Excretion of Citral in the Rat and Mouse', *Fd Cosmet. Toxicol., 14* (1976), pp. 537–540.

349. H. Igimi, M. Nishimura, R. Kodama and H. Ide, 'Studies on the Metabolism of *d*-limonene (*p*-mentha-1,8-diene). 1. The Absorption, Distribution and Excretion of *d*-limonene in the Rat', *Xenobiotica, 4* (1974), pp. 77–84.

350. D.L.J. Opdyke, 'Fragrance Raw Materials Monographs, 1-carvone', *Fd Cosmet. Toxicol., 11* (1973), pp. 1057–1058.

351. D.L.J. Opdyke, 'Fragrance Raw Materials Monographs, *d*-carvone', *Fd Cosmet. Toxicol., 16*, Suppl. 1 (1978), pp. 673–674.

352. G. Kennedy, O.E. Fancher and J.C. Calandra, 'Metabolic Fate of Saccharin in the Albino Rat', *Fd Cosmet. Toxicol., 10* (1972), pp. 143–149.

353. R.L. Anderson, 'Response of Male Rats to Sodium Saccharin Ingestion:

Urine Composition and Mineral Balance', *Fd Cosmet. Toxicol., 17* (1979), pp. 195–200.

354. R.L. Anderson and J.J. Kirkland, 'The Effect of Sodium Saccharin in the Diet on Caecal Microflora', *Fd Cosmet. Toxicol., 18* (1980), pp. 353–355.

355. Anon, 'Cyclamate Still to the Fore', *Fd Cosmet. Toxicol., 10* (1972), pp. 237–243.

356. S. Kojima and H. Ichibagase, 'Studies on Synthetic Sweetening Agents VIII. Cyclohexylamine, a Metabolite of Sodium Cyclamate', *Chem. Pharm. Bull. (Japan), 14* (1966), pp. 971–974.

357. J.S. Leahy, M. Wakefield and T. Taylor, 'Urinary Excretion of Cyclohexylamine Following Oral Administration of Sodium Cyclamate to Man', *Fd Cosmet. Toxicol., 5* (1967), p. 447.

358. J.S. Leahy, T. Taylor and C.J. Rudd, 'Cyclohexylamine Excretors Among Human Volunteers given Cyclamate', *Fd Cosmet. Toxicol., 5* (1967), pp. 595–6.

359. J. Watt and R. Marcus, 'Effect of Degraded and Undegraded Alginates on the Colon of Guinea Pig', *Proc. Nutr. Soc., 30* (1971), p. 81A.

360. A.P. De Groot, H.P. Til, V.J. Feron, H.C. Dreef-van der Meulen and M.I. Willems, 'Two-year Feeding and Multigeneration Studies in Rats on Five Chemically Modified Starches', *Fd Cosmet. Toxicol., 12* (1974), pp. 651–663.

361. T.A. Anderson, L.J. Filer, S.J. Fomen, D.W. Andersen, R.L. Jensen and R.R. Rogers, 'Digestibility of Acetylated Distarch Glycerol-Effect on Growth, Serum Biochemical Values and Body Composition of Pitman-Moore Miniature Pigs', *Fd Cosmet. Toxicol., 12* (1974), pp. 201–207.

362. M.L. Buttolph and P.M. Newberne, 'Sub-chronic Studies in Rats Fed Octenyl Succinate-modified Food Starch', *Fd Cosmet. Toxicol., 18* (1980), pp. 357–362.

363. Joint FAO/WHO Expert Committee on Food Additives, 'Tenth Report – Specifications for the Identity and Purity of Food Additives and their Toxicological Evaluation: Some Emulsifiers and Stabilizers and Certain Other Substances', *Tech. Rep. Ser. Wld Hlth Org., 373* (1967).

364. W.H. Braun, J.C. Ramsey and P.J. Gehring, 'The Lack of Significant Absorption of Methyl Cellulose, Viscosity 3300 cp, from the Gastro-intestinal Tract Following Single and Multiple Oral Doses to the Rat', *Fd Cosmet. Toxicol., 12* (1974), pp. 373–376.

365. G. Feuer, 'Metabolic Rate of ^{32}P-labelled Emulsifier YN in Rats', *Fd Cosmet. Toxicol., 5* (1967), pp. 631–643.

366. J.C. Phillips, I.F. Gaunt and S.D. Gangolli, 'Studies on the Metabolic Rats of ^{32}P-labelled Emulsifier YN in the Mouse, Guinea Pig and Ferret', *Fd Cosmet. Toxicol., 13* (1975), pp. 23–30.

367. J.C. Phillips, J. Kingsnorth, I. Rowland, S.D. Gangolli and A.G. Lloyd, 'Studies on the Metabolism of Sucrose Acetate Isobutyrate in the Rat and Man', *Fd Cosmet. Toxicol., 14* (1976), pp. 375–380.

368. J.W. Daniel, C.J. Marshall, H.F. Jones and D.J. Snodin, 'The Metabolism of Beef Tallow Sucrose Esters in Rat and Man', *Fd Cosmet. Toxicol., 17* (1979), pp. 19–21.

369. R.S. Harris, H. Sherman and W.W. Jetter, 'Nutritional and Pathological Effects of Sorbitan Monolaurate, Polyoxyethylene Sorbitan Monolaurate and Polyoxyethylene Monolaurate when Fed to Hamsters', *Arch. Biochem. Biophys., 34* (1951), pp. 259–265.

370. J.B. Allison, H.L. Rosenthal and A.H. Mills, 'Effects of Non-ionic Surface Active Agents on the Growth of Animals', *Fedn Proc., 11* (1952), p. 435.

371. A.R. Bourke and O.G. Fitzhugh, 'High and Low Levels of Emulsifiers on Rat Intestinal Flora', *Fedn Proc., 12* (1953), p. 302.

372. P. Gyorgy, M. Forbes and H. Goldblatt, 'Effect of Nonionic Emulsifiers

on Experimental Dietary Injury of the Liver in Rats', *J. Agric. Fd Chem., 6* (1958), pp. 139–142.

373. P.H. Elworthy and J.F. Treon, 'Physiological Activity of Non-ionic Surfactants' in M. Schick (ed.), *Nonionic Surfactants Surfactant Science Series* (Marcel Dekker, New York, 1967), vol. 1, p. 923.

374. B.R. Cater, K.R. Butterworth, I.F. Gaunt, J. Hooson, P. Grasso and S.D. Gangolli, 'Short-term Toxicity Study of Sorbitan Monolaurate (Span 20) in Rats', *Fd Cosmet. Toxicol., 16* (1978), pp. 519–526.

375. A.J. Ingram, K.R. Butterworth, I.F. Gaunt, P. Grasso and S.D. Gangolli, 'Short-term Toxicity Study of Sorbitan Mono-oleate (Span 80) in Rats', *Fd Cosmet. Toxicol., 16* (1978), pp. 535–542.

376. W.R. King, W.R. Michael and R.H. Coots, 'Metabolism of Polyglycerol and Polyglycerol Esters', *Toxicol. Appl. Pharmacol., 20* (1971), pp. 334–345.

377. W.R. King, W.R. Michael and R.H. Coots, 'Metabolism of Stearoyl Propylene Glycol Hydrogen Succinate', *Toxicol. Appl. Pharmacol., 17* (1970), pp. 519–528.

378. P.P. Williams, 'Metabolism of Synthetic Organic Pesticides by Anaerobic Microorganisms', *Residue Rev., 66* (1977), pp. 63–135.

379. R.C. Braunberg and V. Beck, 'Interaction of DDT and the Gastro-intestinal Microflora of the Rat', *J. Agric. Fd Chem., 16* (1968), pp. 451–453.

380. F. Matsumura, K.C. Patil and G.M. Boush, 'Formation of Photodieldrin by Microorganisms', *Science, 170* (1970), pp. 1206–1207.

381. J.D. Rosen, D.J. Sutherland and G.R. Lipton, 'The Photochemical Isomerization of Dieldrin and Endrin and Effects on Toxicity', *Bull. Environ. Contam. Toxicol., 1* (1966), pp. 133–140.

382. N.K. Van Allen and T. Kosuge, 'Microbial Metabolism of the Fungicide 2,6-dichloro-4-nitroaniline', *J. Agric. Fd Chem., 22* (1974), pp. 221–224.

383. J.L. Emmerson and R.C. Anderson, 'Metabolism of Trifluralin in the Rat and Dog', *Toxicol. Appl. Pharmacol., 9* (1966), pp. 84–97.

384. M.K. Ahmed, J.E. Casida and R.E. Nichols, 'Animal Metabolism of Insecticides Bovine Metabolism of Organophosphorus Insecticides: Significance of Rumen Fluid with Particular Reference to Parathion', *J. Agric. Fd Chem., 6* (1958), pp. 740–746.

385. J.W. Daniel and J.C. Gage, 'Absorption and Excretion of Diquat and Paraquat in Rats', *Br. J. Ind. Med., 23* (1966), pp. 133–136.

386. I.R. Rowland, P. Grasso and M.J. Davies, 'The Methylation of Mercuric Chloride by Human Intestinal Bacteria', *Experientia, 31* (1975), p. 1064.

387. T. Edwards and G.C. McBride, 'Biosynthesis and Degradation of Methyl-mercury in Human Faeces', *Nature, 253* (1975), pp. 462–464.

388. M. Abdulla, B. Arnesjo and I. Ihse, 'Methylation of Inorganic Mercury in Experimental Jejunal Blind-loop', *Scand. J. Gastroenterol., 8* (1973), pp.565–567.

389. I.R. Rowland and M.J. Davies, *In Vitro* Metabolism of Inorganic Arsenic by the Gastro-Intestinal Microflora of the Rat', *J. Appl. Toxicol., 1* (1981), pp. 278–283.

390. I.R. Rowland, M.J. Davies and P. Grasso, 'Metabolism of Methylmercuric Chloride by the Gastrointestinal Flora of the Rat', *Xenobiotica, 8* (1978), pp. 37–43.

391. I.R. Rowland, M.J. Davies and P. Grasso, 'Volatilisation of Methyl-mercuric Chloride by Hydrogen Sulphide', *Nature, Lond., 265* (1977), pp. 718–719.

392. P.J. Craig and P.D. Bartlett, 'The Role of Hydrogen Sulphide in Environ-mental Transport of Mercury', *Nature, Lond., 275* (1978), pp. 635–637.

393. J.P. Moody and R.T. Williams, 'The Fate of Arsanilic Acid and Acetyl-arsanilic Acid in Hens', *Fd Cosmet. Toxicol., 2* (1964), pp. 687–693.

394. J.P. Moody and R.T. Williams, 'The Metabolism of 4-Hydroxy-3-Nitrophenylarsonic Acid in Hens', *Fd Cosmet. Toxicol., 2* (1964), pp. 707–715.

395. J. Autian, 'Toxicity and Health Threats of Phthalate Esters: Review of the Literature', *Envir. Hlth Perspec., 4* (1978), pp. 3–26.

396. L.G. Krauskopf, 'Studies on the Toxicity of Phthalates via Ingestion', *Envir. Hlth Perspect., 3* (1973), pp. 61–72.

397. S.I. Shibko and H. Blumenthal, 'Toxicology of Phthalic Acid Esters Used in Food Packaging Material', *Envir. Hlth Perspec., 3* (1973), pp. 131–138.

398. D. Brown, K.R. Butterworth, P. Grasso and S.D. Gangolli, 'Short-term Oral Toxicity Study of Diethyl Phthalate in the Rat', *Fd Cosmet. Toxicol., 16* (1978), pp. 415–422.

399. C.P. Carpenter, C.S. Weil and H.F. Smith, 'Chronic Oral Toxicity of Di-(2-ethylhexyl)-Phthalate for Rats, Guinea Pigs and Dogs', *Arch. ind. Hyg. 8* (1953), pp. 219–226.

400. B.G. Lake, S.D. Gangolli, P. Grasso and A.G. Lloyd, 'Studies on the Hepatic Effects of Orally Administered Di-(2-ethylhexyl)phthalate in the Rat', *Toxicol. Appl. Pharmacol., 32* (1975), pp. 355–367.

401. B.R. Cater, M.W. Cook and S.D. Gangolli, 'Zinc Metabolism and Dibutyl Phthalate Induced Testicular Atrophy in the Rat', *Biochem. Soc. Trans., 4* (1976), pp. 652–653.

402. T.J.B. Gray, K.R. Butterworth, I.F. Gaunt, P. Grasso and S.D. Gangolli, 'Short-term Toxicity Study of Di-(2-ethylhexyl)phthalate in Rats', *Fd Cosmet. Toxicol., 15* (1977), pp. 389–399.

403. P.W. Albro, R. Thomas and L. Fishbein, 'Metabolism of Diethylhexyl Phthalate by Rats. Isolation and Characterization of the Urinary Metabolites', *J. Chromatog., 76* (1973), pp. 321–330.

404. P.W. Albro and B. Moore, 'Identification of the Metabolites of Simple Phthalate Diesters in Rat Urine', *J. Chromatog., 94* (1974), pp. 209–218.

405. I.R. Rowland, R.C. Cottrell and J.C. Phillips, 'Hydrolysis of Phthalate Esters by the Gastrointestinal Contents of the Rat', *Fd Cosmet. Toxicol., 15* (1977), pp. 17–21.

406. B.G. Lake, J.C. Phillips, J.C. Linnell and S.D. Gangolli, 'The *In Vitro* Hydrolysis of Some Phthalate Diesters by Hepatic and Intestinal Preparations from Various Species', *Toxicol. appl. Pharmacol., 39* (1977), pp. 239–248.

407. P.W. Albro and R.O. Thomas, 'Enzymatic Hydrolysis of Di-(2-ethylhexyl)phthalate by Lipases', *Biochim. Biophys. Acta, 306* (1973), pp. 380–390.

408. J.W. Daniel and H. Bratt, 'The Absorption, Metabolism and Tissue Distribution of Di-(2-ethylhexyl)phthalate in Rats', *Toxicology, 2* (1974), pp. 51–65.

409. R. Walker, 'Naturally Occurring Nitrate/Nitrite in Foods', *J. Sci. Fd Agric., 26* (1975), pp. 1735–1742.

410. P.F. Swann, 'The Toxicology of Nitrate, Nitrite and *N*-nitroso Compounds', *J. Sci. Fd Agric., 26* (1975), pp. 1761–1770.

411. World Health Organisation, *Environmental Health Criteria* V. *Nitrates, Nitrites and N-Nitroso Compounds* (WHO, Geneva, 1977).

412. World Health Organisation, *International Drinking Water Standards*, 3rd edn (WHO, Geneva, 1971).

413. Z. Knotek and P. Schmidt, 'Pathogenesis, Incidence and Possibilities of Preventing Alimentary Nitrate Methaemoglobinaemia in Infants', *Pediatrics, 34* (1964), p. 78.

414. J. Varady and G. Szanto, 'Untersuchunger über den Nitritgehalt des Speichels, des Magensafts und des Harns', *Klin. Wschr., 19* (1940), p. 200.

415. P.W. Goaz, H.A. Biswell and M.C. Miller, 'On the Addition of a Variable to a Previously Optimized System for the Measurement of Nitrate-Nitrite Reductase Activity in Salivary Sediment', *J. Dent. Res., 43* (1964), p. 380.

416. H.-J. Kuhn, 'Untersuchung über den Nitritgehalt in Speichel, in Abhangigkeit von Zahnzustand und Zahnpflege', Thesis, Medical Faculty; University of Heidelberg, 1964.

417. S.R. Tannenbaum, A.J. Sinskey, M. Weisman and W. Bishop, 'Nitrite in Human Saliva. Its Possible Relation to Nitrosamine Formation', *J. Natl Cancer Inst., 53* (1974), pp. 79–84.

418. S.R. Tannenbaum, M. Weisman and D. Fett, 'The Effect of Nitrate Intake On Nitrite Formation in Human Saliva', *Fd Cosmet. Toxicol., 14* (1976), pp. 549–552.

419. B. Spiegelhalder, G. Eisenbrand and R. Preussman, 'Influence of Dietary Nitrate on Nitrite Content of Human Saliva: Possible Relevance to *In Vivo* Formation of *N*-nitroso Compounds', *Fd Cosmet. Toxicol., 14* (1976), pp. 545–548.

420. M. Harada, H. Ishiwata, Y. Nakamura, A. Tanimura and M. Ishidate, 'Studies on *In Vivo* Formation of Nitroso Compounds. I. Changes of Nitrite and Nitrate Concentration in Human Saliva After Ingestion of Salted Chinese Cabbage', *J. Fd Hyg. Soc. Japan, 16* (1975), p. 11.

421. H. Ishiwata, P. Boriboon, Y. Nakamura, M. Harads, A. Tanimura and M. Ishidate, 'Studies on *In Vivo* Formation of Nitroso Compounds. II. Changes of Nitrite and Nitrate Concentrations in Human Saliva After Ingestion of Vegetables or Sodium Nitrate', *J. Fd Hyg. Soc. Japan, 16* (1975), p. 19.

422. D. Klein, N. Gaconnet, B. Poullain and G. Debry, 'Effet d'une charge en Nitrate sur le Nitrite Salivaire et Gastrique Chez l'Homme', *Fd Cosmet. Toxicol., 16* (1978), pp. 111–115.

423. C.L. Walters, C.S. Dyke, M.J. Saxby and R. Walker, 'Nitrosation of Food Amines Under Stomach Conditions' in E.A. Walker, P. Bogovski and L. Gricuite (eds.), *Environmental N-nitroso Compounds – Analysis and Formation, IARC Scient, Publ. No. 14*, (IARC, Lyon, France, 1976), p. 181.

424. C.L. Walters, F.P.A. Carr, C.S. Dyke, M.J. Saxby, P.L.R. Smith and R. Walker, 'Nitrite Sources and Nitrosamine Formation *In Vitro* and *In Vivo*', *Fd Cosmet. Toxicol., 17* (1979), pp. 473–479.

425. J.W. White, 'Relative Significance of Dietary Sources of Nitrate and Nitrite', *J. Agric. Fd Chem., 23* (1975), pp. 886–891.

426. S.S. Mirvish, 'Kinetics of *N*-nitrosation Reactions in Relation to Tumorigenesis Experiments with Nitrite plus Amines or Ureas' in *N-nitroso Compounds – Analysis and Formation. IARC Scient. Publ. No. 3* (IARC, Lyon, France, 1972), pp. 104–108.

427. P.A.S. Smith and R.N. Loeppky, 'Nitrosative Cleavage of Tertiary Amines', *J. Am. Chem. Soc., 89* (1967), pp. 1147–1157.

428. W. Lijinsky, E. Conrad and R. Van de Bogart, 'Carcinogenic Nitrosamines Formed by Drug/Nitrite Interactions', *Nature, (Lond.), 239* (1972), pp. 165–167.

429. W. Lijinsky, E. Conrad and R. Van de Bogart, 'Formation of Carcinogenic Nitrosamines by Interaction of Drugs with Nitrite' in *N-nitroso Compounds – Analysis and Formation, IARC Scient. Publ. No. 3* (IARC, Lyon, France, 1972), pp. 130–133.

430. W. Lijinsky and M. Greenblatt, 'Carcinogen Dimethylnitrosamine Produced *In Vivo* from Nitrite and Aminopyrine', *Nature, New Biology, 236* (1972), pp. 177–8.

431. K.I. Hildrum, R.A. Scanlan and L.M. Libbey, 'Identification of γ-Butenyl-(β-propenyl) Nitrosamine, the Principal Volatile Nitrosamine Formed in the Nitrosation of Spermidine or Spermine', *J. Agric. Fd Chem., 23* (1975), pp. 34–37.

432. E. Boyland and S.A. Walker, 'Effect of Thiocyanate on Nitrosation of Amines', *Nature (Lond.), 248* (1974), pp. 601–2.

433. D. Lathia and M. Brendebach, 'Influence of Thiocyanate Ions on Starch-Iodine Reaction Used for Estimation of α-amylase Activity', *Clin. Chim. Acta, 82* (1978), p. 209.

434. B.C. Challis and C.D. Bartlett, 'Possible Co-carcinogenic Effects of Coffee Constituents', *Nature (Lond.), 254* (1975), pp. 532–3.

435. E.A. Walker, B. Pignatelli and M. Castegnaro, 'Effects of Gallic Acid on Nitrosamine Formation', *Nature (Lond.), 258* (1975), p. 176.

436. D. Lathia and U. Freutzen, 'Synergistic Effects of Chlorogenic Acid and Thiocyanate on *In Vitro* Formation of *N*-methyl-*N*-nitrosoaniline under Physiological Conditions', *Fd Cosmet. Toxicol., 18* (1980), pp. 463–465.

437. J. Sander, F.U. Schweinsberg and H.P. Menz, 'Untersuchungen über die Entstehung Cancerogener Nitrosamine im Magen', *Hoppe Seylers Z. Physiol. Chem., 349* (1968), pp. 1691–1697.

438. N.P. Sen, D.C. Smith and L.A. Schwinghamer, 'Formation of *N*-nitrosamines from Secondary Amines and Nitrite in Human and Animal Gastric Juice', *Fd Cosmet. Toxicol., 7* (1969), pp. 301–307.

439. C.L. Walters, B.E. Newton, D.V. Parke and R. Walker, 'The Precursors of *N*-nitroso Compounds in Foods', in N-*nitroso Compounds in the Environment, IARC Scient. Publ. No. 9*, (IARC, Lyon, France, 1974), p. 223.

440. J. La Bar and J. Sander, 'Carcinogenic *N*-nitroso-dimethylamine from the Reaction of the Analgesic Amidopyrine and Nitrite Extracted from Foodstuffs', *Z. Krebsforsch., 84* (1975), pp. 299–310.

441. J. Sander, 'Nitrosaminsynthese Durch Bakterien', *Hoppe Seylers Z. Physiol. Chem., 349* (1968), pp. 429–432.

442. W.S.J. Ruddell, E.S. Bone, M.J. Hill, L.M. Blendis and C.L. Walters, 'Gastric-juice Nitrite. A Risk Factor for Cancer in the Hypochlorhydric Stomach?', *Lancet, ii* (1976), p. 1037.

443. M.J. Hill, 'Bacterial Metabolism and Human Carcinogenesis', *Br. Med. Bull., 36* (1980), pp. 89–94.

444. H. Druckrey, D. Steinhoff, H. Beuthner, H. Schneider and P. Klaerner, 'Screening of Nitrite for Chronic Toxicity in Rats', *Arzneimittel-Forsch., 13* (1963), pp. 320–323.

445. M.A. Friedman and H. McClanahan, 'Biosynthesis of Nitrosamines: Reaction of Sodium Nitrite with Dimethylglycine Produces Nitrososarcosine', *Proc. Am. Assoc. Cancer Res., 14* (1973), p. 127.

446. J. Sander, F.U. Schweinsberg, M. Ladenstein, H. Benzing and S.H. Wahl, 'Messung der Renalen Nitrosaminausscheidung am Hund zum Nachweis Einer Nitrosaminbildung *In Vivo*', *Hoppe Seylers Z. Physiol. Chem., 354* (1973), pp. 384–390.

447. B.S. Alam, I.B. Saporoschetz and S.S. Epstein, 'Formation of *N*-nitrosopiperidine from Piperidine and Sodium Nitrite in the Stomach and the Isolated Intestinal Loop of the Rat', *Nature (Lond.), 232* (1971), pp. 116–8.

448. B.S. Alam, I.B. Saporoschetz and S.S. Epstein, 'Synthesis of Nitrosopiperidine from Nitrate and Piperidine in the Gastro-intestinal Tract of the Rat', *Nature (Lond.), 232* (1971), pp. 199–200.

449. M. Greenblatt, V.R.C. Kommineni, E. Conrad, L. Wallcave and W. Lijinsky, '*In Vivo* Conversion of Phenmetrazine into its *N*-nitroso Derivative', *Nature New Biology, 236* (1972), pp. 25–6.

450. J. Sander and F. Seif, 'Bakterielle Reduktion von Nitrat im Magen des Menschen als Ursache einer Nitrosamin-Bildung', *Arzneimittel-Forsch., 19* (1969), pp. 1091–1093.

451. J. Wang, T. Kakizoe, P. Dion, R. Furrer, A.J. Varghese and W.R. Bruce, 'Volatile Nitrosamines in Normal Human Faeces', *Nature (Lond.), 276* (1978), pp. 280–1.

452. D.H. Fine, R. Ross, D.P. Rounbehler, A. Silvergleid and L. Song, 'Formation *In Vivo* of Volatile *N*-nitrosamines in Man After Ingestion of Cooked Bacon and Spinach', *Nature (Lond.), 265* (1977), pp. 753–755.

453. M. Yamamoto, T. Yamada and A. Tanimura, 'Volatile Nitrosamines in Human Blood Before and After Ingestion of a Meal Containing High Concentrations of Nitrate and Secondary Amines', *Fd Cosmet. Toxicol., 18* (1980), pp. 297–299.

454. J. Sander and G. Bürkle, 'Induktion maligner Tumoren bei Ratten Durch Gleichzeitige Verfütterung von Nitrit und Seckundären Aminen', *Z. Krebsforsch., 73* (1969), p. 54.

455. W. Lijinsky, H.W. Taylor, C. Snyder and P. Nettesheim, 'Malignant Tumours of Liver and Lung in Rats Fed Aminopyrine or Heptamethyleneimine Together with Nitrite', *Nature, Lond., 244* (1973), p. 176.

456. R.C. Shank and P.M. Newberne, 'Dose Response Study of the Carcinogenicity of Dietary Sodium Nitrite and Morpholine in Rats and Hamsters', *Fd Cosmet. Toxicol., 14* (1976), pp. 1–8.

457. M. Greenblatt, S. Mirvish and B.T. So, 'Nitroamine Studies: Induction of Lung Adenomas by Concurrent Administration of Sodium Nitrite and Secondary Amines in Swiss Mice', *J. Natn. Cancer Inst., 46* (1971), p. 1029.

458. M.J. Hill, G. Hawksworth and G. Tattersall, 'Bacteria, nitrosamines and Cancer of the Stomach', *Br. J. Cancer, 28* (1978), p. 562.

459. Joint FAO/WHO Expert Committee on Food Additives – Fifteenth Report, 'Evaluation of Food Additives. Some Enzymes, Modified Starches and Certain other Substances: Toxicological Evaluations and Specifications and a Review of the Technological Efficacy of some Antioxidants', *Tech. Rep. Ser. Wld Hlth Org., 488* (1972).

460. T.B. Morgan and J. Yudkin, 'The Vitamin-sparing Action of Sorbitol', *Nature, Lond., 180* (1957), p. 543.

461. N. Hosoya, 'Effect of Sugar Alcohol on the Intestine', IXth Interational Congress of Nutrition, Mexico City, 1972.

462. I.F. Gaunt, A.G. Lloyd, P. Grasso, S.D. Gangolli and K.R. Butterworth, 'Short-term Study in the Rat on Two Caramels Produced by Variations of the 'Ammonia Process', *Fd Cosmet. Toxicol., 15* (1977), pp. 509–521.

463. I.F. Gaunt, F.M.B. Carpanini, I.S. Kiss and P. Grasso, 'Short-term Toxicity of Yellow 2G in Rats', *Fd Cosmet. Toxicol., 9* (1971), p. 343.

464. I.R. Rowland, A.K. Mallett and A. Wise, 'A Comparison of the Activity of Five Microbial Enzymes in Cecal Content from Rats, Mice, and Hamsters, and Response to Dietary Pectin', *Toxicol. Appl. Pharmacol., 69* (1983), pp. 143–148.

465. K. Loeschke, E. Uhlich and R. Halbach, 'Caecal Enlargement Combined with Sodium Transport Stimulation in Rats Fed Polyethylene Glycol', *Proc. Soc. exp. Biol. Med., 142* (1973), pp. 96–102.

466. J.F. Moinuddin and H.W.T. Lee, 'Alimentary, Blood and Other Changes due to Feeding $MnSO_4$, $MgSO_4$ and Na_2SO_4', *Am. J. Physiol., 199* (1960), pp. 77–83.

467. L.J. Filer, 'Modified Food Starches for Use in Infant Foods', *Nutr. Rev., 29* (1971), pp. 55–59.

468. R. Walker, 'Some Observations on the Phenomenon of Caecal Enlargement in the Rat' in C.L. Galli, R. Paoletti and G. Vettorazzi (eds.), *Chemical Toxicology of Food* (Elsevier/North Holland, Amsterdam, 1978), pp. 339–348.

469. R. Walker and E.A. El Harith, 'Nutritional and Toxicological Properties of Some Raw and Modified Starches', *Ann. Nutr. Alim., 32* (1978), pp. 671–679.

470. D.C. Leegwater, A.P. de Groot and M. van Kalmthout-Kuyper, 'The Aetiology of Caecal Enlargement in the Rat', *Fd Cosmet. Toxicol., 12* (1974),

pp. 687–697.

471. E.A. El Harith, J.W.T. Dickerson and R. Walker, 'Potato Starch and Caecal Hypertrophy in the Rat', *Fd Cosmet. Toxicol., 14* (1976), pp. 115–121.

472. Anon, 'Recent Development in the Sulphite Field', *Fd Cosmet. Toxicol., 4* (1966), pp. 187–189.

473. H.P. Til, V.J. Feron and A.P. de Groot, 'Toxicity of Sulphite. I. Long-term Feeding and Multigeneration Studies in Rats', *Fd Cosmet. Toxicol., 10* (1972), pp. 291–310.

474. H.P. Til, V.J. Feron, A.P. de Groot and P. Van der Wal, 'Toxicity of Sulphite. II. Short- and Long-term Studies in Pigs', *Fd Cosmet. Toxicol., 10* (1972), pp. 463–473.

475. V.J. Feron and P. Wensvoort, 'Gastric Lesions in Rats after the Feeding of Sulphite', *Pathol. Eur., 7* (1972), pp. 103–111.

476. M. Glicksman, *Gum Technology in the Food Industry* (Academic Press, London, 1969), p. 214.

477. S. Bonfils, 'Carrageenan and the Human Gut', *Lancet, ii* (1970), p. 414.

478. P. Grasso, M. Sharratt, F.M.B. Carpanini and S.D. Gangolli, 'Studies on Carrageenan and Large Bowel Ulceration in Mammals', *Fd Cosmet. Toxicol., 11* (1973), p. 555.

479. M. Sharratt, P. Grasso, F.M.B. Carpanini and S.D. Gangolli, *Lancet, ii* (1970), p. 932.

480. P. Grasso, S.D. Gangolli, K.R. Butterworth and M.G. Wright, 'Studies on Degraded Carrageenan in Rats and Guinea Pigs', *Fd Cosmet. Toxicol., 13* (1975), pp. 195–201.

481. K.A. Pittman, L. Golberg and F. Coulston, 'Carrageenans: the Effect of Molecular Weight and Polymer Type on its Uptake, Excretion and Degradation in Animals', *Fd Cosmet. Toxicol., 14* (1976), pp. 85–93.

482. J. Watt and R. Marcus, 'Ulcerative Colitis in Rabbits Fed Degraded Carrageenan', *J. Pharm. Pharmacol., 22* (1979), p. 130.

483. J. Watt and R. Marcus, 'Carrageenan-induced Ulceration of the Large Intestine in the Guinea Pig', *Gut, 12* (1971), pp. 164–171.

484. J. Watt and R. Marcus, 'Ulcerative Colitis in the Guinea Pig Caused by Seaweed Extract', *J. Pharm. Pharmacol., 21* (1969), pp. 1875–1885.

485. P. Abraham, R.J. Fabian, L. Golberg and F. Coulston, 'Role of Lysosomes in Carrageenan Induced Caecal Ulceration', *Gastroenterology, 67* (1974), pp. 1169–1181.

486. A.B. Onderdonk and J.G. Bartlett, 'Bacteriological Studies of Experimental Ulcerative Colitis', *Am. J. Clin. Nutr., 32* (1978), pp. 258–265.

487. Anon, 'Carrageenan Under Scrutiny', *BIBRA Bulletin, 13* (1974), pp. 132–134.

488. M.S. Losowki and B.E. Walker, 'Liver Disease and Malabsorption', *Gastroenterology, 56* (1969), pp. 589–600.

7 FOOD ALLERGIES: CLINICAL ASPECTS AND NATURAL ALLERGENS

J.R. Mansfield

Introduction

A chapter on food allergy fits slightly uncomfortably in a book on toxicology because, by its very nature, one is talking of intensely individualistic adverse responses as opposed to more general adverse responses. The concept that foods can produce abnormal reactions has an extremely long history and centuries ago was a subject of considerable medical interest. As interest in pharmacology grew in the past century, this concept was relegated to the back waters of medicine. It is now becoming demonstrated that this was a considerable oversight.

Food reactions can probably be divided into two main categories. There is first, the classic allergic response, usually mediated by IgE and similar circulating antibodies. Examples of this include urticaria, laryngeal oedema and various rashes occurring after the ingestion of slightly exotic and less frequently consumed foods, such as shellfish and strawberries. 'Non-immunological' food reactions may be caused by a wide variety of other possible mechanisms. On the one hand, there may be an immunological-type response, which is as yet not understood. Alternatively, we may be dealing with an enzyme deficiency either relative or total. This enzyme deficiency could possibly result from a genetic deficiency or to an overload of a particular enzyme system produced by too frequent exposure to the food. Although Von Pirquet's original definition of allergy referred simply to an 'altered reactivity' which would cover all these types of reactions, allergy these days has been redefined more narrowly in immunologic terms. Accordingly, the terms 'food intolerance' or 'hypersensitivity' are probably more appropriate to these possibly non-immunologic responses.

Because food allergy has hitherto only been recognised by the medical profession in the rare and easily identified circumstances mentioned above, as a subject it has merited little serious interest. Work which was originated mostly in the United States of America has, however, demonstrated that food intolerance is a very major factor in human disease processes, and recent trials published in Great Britain have supported this view.

Historical Development

In the 1930s Albert Rowe,[1] a physician in California, started to draw medical attention to the significance of food allergy as a factor in a wide range of illnesses. His approach was purely clinical and the technique he used for identification involved the use of a wide range of elimination diets. Although he wrote extensively on the subject, and attained a large following of patients, his work did not receive much credence by the medical profession. In his writing, Rowe distinguished between 'immediate' and 'delayed' reactions to foods, but did not discover that what would normally be a delayed reaction could be converted into an immediate one for diagnostic purposes. If a food which is eaten on a regular basis is totally eliminated from the patient's diet for five days and then deliberately fed, symptoms will occur within two or three hours if allergy exists. This certainly applies to foods other than cereals.

The concept of 'masked' food allergy was originally identified by Dr Herbert Rinkel, a well-known allergist practising in Oklahoma City. His observations were first published in 1944,[2] where masking was defined as: 'If a person ingests a particular food each day, he may become allergic to it and yet not suspect this as a cause of his symptoms. It is usual to feel better after a meal than before. In this case the feelings tend to mask the symptoms of the allergic response.' Rinkel could not explain his observations which occurred in several thousand patients. Since Rinkel's original work, cases of masked food allergy have been reported on many occasions.

Masked food allergy represents an interesting model for addictive behaviour and is, in my opinion, the major basic mechanism behind the addiction to such apparently diverse items as coffee, tea, sugar, alcoholic beverages and tobacco. This concept can be represented graphically as shown in Figure 7.1.

This graph illustrates the results of eating or not eating a masked food allergen in a patient with a single food allergy. Each asterisk represents a feeding of an allergenic food. As can be seen, the second feeding aborts the deterioration of a patient's symptoms as seen after the first feeding. This is followed by an improvement. A similar response is experienced after the third feeding. Because the fourth feed of the allergenic food is delayed, the patient's response is further down the

Figure 7.1: Masking and Unmasking. Symptom Reactions to a Food Frequently Eaten, Then Witheld From the Diet, Then Reintroduced.

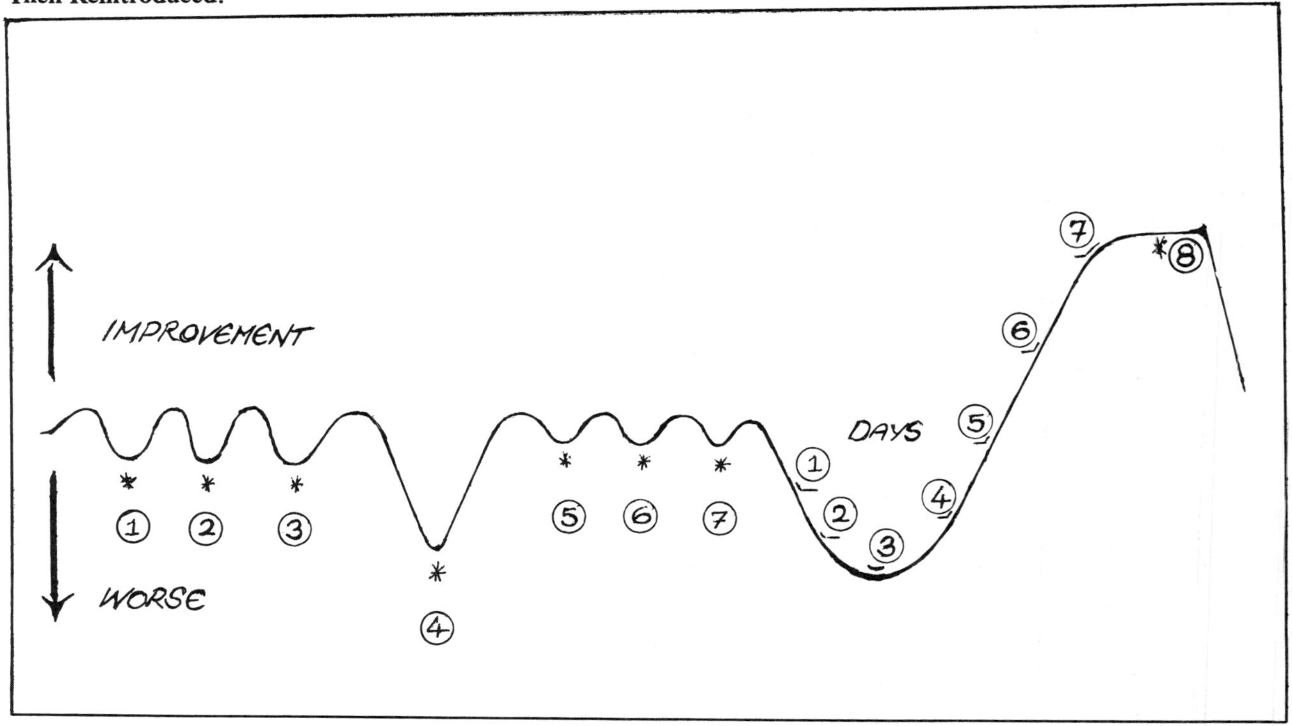

withdrawal curve and is thus worse. Following this fourth feeding the patient's condition usually returns to normal. In migrane patients for example, if this fourth feeding is delayed for too long a stage is reached when further feeding of the allergenic food is to no avail.

After a seventh feeding in this example, a patient has been deliberately told to avoid the allergic food and he or she then exhibits the classic withdrawal phenomenon. This is characterised by a considerable deterioration of symptoms usually by the evening of the first day. The second and third days tend to be quite severe but are followed by a slow improvement until the 6th day. By the sixth day, most patients are symptom free if under 35 years old. In patients older than 35, the symptoms may take another day or so to clear. Feeding at stage 8 represents a deliberate refeeding and the Rinkel Hyper-Acute response. In this, as mentioned earlier, symptoms return frequently quite dramatically.

This graph demonstrates how complicated is the relationship between one commonly eaten food allergen and the symptomatology that it creates. Imagine therefore one is dealing with a patient who has allergies to wheat, corn, milk and yeast. There will be a different curve of symptomatology for each food which depends on the frequency at which it is eaten. These curves of response will then become superimposed on each other. Any relationship between food and symptoms will in these circumstances become far from obvious. I have emphasised this point to some extent as many physicians believe that if food allergy is present, it will be obvious to all those concerned. It is this simplistic view of the subject which has partly led to its neglect.

Food intolerance can exist by itself. Occasionally this may be due to a single food but more commonly several foods are involved. Sometimes the food intolerance co-exists with other allergies, e.g. house dust, house dust mite, various moulds and pollens. These inhalant reactions can cause a proportion of a patients' symptomatology and the foods can cause the rest.

Sometimes reactions to environmental contaminants such as inhaled hydrocarbons from gas, petrol, diesel, paint, perfumes and glues can further complicate the picture. Like food reactions, hydrocarbon responses have not been well understood in the past, but the subject received a considerable impetus after the publication in 1962 of *Human Ecology and Susceptibility to the Chemical Environment* by Dr T.G. Randolph.[3] Because most physicians who work with food allergy have realised the importance of concurrent inhalation problems, they have tended to call themselves 'Clinical Ecologists' rather than food allergists.

Clinical Ecology

Clinical ecology is concerned with adverse reactions to environmental insults modified by individual susceptibility in terms of specific adaptation. Ecological illness is manifested by a wide variety of physical and mental symptoms. These are often chronic in nature and cyclic in occurrence producing conditions that are too frequently undiagnosed or poorly identified. The excitants found in the environment may be associated with air, water, food and drugs. The outcome of this interreaction of excitants versus host is modified by the host's individual susceptibility which is governed by his genetic background, current state of physical and mental wellbeing and his ability to adapt. This definition of Clinical ecology is accepted now by most of the physicians who practice it.

Clinical Conditions Related to Clinical Ecology

These syndromes can be classified according to the various systems involved.

1. Cerebral	Headaches
	Migraine
	Endogenous depression
	Hyper-activity in children
	Chronic anxiety states
	Some cases of thought disorder
	Alcoholism
2. Abdominal	Spastic or irritable colon
	Colitis
	Ulcerative colitis
	Regional ileitis (Crohn's disease)
3. Metabolic	Obesity
4. Musculo-skeletal	Rheumatoid arthritis
	Osteoarthritis,
	Ankylosing spondylitis
	Non-specific arthritis
	Polymyalgia rheumatica
	'Fibrositis'
5. Genito-urinary	Recurrent cystitis without evidence of infection

	Menorrhagia and dysmenorrhoea
6. Cardio-vascular and respiratory	Tachycardias
	Idiopathic oedema
	Vasculitis in all its forms
	Asthma
7. Skin	Urticaria
	Eczema
	Acne
8. Ear, nose and throat	Perennial rhinitis and syndromes which follow as a consequence such as sinusitis.

Most readers will, I expect, view the above list with almost total disbelief. Certainly I did when I first became acquainted with this field of medicine.

Clearly, many physicians in various parts of the world originally viewed the subject with the same scepticism but were amazed and delighted when they found that they could successfully treat many of their patients with the above conditions. The significance of the clinical results is sometimes difficult to convey in print, but in contrast to the drug treatment to depressive illness, the results are frequently very dramatic. Symptoms which have been present in some cases for over 40 years can completely abate after a few days. The appearance of the patient can improve dramatically.

Quite aside from these thousands of observations by many different physicians which are admittedly anecdotal, the relationship of some of the above disease processes to food and inhalant allergy has been the subject of a number of learned papers in various medical journals. The conditions surveyed include migraine, rheumatoid arthritis, vasculitis, hyper-activity in children and tachycardias. These papers will be briefly reviewed later when I survey the current evidence that food allergy is a major factor in human disease.

Investigation of Food Allergy

How then can a patient with a suspected food allergy be investigated? Clearly, the existence of any one of the above diseases is enough to make a clinical ecologist suspicious of the existence of food and other allergies. Secondly, evidence of repetitive eating patterns would re-inforce this probability. Thirdly, and very importantly, there are a number of minor symptoms quite aside from the original disease entity which the patient may or may not even mention but which are particularly suggestive of food allergy. These symptoms are:

1. Persistent fatigue, only partly relieved by rest.
2. Idiopathic oedema of the fingers or ankles (often mild).
3. Bouts of racing pulse, unrelated to exercise or specific anxiety.
4. Bouts of sweating for no apparent reason.
5. Abdominal distension for no apparent reason.
6. Marked fluctuations in weight.

Two or three of these symptoms in addition to the original symptomatology would reinforce the view that food allergy was present.

Having decided that the patient is a suitable candidate for investigation, a physician can choose various methods. It must be emphasised at this point that it is assumed in this presentation that the patient has been extensively investigated along conventional lines for the possibility of serious pathology such as neoplasia.

The most popular methods of investigation are:

1. Dietary exclusion followed by deliberate re-introduction of foods in a certain sequence.
2. The intracutaneous provocation method of food allergy management.
3. The Bryan-Cytotoxic test.

The Dietary Exclusion Method

This is delightful in its basic simplicity. If a patient who has very constant symptoms is fasted for six days on spring water alone, his symptoms should clear completely if they are related to food allergy alone. Exceptions to this are: (a) older patients, especially over 50, may take 7 to 8 days to clear; (b) patients who have conditions which considerable tissue damage such as ulcerative colitis or eczema. In cases of eczema, the irritation may have gone by the sixth day but the rash may take two or three weeks to disappear completely. Obviously without any allergen at all present, epidermal cell turnover takes this order of time to occur.

In addition to fasting, the patient should take Epsom Salts on the first morning of the fast to clear his bowels. If he is a cigarette smoker he should completely stop smoking as cigarette smoke contains the sugar with which tobacco is cured. There should be no taking of medications as all medications contain food stuffs. For example, most tablets contain corn starch as a filler and CORN is the second most common food allergy seen. Other food constituents, such as sugar, are present in tablets and syrups.

On an out-patient basis, a fast is often difficult to perform, particularly with patients who have grown up in a Western culture. Accordingly, I devised a few years ago the Lamb and Pear Diet which a number of other physicians in the United Kingdom now also use. Lamb and pears represent two foods to which the patient is unlikely, on a statistical basis, to be allergic. They tend to complement each other, are very nutritious and are available all year round. I find this a very practical approach, particularly in mild and moderately severe cases. There are many patients who will not tolerate a fast as an out-patient but will tolerate a lamb and pear diet. Occasionally one may use this approach with someone who is allergic to one of these foods, but the response to the diet is usually so disastrous that one can, with experience, detect it. Other foods with a low risk of producing allergic responses are cod, plaice, rabbit, runner beans and pineapple.

The withdrawal phase on a fast or hypo-allergenic diet is a fascinating time and a whole book could be written on it. As mentioned earlier when discussing Rinkel's concept of masking, when one stops eating a masked food allergen, there is at first a considerable increase in symptomatology before the improvement occurs. For example, in a patient with a migraine problem, one sees a severe migraine developing usually during the afternoon of the first day of the diet. The second day precipitates as bad a migraine as the patient can remember. By the afternoon of the third day only a mild headache usually remains and this gradually clears completely on the fourth and fifth days. On the sixth day the patient usually reports no symptoms at all and his head feeling much clearer than normal. In a patient with endogenous depression his feelings can be quite suicidal on the second and third day and it is important that he or she should be accompanied. In many patients a withdrawal myalgia appears usually on the third day. Most commonly, the patient reports generalised pain and stiffness, particularly in the lower back, buttocks and thighs. This can frequently be a totally new symptom to the patient. It usually persists with decreasing intensity until the evening of the sixth day. The onset of nasal catarrh on the third day is another interesting symptom observed sometimes in this phase.

Quite obviously many patients, especially if they are very thin initially, will feel rather weak on the sixth day of the hypo-allergenic diet, but if their constant symptomatology has completely or near completely disappeared, the physician can be confident he is dealing with a food allergy problem. To simplify matters, I then give the patient a sheet detailing foods to test in a specific sequence. At first, to build up the patient's diet as quickly as possible, I suggest he or she tries reintroducing foods that are not common offenders. Thus in the first few days patients test such food as plaice, green beans, pineapple, cod, rice, beef and carrots. One food is tested at each meal with three meals per day, so that the patient's diet should expand by three foods a day. If the patient notices no symptoms appearing in the four hours subsequent to the test, he passes the food as safe and can eat it in subsequent meals with the foods he is testing.

The diet in most Western civilised countries is dominated by a few foods. These foods are wheat, corn, milk, eggs, yeast, cane sugar, beet sugar and potatoes, and these are also the most common allergies we deal with. It is generally considered that food allergy is caused by the frequency of contact and these items occur with monotonous regularity in most people's diet. Some of these foods are taken more than six times daily by many people. Although most of these items are usually eaten in combination, e.g. cakes, biscuits, bread, sweets, alcoholic beverages and soft drinks, they are tested individually in pure form in our test routine. I normally assess about 70 commonly eaten food items and this takes about six weeks on average. I consider that I will include at least 97 per cent of what most people commonly eat during this period.

The problems with this method of treatment lie first, in the high degree of co-operation which is needed from the patient. Many patients are excellent in this respect, but some are less co-operative. Some, particularly elderly patients, find the reactions rather insidious and become confused with their results. Patients with a very high number of reactions find the continual going into one reaction and then into another reaction both very exhausting and dispiriting. However, the most common problem is that at the end of the day, a patient has discovered that he reacts to a spectrum of foods such as WHEAT, CORN, OATS, RYE, YEAST and EGGS. They are symptom free while avoiding these foods, but the problems of avoidance are immense, particularly if their occupations necessitate them eating out in hotels and restaurants on a frequent basis.

Intradermal Provocation Neutralisation Test

For patients who have multiple food allergies, or who are sensitive to allergens which are difficult to avoid, the Intradermal Provocation Neutralisation Test is highly beneficial as it serves both as a test and as a treatment. Skin tests for food allergy have, prior to the discovery of this test, been highly unsatisfactory. Previously, skin testing for foods had been limited to scratch or prick tests, or single intradermal tests. These tests rarely demonstrated positive results, even in known food allergies. The fact that scratch tests are useless in the evaluation of food allergy is largely responsible for the fact that interest in food allergy has lagged about half a century behind interest in inhalant allergy.

The provocation neutralising technique discovered in 1957 by Carleton H. Lee of St. Joseph, Missouri.[3] He had been studying variations in wheal sizes produced by intradermal injections of varying dilutions of given food extracts. When one injects 0.05 ml of any fluid between the layers of the skin, the fluid distends the skin and produces a bleb or wheal of up to 7 mm in diameter. What Carleton Lee observed was that at certain dilutions of the food extracts there would be no increase in the size of the wheal as measured ten minutes after the original injection, and at other dilutions there would be an increase. He also noticed that he was making some patients temporarily worse with certain dilutions and making them better with other dilutions of the same extract. Patients furthermore reported to him that after the injections which made them better, they were able to eat the allergenic food for the next two days without symptoms recurring. The eminent American allergist, Dr Joseph B. Miller of Mobile, Alabama, subsequently organised, clarified and extended many points of the technique, which then became possible to teach. The whole technique is admirably described in his book *Food Allergy – Provocative Testing and Injection Therapy*.[4] At present in the United States there are well over a 100 clinics using this technique and in the United Kingdom about eight.

In a classic case, one sees a considerable increase in wheal size combined with symptoms when the higher concentration of the food extract is given. On giving progressively weaker dilutions of the food extract, a level is reached at which the wheal does not grow. When this point is reached the symptoms normally disappear, and this confirms the neutralising point. If a still weaker injection is given, the wheal commonly remains negative but the symptoms recur. This is called under-dosage and the symptoms can be alleviated by re-giving the preceding neutralising dose. For therapy the neutralising doses of all the allergenic food which the patient finds difficult to avoid are given together, either by sublingual drop or subcutaneous injection. We find that by converting the information we discover by intracutaneous injections into appropriate sublingual dosage, we give excellent symptomatic relief to about 90 per cent of our patients. The sublingual drops need to be taken before each meal containing the allergic food. The remaining 10 per cent do much better by injections, which initially are given on alternate days but can be usually reduced to twice weekly within a month or so.

Apart from its dramatic effect of enabling a patient immediately to eat food to which he is allergic without any adverse effect, the intradermal provocation test is quite unique in that the procedure re-creates the patient's condition in miniature. Headaches, nausea, depression, abdominal pain, joint pain, fatigue, wheezing, nasal catarrh and pruritis are frequently turned on and turned off within minutes in the intradermal testing room. As a demonstration it is highly convincing to the patient. I have seen highly sceptical patients change to total believers within a few minutes. Medical observers are more difficult to convince as most of these symptoms are fairly subjective. However, medical observers have, in my experience, been highly impressed by the production of more objective physical signs such as urticaria (hives), profuse nasal discharge, oedema of the face, hands or ankles and wheezing as assessed by a peak flow meter.

When writing a scientific paper on this subject, the most easily studied parameter and the most relevant clinically is therapeutic relief. This is observed when the patient is given subcutaneous food injections over a period of time, using protective doses based on intradermal provocative testing, as opposed to the recurrence of symptoms on the same diet when given subcutaneous placebo injections over an equal period of time in a double-blind fashion. It is of course essential that any investigator be competent in first determining the appropriate treatment levels.

In 1977, Dr J.B. Miller reported a double-blind study of food extract injection therapy.[5] This study illustrated the superiority of neutralising food injections over placebo at a 99.8 per cent level of confidence. Most of the cases responded quickly and dramatically to food neutralising injections and regressed equally rapidly when given the placebo injections. This technique, which has been found to be such a valuable tool by all those who have gained proficiency in its use, has been subjected to a double-blind controlled study and its value has been demonstrated.

Cytotoxic Food Test

This test was originally introduced by Black[6] and later modified and improved by Bryan and Bryan.[7] The test is based on the fact that polymorphonuclear leucocytes will undergo certain morphological changes when incubated with the antigen of a particular food that is present in a patient's diet. The procedure has a number of problems with interpretation and quality control but technicians who have had special training in its use, particularly under the guidance of the Bryans, seem to acquire reasonable results. The method has the superficial advantage, from the patient's point of view, of a relative lack of personal involvement in the procedure. All he needs to do is to give a sample of his intravenous blood and await results. Thereafter, however, a teaching process is required in order to instruct the patient to avoid those foods to which he is apparently allergic. Experience with this method, together with the problems of unreliability and expense, have led many clinical ecologists to abandon this test. However, the main problem is that at the very best, it is only a means of identifying allergies. Most patients, once their allergies are identified, need to be desensitised and intradermal testing would be needed therefore in any case.

Another *in vitro* test that can be performed on a patient's serum is the radio-allergosorbent test, 'RAST'. As a practical tool for day to day management of patients with food allergic disease, this has also been disappointing so far, though as a research tool it has recently supplied important *in vitro* confirmation of elimination diet results. Basically it supplies a laboratory report of specific IgE Levels, not a judgement as to the allergenicity of the food in the patient. The original test kit is very expensive, the test is complex and the number of foods at the moment which can be tested is very limited. Its accuracy is certainly no greater than the intradermal provocation test and, like the cytotoxic test, it has no treatment possibilities other than avoidance.

General Discussion

A number of techniques currently used in the identification of food allergens have been detailed and means of overcoming the symptoms discussed. It is of interest at this point to consider to what extent these concepts have been scientifically validated.

There is some problem in comparing a large group of people with complex allergic phenomena. If one takes a superficially simple condition such as migraine, it quickly becomes apparent that one migraine sufferer may also have an associated fatigue problem, another a hypertensive problem, another an idiopathic oedema problem and another a depressive problem. Clinical ecologists realise that these apparently unrelated problems are in fact related and are just varying individual facets of an underlying allergic problem. There is therefore a tendency for clinical ecologists to present their data in the form of individual case reports with details of clinical improvement followed by the recurrence of symptoms with exposure to known allergens. This re-exposure can be arranged to be double blind of course by the use of such devices as naso-gastric tubes or by the technique mentioned earlier in Miller's double-blind study.

The use of individual case reports, however dramatic, has fallen into considerable disfavour with the medical profession and is regarded as purely anecdotal. This is despite the fact that most of the early discoveries in medicine emanated from this source, and that case law is still held in very high esteem by the legal profession. There have been countless papers on this subject produced in medical journals all round the world. I intend only to mention in detail the British papers on this subject.

One of the best British papers was published in the *Lancet* on February 25th, 1978, by Finn and Cohen.[8] Dr Ronald Finn, a Consultant Physician at the University of Liverpool, was the discoverer of Anti D Serum which revolutionised the treatment of Rhesus incompatibility in the new-born baby. The paper described six cases of patients with long-standing physical and mental symptoms who had not been helped by many years of conventional treatment. They all experienced immediate relief of symptoms on avoidance of certain foodstuffs. Symptoms included chest pain, palpitations, dyspnoea, severe throbbing headaches, severe aphthous ulcers, persistent abdominal pain, nausea and vomiting. There were also a number of psychiatric symptoms such as panic attacks, anxiety and depression. All these symptoms were absent while the incriminated foods were avoided. Some of these patients were then subjected to refeeding of the incriminated foods or a placebo food via naso-gastric tubes. The food was administered by a nurse from a syringe which was masked and given a code number only. Foods thought to be allergic and placebo foods were given in a random sequence and the chance of the patient identifying the allergic foods by chance was 1 in 1024. The patients all managed to identify correctly the adverse reacting foods.

Although clinical ecologists are reluctant to look at large numbers of patients with a single disease entity, for the reasons explained above, in order to have some common basis with the rest of the profession, two trials have recently been published on the relationship of migraine and food allergy. In the first of these, Dr Ellen Grant of Charing Cross Hospital reported that in her trial 60 migraine patients completed elimination diets after a 5-day period of withdrawal from their normal diet.[9] Of these, 87 per cent had been using oral contraceptive steroids, tobacco, and/or ergotamine for an average of three years, 22 years and 7.4 years respectively. The commonest food causing reactions were wheat (78 per cent), orange (65 per cent), eggs (45 per cent), tea and coffee (40 per cent each), chocolate and milk (37 per cent each). When an average of ten common foods were avoided there was a marked fall in the number of headaches per month, 85 per cent of the patients becoming headache-free. The 25 per cent of patients with hypertension became normotensive. Chemicals in the home environment can make this testing difficult for out-patients. Both immunological and non-immunological mechanisms may play a part in the pathogenesis of migraine caused by food intolerance.

The reader has probably noted that the common allergens noted in this study, which were wheat, orange, eggs, tea and coffee, tend to contrast rather sharply with the conventional opinion that the predominantly incriminated foods are cheese, chocolate, citrus fruits and alcohol. Cheese, chocolate and citrus fruits tend to be intermittently consumed foods, and so if patients are allergic to them they tend to notice it as the reaction is naturally unmasked. In contrast, once the patients have been through a 5 day elimination of all continuously consumed foods, wheat becomes the most common allergen noted.

On July 5th, 1980, a further paper appeared in the *Lancet* entitled 'Food Allergy in Migraine'.[10] This was a study of dietary exclusion and RAST. The authors were from the National Hospital for Nervous Diseases and the Department of Immunology at the Middlesex Hospital.

Summarising these results, the authors stated that they found 'Two-thirds of severe migraine sufferers were allergic to certain foods, shown by dietary exclusion and subsequent challenge. Radio-allergosorbent tests confirmed the relevance of these foods. Oral sodium cromoglycate protected that patients from food challenges. The initial specific allergic reaction in the gut may result in increased mucosal permeability which allows food antigens, complexes, or mediators to be absorbed and cause symptoms.'

The authors also stated that when the RAST tests were compared with the foods in those 23 patients in whom provoking foods were identified, a good correlation was found between high RAST titres and clinically relevant foods. The six patients in whom dietary factors were not identified had low RAST assays.

Sodium cromoglycate in its inhalant form is the well-known anti-asthma drug Intal. This works by affecting the mast cells in the respiratory mucous membrane. The oral form of course affects the mast cells in the alimentary canal and in sufficient dosage is quite an effective agent against single challenges of allergic food. Physicians working in this field have generally found that it is ineffective against continuous challenge from multiple foods.

In the field of rheumatology, excellent results have been reported by many physicians in the treatment of both rheumatoid and osteoarthritis using these techniques. Dr T.G. Randolph summarised these results in the book *Clinical Ecology*[11] and goes on to describe in detail many of his own results. In the United Kingdom in 1981 the *British Medical Journal* published a paper entitled 'Rheumatoid Arthritis and Food: a case study'.[12] This work originated from the highly respected Rheumatology Unit at the Hammersmith Hospital in London. They concluded that 'clinical and laboratory studies in a patient whose rheumatoid arthritis appeared to be exacerbated by dairy produce showed that challenge with milk and cheese resulted in a pronounced increase in synovitis and changes in immune complexes, IgE antibodies, and heat-damaged red cell clearance rates. Exclusion of dairy produce from the diet produced a considerable improvement in her previously aggressive disease.'

Rheumatoid arthritis is a condition in which progress can be monitored by a whole range of objective clinical measurements such as grip strengths, morning stiffness, ring size of joints and visual pain analogue. There are also a number of laboratory measurements which can be made such as circulating immune complexes, the RAST test and clearance of heat-damaged red cells. All these clinical and laboratory parameters increased considerably after challenging the patient with milk and cheese and returned to normal later.

In January, 1961, Professor Truelove of the Radcliffe Infirmary in Oxford described how ulcerative colitis was provoked by milk.[13] He noted that Andressen in the USA in 1942 had concluded that ulcerative colitis was due to food allergy in at least 66 per cent of cases.[14] The food item which was regularly incriminated was milk. Truelove summarised that during the last few years it had proved possible to recognise a group of patients with ulcerative colitis in whom removal of milk and protein-containing milk products from the diet has been followed by marked improvement in the clinical course of the illness. In five of these cases, milk had been re-introduced into the diet and on every occasion this has been followed by a frank attack of ulcerative colitis. Despite the fact that this had been recognised nearly 20 years ago, no-one else has seen fit to investigate this observation further in the United Kingdom. It is also rare for a hospital even to try a milk free diet in cases of ulcerative colitis. This emphasises the huge gap in outlook between the conventional anthropocentric view of medicine and the ecologic orientation I have described.

Clearly, food allergy presents an enormous problem in clinical medicine and may underlie many of the medical problems encountered in a doctor's waiting room. In recent years, very significant progress has been made in devising critical tests for the identification of offending food constituents and in treating affected patients. A large number of physicians throughout the world have become actively interested in food allergy and through experience gained and by using existing technology have attained commendable results in what still remains a challenge in general medicine.

Summary

Recent studies have indicated that food allergy is a more significant factor in human illness than had been previously suspected. It is now appreciated that certain foods, either alone or in combination with inhaled allergens, can be responsible for a wide range of chronic disease processes. Study of the ability of foods and environmental allergens to provoke chronic disease in susceptible persons is now referred to as clinical ecology. The present chapter discusses the historical development of clinical ecology, its basic concepts and techniques currently used in diagnosis and management with reference where necessary to pertinent published work.

Further Reading

Over 100 books have been published on the subject of food allergy, most of these being produced in the United States. Space here does not permit discussion of much of this work but for further reading the following publications are recommended:

1. R. Mackarness, *Not all in the Mind* (Pan Books, 1976).
2. R. Mackarness, *Chemical Victims* (Pan Books, 1980).
3. L. Dickey, *Clinical Ecology* (Charles C. Thomas, Springfield, Illinois, 1976).

4. T.G. Randolph and R.W. Moss, *Allergies: Your Hidden Enemy* (Turnstone Press Ltd, 1981).

Richard Mackarness deserves credit for introducing the subject of food allergy to Great Britain and his books were written for general interest. Lawrence Dickey has edited contributions from 40 separate physicians and this publication presupposes a rudimentary knowledge of the subject.

References

1. A.H. Rowe, *Food Allergy – its Manifestations, Diagnosis and Treatment with a General Discussion of Bronchial Asthma* (Lea and Febiger, Philadelphia, 1931).
2. J.H. Rinkel, 'Role of Food Allergy in Internal Medicine', *Ann. Allerg., 2* (1944), p. 115.
3. T.G. Randolf, *Human Ecology and Susceptibility to the Chemical Environment* (Chas. C. Thomas, Springfield, Illinois, 1962).
4. J.B. Miller, *Food Allergy, Provocative Testing and Injection Therapy* (Chas. C. Thomas, Springfield, Illinois, 1972).
5. J.B. Miller, 'A Double Blind Study of Food Extract Injection Therapy – a Preliminary Report', *Ann. Allerg., 38* (1977), p. 185.
6. A.P. Black, 'A New Diagnostic Method in Allergic Diseases', *Paediatrics, 17* (1956), p. 716.
7. W.T.K. Bryan and M.P. Bryan, 'Cytotoxic Reactions in the Diagnosis of Food Allergy and Otolaryngology', *Clin. N. Am., 4* (1971), p. 523.
8. R. Finn and H.N. Cohen, 'Food Allergy – Fact or Fiction', *Lancet, i* (1978), p. 426.
9. E.C.C. Grant, 'Food Allergies and Migraine', *Lancet, ii* (1979), p. 966.
10. J. Munro, C. Carini, J. Brostoff and K. Zilka, 'Food Allergy in Migraine', *Lancet, ii* (1980), p. 1.
11. L. Dickey, *Clinical Ecology* (Chas C. Thomas, Springfield, Illinois, 1976).
12. A.L. Parke and G.R.U. Hughes, 'Rheumatoid Arthritis and Food: a Case Study', *Br. Med. J., 1* (1981), p. 2027.
13. S.C. Truelove, 'Ulcerative Colitis Provoked by Milk', *Br. Med. J., 1* (1961), p. 154.
14. A.F.R. Andresson, 'Ulcerative Colitis – an Allergic Phenomenon', *Am. J. Dig. Dis., 9* (1964), p. 91.

INDEX

accidental poisoning 2, 10, 75, 94
adaptation, metabolic 209–10
additives, food 1, 14–15, 96–9, 133–6, 160, 235–42
administration, methods 35, 134, 197
AF2 food preservative 160, 163
aflatoxins 13–14, 17, 105, 124–6, 140, 147, 159
age factors 192, 246
agent orange 109–10
agrochemicals 1, 10, 15–16, 106–11, 136–8, 161–3, 242–43
alcohol, toxicity 5, 6, 77, 100–2
alcoholic drinks 91, 131, 162, 164–5, 279, 286
alcohols, dehydroxylation 203–4
aldrin 108
alkaloids 12, 103, 105, 147, 160, 202
allergic responses 35–7, 282–3
allergy, defined 23, 57, 278
amaranth 96, 98, 160, 231, 236
amines, toxicity 220–1
amino acids 220–1
anaemia 7
anaerobes 187, 188, 190, 194–5
anaerobiosis 198, 217
analytical methods, sensitivity 126
anaphylaxis 29–30, 36, 37–45
anencephaly 102–3
animal studies, relevance to man 2, 62, 65, 78, 136, 165–6
antibiotics 197–8
antibodies 29–31, 38–9, 59–60, 278
antigenic potential 65–7
antigenic processes, defined 22–4
antioxidants 1, 58, 63, 81, 99, 136, 140
arsenic compounds 88, 200, 204, 244–5
arthritis 36, 47, 49, 50, 282, 292
arthus reaction 36, 45–50
Aspergillus 9, 13–14, 39, 126, 159
aspirin 58–61
assay methods 30–1, 39, 150–2, 153–6
asthma 28, 30, 33, 36, 38–9, 42, 58–61, 283, 292
atopic disorders 28, 30, 33, 36, 38–9, 42, 54, 58–61, 283, 292
avoidance diets 39, 281-3, 291
azo compounds 3, 202–3, 207–8, 235–8

B cells 25–8, 34, 35, 50, 52
bacteria, gastrointestinal 183-9, 229–30
baseline risk testing 135–6
base pair substitution (DNA) 148
basophils 31, 37–8, 55, 68,
beers, antifoaming agents 5
benzanthracene 129
benzoates 59–61
benzopyrene (BP) 128–9, 164
beryllium 55
bile, 33
acids 215–17, 230–1
bladder cancers 135, 221–2
bracken 106, 123–4, 139, 159
brain damage 94, 95, 98–9, 100–1
Brassica vegetables 164, 247
Bryan cytotoxic tests 284, 288–9
butylated hydroxyanisole (BHA) 58, 63, 99
butylated hydroxytoluene (BHT) 99, 136, 140

cadmium 6-7, 17, 92–3
caecum, enlargement 223–7, 241, 249–51
caffeine 77, 104, 164
calcium 81, 84–5
calorie intake, reduced 138–9
captan pesticides 161, 167
carbaryl 107, 108
carbohydrates 79, 213, 223–7
carbonyl compounds, reduction 203
carcinogens, food 122–40, 147, 160, 163, 195, 208, 209, 215–20
cardiovascular diseases 5-7, 230
carrageenans 106, 241, 252–3
casein 43, 139
case reports 290–3
cattle 199, 246

cell-mediated immune (CMI)
responses 36, 50–7
celluloses, modified 241
central nervous system 7–12, 100
cereal products 43, 49, 94, 222, 285–6, 291
contaminated 16-17, 94, 104-5, 109, 112
cheese 45, 131, 291–2
chemicals, screening 148, 150, 153, 158
chemostats 199
children, lead poisoning 10, 89–90
Chinese restaurant syndrome 11, 98
chlorine-containing contaminants 15–16, 107–9, 136–7, 161, 242–3
chocolate 43, 45, 60, 291
cholesterol 196, 216, 218–22
choline 83
chromosomal mutations 148, 152, 155–6
citrinin 17
citrus fruits 162, 164–5, 291
cleft palates 92, 95, 110
cobalt 5–6
coeliac disease 36, 55, 56, 222–3
coffee 78, 104, 129, 162, 164–5, 279, 291
colon
bacteria 189, 190
cancer 138–40, 194–5, 215–20, 222, 228–30
colours, food 11–12, 98, 133–4, 160–1, 163, 232-5; *see also* amaranth, indigo carmine, tartrazine
conjugation reactions 207, 208–9, 212
contact dermatitis 55–6
contaminants, food 2, 102–12, 161, 242–9
copper 87
costs, testing 62, 154, 158
coumarins 205, 238–9
Crohn's disease 36, 50, 55, 193, 214, 282
cromoglycate, sodium 291–2
cultures, cell 52–3, 62, 152, 199
cyanogenic glycosides 201, 232–3
cycasin 8, 12, 122–3, 160, 162, 201, 208, 209, 231–2
cyclamate 63, 96–8, 135, 160, 197, 201, 209–10, 240
cyclochlorotine (CC) 126–7
cyclohexylamine 97, 201, 209–10, 240
cytotoxic reactions 36, 56–7
cytotoxic tests 288–9

dealkylation reactions 200, 205, 212
deamination reactions 209, 220–1
decarboxylation reactions 200, 204–5, 220
decay processes 127
degradation reactions 200, 204–5, 243
dehydroxylation reactions 200, 203–4
delayed hypersensitivity (DH) 26, 36, 50–7, 279
deoxyribonucleic acid (DNA) reactions 81, 134, 148, 156–7
diagnostic aids 43, 53
diazinon 108
dichlorodiphenyltrichloroethane (DDT) 107, 108, 137, 204, 242–3
dieldrin 108, 137, 243
dietary exclusion method 284–6
dietary factors 139, 193–7, 210, 216
dioxin 110
diquat 243
disorders
gastrointestinal 36, 50, 55, 192–3, 214, 222–3, 228, 253, 282–4, 292–3
intolerance 36, 41–3, 49–50, 55–8, 278, 282–4
diverticular diseases 190, 228
L-dopa 204–5
dosage rates, feeding trials 5, 11, 77–8, 146, 153, 251
double-blind tests 57, 288, 290
Drosophila 151, 153–4, 155–6, 159–61
drugs, effect 6, 24, 46–7, 60–1, 210
duodenum, bacteria 189

ecology, clinical 281–3, 293–4
eczema 13, 30, 36, 39, 42, 55–6, 58–9, 278, 283–4
eggs 39, 49, 60, 65, 286, 291
egg white 43–5
emulsifiers, food product 106, 242, 252
end-points, genetic 148, 151–6, 158

entero-hepatic circulation 207, 208–9, 239
enterokinase 212, 213
Environmental Protection Agency (EPA) 109, 157
enzyme-linked immunosorbent assays (ELISAs) 39, 43, 49
enzymes 77, 209–14
  inhibitors 233–5
ergotism 9, 105
ethylenediaminetetra-acetic acid (EDTA) 87, 88
ethylenethiourea (ETU) 137–8
eurucic acid 7

faeces
  bacteria 186-8, 191-4
  transit times 229
fasting 284–5
fats, dietetic 6, 138–9, 195–6, 215–16, 222
fatty acids, unsaturated 63–4, 214–15
favism 7
fertility, reduced 74, 84, 91, 98, 101–2, 107–8, 112
fetal alcohol syndrome 100–1
fetal skin defects 108
fibre, dietetic 138, 228–31
fish, sensitivity 8, 43, 45
flatulence 225, 228
flavonoids 160, 162, 165, 204, 205
flavours, food 238–9
foodstuffs
  amount consumed 145
  listed 43, 129, 131, 159–62, 291
  low-risk (allergic) 285, 286
fungal toxins 103, 124–7, 159
fungicides 110–11

gallate esters 63
gene mutations 148, 151–2, 153–4
genetic factors 6, 7, 39, 56, 147–9, 213
genetic markers 150–2, 153–6
germ-free animals 197–8, 201
glucuronides 200, 201, 209, 212
gluten 30, 50, 56, 222–3
glycosides, hydrolysis 200, 201, 208, 232
glycyrrhinic acid 6
gut-associated lymphoid tissue (GALT) 34–5

haptens 22–3, 46–7, 55, 64
hay fever (allergic rhinitis) 33, 36, 38, 42, 58–9, 283
herbicides 109–10
heritable translocation tests 156
histamine, release 25, 45, 68, 220
histological changes 40–1, 47–9, 54–5
house dust mites 39, 281
humans, testing 68, 157
hydrocarbon responses 281
hydrolysis reactions 200–2, 207, 212
hyperactivity syndromes 11, 97
hypersensitivity, defined 11, 278
hypoglycin A 106

idiosyncrasies 25
IgE antibodies 30–1, 38–9, 59–60, 278
ileum, bacteria 189
immediate-response hypersensitivity 37, 279
immune complex mediated disorders 24, 36, 49–50
immunoglobins 26, 28–31
immunological responses 59–60, 223
immunosuppression 61–4
imports 1–2
impotence 101–2
in vitro tests 60–4, 66–7, 148, 150–2, 198–9, 214, 248, 289
in vivo tests 62–4, 152–6, 155, 158, 248–9
India, diet 3, 194
indigo carmine 205, 206, 238
industrial chemicals 2, 55, 128
infection, prevention 30, 33, 61–4
inhalent allergies 39, 281, 283
injection therapy 287–8
inoculation 28; *see also* serum sickness
insect bites 55
intestinal metabolism
  enzyme activity 211–14
  microbiological action 132, 186–211
  toxicology 214–49
intestinal reactions 31–5, 40–1, 48, 200, 212
intolerance, food 41–2, 57–61, 213–14, 222–3, 278
iodine 87
iron 87
irradiation damage 108, 161, 164,

164
isomerisation reactions 206
itai-itai disease 92

J (junction) chains 30, 32
Japan 6, 17, 94, 109, 128, 160, 163

K (killer) cells 27, 29
Kenya 125–6
kidney
  cancers 134
  toxicity 16–18

β-lactoglobulin 40, 43, 53
lactose, intolerance 213–14, 225
lamb-and-pear diet 285
lathyrism 8, 106
lead 7, 9–10, 17–18, 89–92, 100
lead chromate 3
lectins 232
legumes 7, 8, 43, 131, 225, 228, 235
lethal synthesis, defined 207–8
lethal-assay methods 153–4, 155–6
libido, loss 91, 102
lipase, deficiency 211–12
lipids 79–80, 214–20
literature
  allergies 290–3
  food additives 74
liver
  cancer 123–6, 132, 134, 139
  reactions 12–16, 124–6, 206–9, 239
locoweed 8
luteoskyrin (LS) 126–7
lymphocytes 25–8, 34, 51–4
lymphokines 50–1

magnesium 85–6
maize starch 227, 250, 285
malabsorption syndromes 36, 56–7, 213–14, 222–3
malathion 108, 243
manganese 88
masked food allergies 279–81
mast cells 31, 37–40, 292
maternal diets 73–4, 78, 81, 86
meat-based diets 138, 194–7, 215–16, 222
mental retardation 10, 11, 79, 82, 90, 94
mercury 10–11, 18, 75, 76, 93–6, 206
mercury compounds 244
metabolic reactions 200, 207, 212
metabolic systems 150, 153
metals 5–7, 17–18, 63, 84–96, 244–5
metaphase analysis, chromosome 155
micronucleus assay 155, 217
migraine 282, 285, 289, 291
milk 39, 43, 49, 60, 65, 67, 162, 164–5, 291–3
  mercury content 94–5
Mishell-Dutton technique 63
mitogens 52, 54
monosodium glutamate (MSG) 11, 58, 98–9
mould infestations 9, 13, 16–17, 103–5
mucosal immune systems 30, 31–5
mushrooms 43, 159, 162
mutagenicity 145–67
mutagens 158–65, 166–7
mycotoxins 8–9, 13–14, 16–17, 104–5, 124, 159, 162

N (null) cells 27, 52
N-oxides, reduction 200, 203
natural chemicals 2, 8–9, 12–17, 59, 78–88, 105–6, 138–40, 214–35
neural tube abnormalities 103–4
niacin 12–13, 83
nitrates 130, 132–3, 140, 204, 242, 245–9
nitrites 15, 130, 132, 140, 161, 163
  reactions 202, 206, 242, 246–9
nitro-compounds 200, 202, 203, 243
nitrofurans 150, 160
nitrosamines 15, 130–3, 140, 161, 163, 204, 207, 242, 247–9
nitrosation reactions 163, 247–9
nuts 43, 130, 162

obesity 6, 140, 282
occupational exposure 56, 63, 89, 91, 93
ochratoxins 14, 16–17, 105
oestrogens 81, 233
oligosaccharides 213-14, 225, 228
oxalic acid 16
oxidative reactions 206, 207, 212, 218

packaging materials 15, 161, 163, 245
pancreas, enlargement 234
pantothenic acid 83
paper disc radio-immunoassay (PRIST) 31
paraquat 243
parathion 202, 243
pathology, gastrointestinal 249–53
Penicillium 9, 14, 16–17, 126–7, 159
pesticides
  advantages 1, 16
  toxicity 15, 107–8, 136–8, 161–3, 242–3
Peyer's patches 32, 34, 35
pH, gastrointestinal 188, 198, 212, 246, 247, 248
pharmacological intolerance 60–1
phenols, reduction 200, 203–4
phosphates, dietetic 85
phosphorus-containing pesticides 108, 202, 243
phthalate diesters 245
placental transfer 84, 94, 95, 100, 108, 123
point mutations 148, 151–2, 153–4
polychlorinated biphenols (PCBs) 108–9
polycyclic aromatic hydrocarbons (PAHs) 127–9, 161
polyploidy 161, 162, 164
polyunsaturated fatty acids (PUFA) 63–4
potassium 6
potato
  blight 102–4
  starch 223–7, 250–1
Prausnitz-Kustner transfer tests 39
pre-clinical studies 75–6
preservatives 127–33, 242, 251
processing, food 14, 127–33, 161, 163
proteins 79, 213, 220–3
provocation neutralisation tests 286–8
pseudo-allergies 24–5, 43, 45, 57
pyrolysis products 128–9, 161, 163–4
pyrrolizidine alkaloids 12, 147, 160, 202

quinic acid 209

radiation equivalents 167
radio-allergosorbent tests (RASTs) 31, 39, 43, 66, 289, 291–2
radioactive compounds 85
reagins 30–1, 38–9
reduction reactions 200, 202–3, 207, 212, 239, 247
regulatory bodies 2–3, 134, 137, 167
repair mechanisms, cell 133, 156–7, 164, 166
retinol 80–1
rheumatoid factors 30, 47
riboflavine 83
ring fission reactions 200, 205
Rinkel responses 279–81
risks, mutagenic 166–7
rubratoxin 105
ruminants 105, 202–23, 246

S9-fractions 150
saccharin 96–7, 134–6, 140, 160, 197, 240
salicylates 58–61
saliva, microflora 188
saliva compounds 132, 247
Salmonella mutagenicity tests 150, 151–2, 159–60, 165, 217, 222
salt consumption 6, 77
sassafras, oil 123
seafood, contaminated 94, 112
secretory antibodies 30, 32–3
Senecio alkaloids 12
sensitivities, anaphylactic 66–7
sequestration 87, 88, 230
serum sickness 24, 36, 49
sheep 14, 87, 90–1, 202–3
shellfish 8, 43, 129, 278
shock reactions 37–8
smoking, food 128
solanine 103
solvents 148, 150
sorbitan esters 241–2
species differences 14, 76–8, 83, 108–9, 111, 166, 188–92, 209
specific locus (mutagenicity) tests 154
sperm abnormalities 91, 102, 157
spices 123, 160
spina bifida 102–3
spot tests 154
sprue, tropical 193
stabilisers, food product 106, 241–2
starches 223–7, 250–1

chemically modified 241, 250
sterility 81
steroids 208–9, 212, 215–20
stilboestrol 208–9
stomach, microflora 188–9
strategies, testing 62–3, 158
strawberries 43, 278
strontium 85
sugar 96, 279, 284–6
sulphamates, hydrolysis 201
sulphites 58, 242, 251–2
sulphonated azo dyes 236–7
sulphur dioxide 58, 242, 251–2
susceptibility, allergic 37
sweeteners 96–9, 134–6, 239–40
symptoms, allergic 42, 284
synergistic interactions 126, 129, 247
syntheses 200, 205–6, 207

T cells 25–8, 34–5, 50–1, 52, 56–7
tartrazine 11, 57–61, 205
tea 104, 129, 162, 165, 279, 291
teratogenicity studies 74–112
teratology, defined 73
terpenoid flavours 239
tests
food allergy 283–9
immunological 62–3, 64–8
mutagenicity 146–58, 217, 222
skin 35–7, 39, 55, 60, 287–8
tissue 62, 152, 199
Thailand 124–5, 214
thiamine deficiency 6, 83
thickening agents 241, 252
threshold limit values 89–90, 93, 109
times, reaction 36, 37, 41–2, 47, 55
tobacco 247, 279, 284
tolerance, defined 23–4
tomatoes 43, 45
trace elements 87–8
tremorgans 9
2,4,5-trichlorophenoxyacetic acid (2,4,5-T) 109–10
tryptophan 164, 220, 221–2
tuberculin reaction 54–5
tyrosine 45, 220, 221

Uganda 124–6, 194, 214
ulceration, gastrointestinal 251–83
ulcerative colitis 50, 55, 193, 253, 284, 292–3
United Kingdom (UK) 17, 128, 194
urticaria (hives) 36, 38, 42, 58–60, 278, 283, 288

vanillin 204
vegetable oils 7, 129, 131
vegetarian diets 195
vinyl chloride 161, 163
vitamins 12–13, 16, 80–4, 102, 160, 164

water, analysis 6, 131, 246
whole food testing 153, 158
wine flavouring 91, 164, 165
withdrawal symptoms 281, 285

yeasts 43, 152, 187, 286
Yushio oil disease 108

zinc 86–7
zirconium 55